From Atoms to Quarks

BOOKS BY JAMES TREFIL

James Trefil

From Atoms
to Quarks

An Introduction
to the Strange World
of Particle Physics

Illustrations by Judith Peatross

Anchor Books

DOUBLEDAY

New York London Toronto Sydney Auckland

AN ANCHOR BOOK
PUBLISHED BY DOUBLEDAY
a division of Bantam Doubleday Dell Publishing Group, Inc.
1540 Broadway, New York, New York 10036

ANCHOR BOOKS, DOUBLEDAY, and the portrayal of an anchor are
trademarks of Doubleday, a division of Bantam Doubleday Dell
Publishing Group, Inc.

Library of Congress Cataloging-in-Publication Data

Trefil, James S., 1938–
    From Atoms to quarks: an introduction to the strange world of particle
physics/James Trefil; illustrations by Judith Peatross.
    p. cm.
    Originally published: New York: Scribner, © 1980.
    1. Nuclear physics. 2. Particles (Nuclear physics) I. Title.
QC777.T73   1994
539.7—dc20   93-45263
    CIP

To my mother and to Flora

Contents

x   Contents

Introduction
to the Revised Edition

IT IS a somewhat disquieting experience to return to a book that one wrote fifteen years ago, particularly a book that deals with a subject like particle physics that has been through revolutionary changes. Nevertheless, the success of the original edition of *From Atoms to Quarks,* the kind comments of colleagues and readers, and the phone calls I started getting when the last copy of the original paperback disappeared from the publisher's warehouse all convinced me that it was time to look at the field again.

I wrote the original version of the book in the late 1970s, a time when many particle physicists were going through a period of disillusionment. The original promise of the quark model seemed to have led nowhere, and the blinding insights of the unified field theories were still the property of relatively small groups of workers. It seemed that progress on understanding the basic structure of the universe—a quest that can be traced back two millennia to the Greeks—had stalled. This led to a somewhat negative tone in some parts of the original, a tone that, in retrospect, was totally unjustified.

The last decade and a half have seen unprecedented progress in our understanding of the basic structure of the universe. They have also seen a coming together of the study of elementary particles and cosmology, so that now astrophysicists talk with confidence about how the universe behaved during its first second of existence. From a purely scientific point of view, the world couldn't be rosier. As I point out in Chapter XV, we stand at the brink of developing a true "Theory of Everything"—a single equation that, in some sense, contains within it all the knowledge of the universe.

Yet despite all this, I find myself doing this revision at a time when particle physicists are once again discouraged. The problem is that the field has gone as far as it can without new infusions of data, and this data can come only from new particle accelerators like the Superconducting Supercollider, the massive project in Texas that was recently killed in midconstruction by the House of Representatives. That we should have come so close to the end of the quest, only to be stopped just short of it, has cast a pall of gloom and discouragement over the whole field.

It seems worthwhile, therefore, to look again at the historical context of the physics of elementary particles, to trace once again the progression of our knowledge from atoms to nuclei to elementary particles to quarks, to remind ourselves about how far we have come on this road. With this goal in mind, I offer this updated version of that story, with a special emphasis on what we have learned in the last fifteen years and what we still have to understand.

JAMES TREFIL
Red Lodge, Montana

From Atoms to Quarks

I

The Quest for
Ultimate Simplicity

Gaily bedight,
A gallant knight,
In sunshine and in shadow,
Had journeyed long,
Singing a song,
In search of Eldorado.
—EDGAR ALLAN POE, "Eldorado"

INTRODUCTION

THERE IS SOMETHING in people that makes them want to know. They look at the world and ask questions. What is it made of? Why is it there? How does it work? Some of this yearning to know is expressed in religion, some in art, and some in philosophy. But for those who share the general background usually called Western culture, the yearning is most completely expressed in the institution called science.

Physics is the science that is concerned with matter and motion, and it is to physics that some of the most interesting of these basic questions apply. In particular, the questions about how the material world is put together and what it is made of have long occupied the thoughts of men who would now be called physicists. Although these men lived thousands of miles and thousands of years apart, a single concept runs through their ideas like a unifying thread. Perhaps concept is the wrong word. Perhaps we should call it a belief, a hope, a dream. From the time people first started asking scientific questions, they have assumed that when they knew enough to answer the questions fully, they would find that the world was really a simple, uncomplicated sort of place.

When you consider the overwhelming complexity that we see around us, you have to appreciate what a tremendous leap of faith this was. Yet in about 585 B.C., Thales of Miletus, the Ionian Greek who is called the first scientist, noticed that matter comes in three forms—liquid, solid, and gas—and suggested that the entire world might be nothing more than water (which is also seen as liquid, ice, and steam). From this beginning his students developed the familiar Greek system in which everything known is some combination of just four elements—earth, fire, air, and water. This theory, containing as it does four basic constituents, is the first serious attempt to find a scheme that is both simple and explains the observed complexity of the world. If we follow modern usage, we would call earth, fire, air, and water elementary or fundamental, and if we believed that they came in small pieces, we might even call them elementary particles.

In the late fifth century B.C., the philosophers Leucippus and Democritus, trying to resolve the conflict between the observed complexity and transience of the physical world with the Greek idea that Truth must be eternal and unchanging, suggested that matter might actually be composed of small particles. They called these particles atoms (literally, that which cannot be split), and pointed out that while atoms could be unchanging, the relationship between atoms could change. In this way, both of the seemingly contradictory observations could be satisfied in a single philosophical system. Although this idea appeals to us a great deal in hindsight, it enjoyed little currency among Greek and Roman philosophers (except for a brief revival by the Epicureans), and it remained something of a philosophical sidelight until modern times.

The first statement of the modern atomic theory was put together in the early nineteenth century, when John Dalton, an English chemist, published his two-volume book called *A New System of Chemical Philosophy* (1808). In this book Dalton pointed out that many of the laws of chemistry that were known in his time could be explained easily if it were assumed that to each chemical element there corresponded an atom of matter. The atom for hydrogen was different from the atom for sulphur, of course, but the basic idea was clear. Every substance in the world must be made up of different combinations of a few different kinds of atoms, so that in this picture it is the atoms themselves that would be the fundamental building blocks of matter. Consequently, it is to the atoms that we would award the title "elementary particle." And even if today we know of over 100 chemical elements (as opposed to 26 in Dalton's time), it can still be argued that the modern atomic theory does impose a kind of simplicity on nature.

To Dalton, as for Democritus, the atoms were indivisible. Indeed, they were represented as featureless spheres in most of his books. Consequently, there was little in the atomic theory to prepare the scientific mind for a rather astounding series of discoveries that took place between 1890 and 1920.

THE DISCOVERY OF THE ELECTRON

By the middle of the nineteenth century, physicists were experimenting with a new phenomenon that was eventually to show that the simple atomic picture of matter had to be drastically changed. By that time it had become possible to produce an experimental apparatus in which a glass tube had a metal plate (called an *anode*) at one end, a wire capable of carrying electrical current (called a *cathode*) at the other, and a pretty good vacuum in the tube itself. By passing electrical currents through the cathode, physicists hoped to study the electrical properties of the rarefied gas in the tube.

When current was introduced in the cathode of such a device and the anode was held at a high positive voltage, a strange phenomenon was seen. A thin line of glowing gas formed near the cathode and extended toward the anode. From an analysis of the light emitted by the tube, it was clear that this thin line was made up of residual gas that had been heated when something passed through it. The unknown something was called *cathode rays,* and the consuming question then centered around their identity.

At that time, physicists knew that an object that carried an electric charge could be affected by two kinds of forces. If it were brought near the pole of a battery, it would be either attracted or repelled, depending on whether the positive or negative pole of the battery was involved. In technical language, it was said that such a particle was acted on by an electrical force.

In a similar way, if an electrically charged object was moving and a magnet were brought near its path, the charged object would change direction. This was taken as evidence that a magnetic force existed in addition to the electrical force. These two are sometimes referred to as *electromagnetic* forces.

The electric and magnetic forces will act on any charged object, but will not act on something that carries no charge, such as light. Thus, the debate on cathode rays eventually came down to the question of whether they possessed an electrical charge.

By the early 1880s, there were two definite schools of thought on this topic. In Germany the general feeling was that the cathode rays were a new type of radiation, something like light. To support this view, they pointed out that light was capable of transmitting energy in the form of heat. Isn't sunlight, after all, the basic source of warmth on the earth? In England it was felt that the cathode rays had to be particles of some type. The primary reason for this belief was that it was known that the glowing line that marked the passage of the cathode rays could be deflected and moved by bringing magnets near the apparatus, and this could only mean that the cathode rays were affected by the magnetic force. Since this had never been observed for light, the thinking went, the rays must be something else. The primary theory in England at the time was that atoms in the gas, upon striking the cathode, somehow became negatively charged and were then pulled through the gas toward the anode.

In 1897, J. J. Thomson, a young English physicist, performed a series of experiments that seemed to settle the matter once and for all. He reasoned that if the cathode rays were really particles, they would not only be affected by magnets but by large electrical charges as well. He knew that by measuring the amount by which a particle of known velocity is deflected by a magnetic field, it would be possible to determine the ratio of the particle's charge to its mass. (We will talk about the details of how this is done in Chapter VI.) If the charge of the cathode ray were called e and the mass m, then the charge to mass ratio would be e/m. The problem was that no one knew how to find the velocity of the cathode rays. If a way could be found to measure it, scientists would know whether the cathode rays were really radiation, because if they were, their velocity would be the same as that of light. As an added benefit, scientists would also know the value of e/m of the particle.

In Thomson's apparatus, the cathode rays were directed across a region between two charged plates, and in this region there was also a magnetic field. What he did was to adjust the charge (or, equivalently, the voltage) on the plates until it exactly canceled the deflecting effects of the magnetic field. In other words, if the magnetic field would cause the cathode rays to move downward, it would mean that the plates were charged in such a way as to move the beam back up by an equivalent amount. By measuring the electrical field that will exactly cancel the deflection due to the magnetic field, the velocity of the particles can be found. A more detailed description of this process will be given in Chapter VII.

The important result that Thomson obtained was that the velocity of the cathode rays was about 3×10^7 kilometers/second, which is about a tenth of the velocity of light. Clearly, the cathode rays were particles. From the fact that they were attracted toward a positively charged anode, he could conclude that the particles carried a negative electrical charge. The particle is now known as the *electron,* and its mass is 9.1×10^{-28} gram—a very small mass.

Once it was properly identified and labeled, it was realized that the electron is a very important particle indeed. Every electrical current, whether it is a man-made circuit or is in a nerve in the body, is simply the flow of electrons. For example, it takes about 6,000,000,000,000,000,000 electrons *per second* to keep a 100-watt light bulb going. Moving electrons are, in fact, the "electrical fluid" that Benjamin Franklin first postulated to explain electricity in his time.

From the point of view of the search for simplicity, however, the discovery of the electron was rather ominous. After all, the only place the particle could come from was the interior of some "indivisible" atom. The existence of a negatively charged particle that can be taken from the atom implied that there must be a positively charged segment left behind, and this, in turn, implied that the atom must have structure. If this was so, then there must be a type of matter more fundamental than the atom. The electron was the first example of matter from this deeper level.

THE PHOTON

Light has been known to be a wave since the eighteenth century, but the discovery that its origins are tied to electrical charges and that it is only one example of electrically generated waves was one of the great triumphs of nineteenth century physics. From the discovery have flowed some of the most important artifacts of our culture—radio, television, radar, and microwave technology, to name but a few. In this section, I shall talk about the idea of light as a wave and then show how some developments early in the twentieth century suggested another way of looking at it.

A classical example of a wave is surf coming into a beach. The wave is characterized by three numbers—the wavelength, or distance between crests of the wave; the frequency, or the number of crests that go past each second; and the amplitude, or maximum height of the wave. These quantities, in general, are independent of each other. In fluids it is possible, in principle, to have any combination of wavelength, frequency, and amplitude in a wave.

If you have ever seen surf coming into a rocky coast, you realize that waves can carry tremendous amounts of energy. The energy carried by a wave always depends on the amplitude, and for some waves it can depend on the other two parameters as well. Thus, we could just as well characterize a wave by frequency, wavelength, and energy, as by frequency, wavelength, and amplitude. This is the most useful way to think of light waves.

Finally, we note that if the frequency is the number of crests passing a point per second and the wavelength is the distance between crests, the speed of the wave must be $V = \lambda\nu$, where we have followed the usual convention and represented the wavelength by the Greek letter λ (lambda) and the frequency by the Greek letter ν (nu). Remember that this is the velocity of the wave and not of the water. As a particular wave moves by a point, the water moves up and down—it does not move along with the wave.

The identification of light with this phenomenon rests on arguments too detailed to go into here, but essentially it depends on the fact that light seems to behave in ways that are very similar to water waves. The discovery about light alluded to above had to do with how the waves are generated. It turns out that whenever electrically charged objects are accelerated, a wave is generated that moves outward from the objects in an ever expanding sphere. It is a three-dimensional analog of the circular wave that moves away from the point where a rock is thrown into a still body of water. Because the wave moves away from the charged object, it is called *radiation,* and because it is electromagnetic in origin, it is called *electromagnetic radiation.* In modern terms, the phenomena that correspond to the displacement of the water in surf are oscillating electrical and magnetic fields through moving space.

In classical physics, there is no reason why these electromagnetic waves cannot have any combination of wavelength and energy. Theory requires, however, that all the waves move at the same velocity. This velocity is so important that it is given a special letter—c—and it has the experimentally determined value

$$c = 3 \times 10^8 \text{ m/sec} = 186{,}000 \text{ mi/sec}$$

This means that if we know either the frequency or wavelength of a particular radiation, we can determine the other from the relation $c = \lambda v$. The most common examples of radiation at different wavelengths are the colors that we can see with the unaided eye. Each different color corresponds to a different wavelength and frequency of light. Red has the longest wavelength and lowest frequency, while violet has the shortest wavelength and highest frequency.

In more quantitative terms, visible light corresponds to electromagnetic radiation, for which the wavelength is between 3.8×10^{-5} centimeters (violet) and 7.8×10^{-5} centimeters (red). Clearly, this rather narrow band of wavelengths does not exhaust the possibilities for electromagnetic radiation. Even subject to the constraint in the equation, there is a continuum of possible wavelengths. Some typical familiar electromagnetic waves are listed in the table, together with the names usually assigned to each wavelength interval.

v	λ (cm)	
10^6	3×10^4	AM radio
10^7	3×10^3	shortwave radio
10^8	3×10^2	FM radio and TV
10^{10}	3	microwave
3×10^{13}	10^{-3}	infrared
3×10^{16}	10^{-6}	ultraviolet
3×10^{18}	10^{-8}	X ray
3×10^{20}	10^{-10}	gamma ray

Once the infinite variety of possible electromagnetic radiation was realized, some problems arose. The central problem was that if the radiation of a given frequency had an energy independent of the frequency, the principles of classical physics seemed to predict that the universe would be full of high-energy radiation, which is not the case. To get around this difficulty, the German physicist Max Planck suggested in 1900 that the energy of radiation emitted by an atom could not be just any number, but had to be an integral multiple of a basic energy given by $E = hv$. In this equation, E is the energy of the wave, v its frequency, and h is a constant, now called *Planck's constant.* We shall encounter this quantity again, since it plays an important role in all interactions on the atomic level. For our present discussion, the important point is that this equation seems to suggest that light is emitted in little bundles rather than in some kind of continuous spectrum. Planck called these little bundles *quanta,* but he never really accepted the idea that light was not an ordinary classical wave.

Albert Einstein, however, saw that Planck's quantum postulate could have far-reaching consequences if it were taken to its logical conclusion. For example, there was an experimental phenomenon known as the photoelectric effect, in which it was observed that when light was allowed to shine on one side of certain metals, electrons were emitted from the other side. This phenomenon could be understood in the classical wave theory—the electromagnetic waves would move electrons out of the metal in much the same way as the surf brings driftwood to a beach. The problem was that it would take a classical wave a long time to work an electron out of an atom, and the experimental fact was that electrons were emitted from the metal as soon as the light was turned on.

In 1905 Einstein suggested that if light were really composed of particlelike bundles rather than being a classical wave, this experimental fact could be understood. If Einstein's theory was true, then the interaction between light and electrons would be analogous to a collision between billiard balls, and the electrons would come flying out immediately after the collision. For this work, Einstein was awarded the Nobel Prize in 1921. (The theory of relativity was then too "risky" for the conservative Nobel committee to honor.)

The particles that make up light and other electromagnetic radiation are now called *photons.* They can be thought of as the constituents of radiation, and in this sense they would have to be included in any list of elementary particles that are supposed to give a simple explanation of matter.

In the early 1900s, a rather extraordinary figure appeared on the stage of world science. The son of a New Zealand wheelwright, Ernest Rutherford became the leading figure in the exploration of atomic structure and is largely responsible for the picture of the atom that we have today.

At this time, a great deal of work had been done on the chemistry of radioactive elements. It was known, for example, that there were three types of radiation given off by these elements, and one of the important tasks was to identify them. They were called α (alpha), β (beta), and γ (gamma) radiation, respectively, and each had different properties. In modern terms, the beta rays are electrons and the gamma rays are high-energy photons. Working at McGill University in Montreal, Rutherford was able to show that the mysterious alpha rays were, in modern terms, the nuclei of helium atoms. He was able to show this by performing experiments in which small samples of naturally radioactive elements that were known to emit alpha radiation were placed near sealed, evacuated tubes. After a period of time, sensitive chemical analysis showed the presence of helium in the tube where none had existed before. Since only the alpha radiation could have entered the tube, the connection between these rays and helium was established. For this experiment, and for other work in sorting out the behavior of radioactive elements, Rutherford was awarded the Nobel Prize in chemistry in 1908.

Contrary to the usual rule, Rutherford actually did his most important work *after* he received the Nobel Prize rather than before. When he went to Manchester University in England, he continued his experiments with alpha particles. One of the "hot" topics in those days was the study of the way in which these particles passed through thin metal foils. The idea was that you could learn something about an atom by seeing how the alpha particles bounced off of it. Almost as an afterthought, he suggested to some co-workers in 1911 that it might be interesting to see if any particles were ever bounced back (i.e., suffered a change of direction of 180°) when they hit a nucleus.

According to the ideas current at the time, the atom was a

large, diffuse, positively charged material in which electrons were embedded like raisins in a bun. Such an atom could not deflect an alpha particle, any more than a cloud of gossamer could deflect a bullet. When Rutherford's experiment was carried out, however, a surprisingly large number of alphas—perhaps one in a thousand—were scattered through angles close to 180°. This could only be understood if it was assumed that the atom, far from having its mass smeared out over a fairly large area, actually had virtually all of its mass concentrated in a small spot in the center. This concentration of mass Rutherford called the *nucleus* of the atom. Since the nucleus repelled the positively charged alpha particles, it must also be positively charged.

The picture of the atom that Rutherford evolved would be familiar to us (Illus. 1). The atom consisted of a small, dense, positively charged nucleus in which most of the mass resided and around which the electrons orbited. The analogy between the nuclear atom and the solar system was remarked upon by many observers at the time, and served as the plot for innumerable science fiction stories during the 1930s and 1940s. In that sense, it has become part of the accepted folklore of modern culture.

Once the existence of the nucleus became established, scientists could begin to ask about its composition. Was it a uniform blob of positively charged matter analogous to Dalton's atom, or were there constituents inside, waiting to be discovered? In 1919, using techniques similar to those that had allowed him to identify the alpha particle, Rutherford was able to show that a certain particle emitted from nuclear collisions of alphas with nuclei was itself the nucleus of hydrogen. Since hydrogen is the lightest atom, it follows that its nucleus must play some special role in nature. Rutherford recognized this fact by naming the new particle the *proton* (the first one).

The proton is a particle that has a positive electrical charge. The magnitude of this charge is precisely equal to the magnitude of the charge on the electron, although the signs of the charges of the two particles are opposite. The hydrogen atom must therefore be a system in which a single electron orbits a single proton, making hydrogen the simplest possible atom that can be imagined. Since protons are found in the debris of nuclear colli-

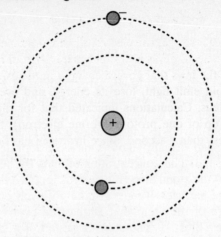

1. The atom as Rutherford envisioned it.

sions, they must exist in heavier nuclei as well. In fact, each unit of positive charge in a nucleus is supplied by a proton, so that helium (the next heavier element after hydrogen) must have two protons in its nucleus, lithium three, and so forth. The heaviest natural element, uranium, has a total of ninety-two protons in its nucleus.

As we might expect from the Rutherford experiment, the proton is much heavier than the electron. Its mass is now known to be

$$m_p = 1.67 \times 10^{-24} \text{ g}$$

so that the ratio of the proton to the electron mass is 1,836.

In passing we should note that in 1920 Rutherford suggested that there was probably another constituent to heavier nuclei; that is, an electrically neutral particle about as massive as the proton. He called this hypothetical particle the neutron. He came to this conclusion by noting that most atoms seemed to weigh about twice as much as one would expect them to if one added up the masses of the protons and electrons in them. The neutron and its eventual discovery in 1932 will be discussed in Chapter II.

A REMARK ABOUT THE BOHR ATOM

Once it was realized that the electrons in an atom move in orbits around the nucleus, physicists found themselves in a quandary. According to the laws that govern charged particles, an orbiting electron should emit light, lose its energy, and eventually fall into the nucleus. Calculations indicated that for the hydrogen atom (made up of one proton and one electron), this process would take less than a second. Since hydrogen has been around since the beginning of the universe, there was clearly something wrong with these arguments. In 1913 a young Danish physicist by the name of Niels Bohr suggested a way out of this dilemma.

The essential difference in Bohr's picture of the atom is that electrons circle the nucleus only at certain well-defined distances. For example, in the hydrogen atom sketched in Illustration 2, the electron can be in an orbit of radius r_2 or r_1, but it cannot be at any orbit between the two. The orbits at radius r_2 and r_1 are known as *allowed orbits*. Why the atom should arrange itself in this way did not become clear until the advent of quantum mechanics in the next decade, but the theoretical difficulties with the classical nuclear atom do not appear in the Bohr atom. In addition, the Bohr atom provides a very simple visualization of the process by which atoms emit light. If an electron happens to be in the orbit at radius r_1, the only way it can move to a lower orbit is to make an instantaneous jump, called a *quantum jump*. For example, the sketch shows the electron making a quantum jump from orbit r_1 to orbit r_2. But these two orbits have different energies, so in order for things to balance, the lost energy must be radiated away from the atom. It is, in fact, carried away by a photon whose energy is equal to the differences in energy between the two orbits. If the energy of the photon is right, we will see the light emitted by the electron transition, and we say that the atom is in a piece of material that is "red hot."

In the orbit r_1 the same process can occur. The electron could jump to the next lower orbit and another photon could be emitted. Alternatively, the electron could jump from the orbit r_2 to the orbit below r_1 and emit a single higher energy photon. The complete working out of the connection between an atom's

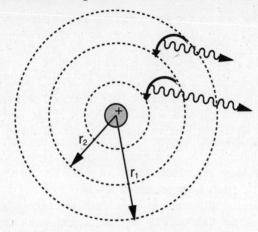

2. The electron's allowed orbits and quantum jumps.

structure and the light it emits is the domain of atomic physics. For our purposes, the important point to note is that Planck's assumption of the existence of quanta is mirrored in the way electrons are arranged in an atom.

DETECTION OF CHARGED PARTICLES

If elementary particles are going to be searched out and studied, it is clear that some way of detecting them has to be found. Thus far, we have glossed over the details of how this is done. Particles that carry an electrical charge are the easiest to detect; hence, they were the first elementary particles found. The reason for this easy detection is that when a charged particle approaches an atom, the electrical force acts between the particle and the electrons in the atom, so that the atom is changed during the passage. In one common occurrence, the particle actually tears an outer electron from the atom, leaving in its wake a free electron and a residual atom that has a net positive charge. An atom that is missing an electron is called an *ion,* and the process described above is called *ionization.*

Alternatively, the passage of the charged particle could simply leave behind an atom in which one or more electrons were lifted to orbits far away from the nucleus. In that case, the electrons would eventually jump back down to lower orbits, and radiation would be emitted. This radiation could be perceived as visible light. A material that emits light when a charged particle enters it, either through the process we have described or a more complicated one, is called a *scintillator*. Many of the early experiments on alpha particles were done by people sitting in darkened rooms watching for the flashes of light that indicated that a particle had struck a sheet of scintillating material, which had, in turn, given off light.

It has been the detectors that depend on ionization that have played the major role in the exploration of the world of the nucleus. Perhaps the most commonly known detector is the Geiger-Müller counter (usually just called the Geiger counter). This instrument consists of a rarified gas in a long metal tube, with a wire running down the center. By connecting this wire to one side of a voltage source (such as a battery) and the metal walls to another side, it is possible to create a voltage between the wire and the wall of the counter.

Suppose that a charged particle now passes through the chamber. It will leave behind it some free electrons and positive ions, as we discussed above. If the central wire has been connected to the positive pole of the voltage source, then the positive ion will be attracted to the wall and will start to move in that direction. Similarly, the electron will start to move toward the wire. If the voltage is high enough, the electron will soon be moving fast enough so that it, too, can ionize atoms of the gas and create new electrons that, in turn, will ionize still other atoms and create still more electrons. This process then builds up an *avalanche* in the gas, and when the large numbers of electrons produced in this way strike the central wire, a large current will appear. This signal, suitably amplified and fed through a loudspeaker, produces the familiar "click" that the Geiger counter emits in the presence of radioactive material.

The principle of the Geiger counter (creating extra electrons by applying a voltage and then detecting the enhanced signal) is used in many different types of counters in modern physics. This short sketch of its operation, however, should give you some idea of the processes involved when a physicist says "A particle was detected. . . ."

In familiar units, a typical nuclear diameter might be, say, 0.000000000000006 meter, and the speed of light is 300,000,000 meters/second. Rather than go through the bother of writing all of those zeros, physicists use a notation that makes handling both large and small numbers more convenient. I have seen it called scientific notation and powers-of-ten notation, as well as other titles. I see no point in quibbling over the name we use for it; we shall just use it without worrying about what it is called.

The key point is that every number, no matter how large or how small, is written as a number between 1 and 10 times a power of 10. For example, the number 250 would be written 2.5×10^2 and the number 0.003 as 3.0×10^{-3}. A useful mnemonic for this notation is to think of a positive exponent in the power of 10 as a direction to "move the decimal point to the right," and the magnitude of the exponent as saying how many places it should be moved. Similarly, negative powers of 10 imply that the decimal point should be moved to the left. The number 2.5×10^2 would therefore be interpreted as the direction "move the decimal point in the 2.5 two places to the right," while the number 3.0×10^{-3} would mean "move the decimal point three places to the left."

In this notation, the nuclear size noted above would be 6×10^{-15} meters, and the speed of light would be 3×10^8 meters/second.

Arithmetic manipulations in this notation are also quite simple. To multiply two numbers, you multiply the numerals in the usual way and add the powers of 10. For example, to multiply 250 by 250, we write:

$$2.5 \times 10^2 \times 2.5 \times 10^2 = 6.25 \times 10^4 = 62,500$$

To multiply 250 by 0.003, we write:

$$2.5 \times 10^2 \times 3 \times 10^{-3} = 7.5 \times 10^{-1} = 0.75$$

The rule for division is that the exponent of the denominator is subtracted from the exponent of the numerator, so that 250 divided by 0.003 would be

$$\frac{2.5 \times 10^2}{3 \times 10^{-3}} = 0.83 \times 10^5 = 8.3 \times 10^4 = 83,000$$

while 0.003 divided by 250 would be

$$\frac{3 \times 10^{-3}}{2.5 \times 10^2} = 1.2 \times 10^{-5} = 0.000012$$

As an example of how this notation is used, let us see how long it would take light to cross a typical nucleus. This should be the distance traveled divided by the velocity. Using the numbers above, we find the transit time to be

$$t = \frac{6 \times 10^{-15}}{3 \times 10^8} = 2.0 \times 10^{-23} \text{ sec}$$

This very short time will acquire a great deal of significance later, since we shall see that events inside a nucleus seem to take about this long to happen.

We can use the notation to discuss the size of some of the things we have talked about so far. If we consider a typical nucleus to have 10–20 objects the size of a proton in it, and a typical atom to have 5–10 electrons, then the table below lists some typical sizes. (For reference, oxygen has sixteen proton-sized objects in its nucleus and eight electrons in orbits.) The length 10^{-15} meter is an important one, and is therefore given a special name. It is called the fermi (after Enrico Fermi, whose work we will discuss later).

From the following table an important quantitative fact emerges. While the proton and the nucleus do not differ greatly from each other in size, the nucleus is much smaller than the atom. Rutherford had conjectured that the atom was mostly empty space, but with the information in the table we are now in a position to see just how empty it is.

QUANTITY	LENGTH (m)
Radius of the proton	8×10^{-16}
Radius of a typical nucleus	3×10^{-15}
Radius of a typical atom	3×10^{-10}

We can start by comparing the volume of the nucleus to the volume of an atom, assuming that both are spheres. The formula for the volume of a sphere of radius R is

$$V = \frac{4}{3} \pi R^3$$

so the volume of the proton is

$$V_P = \frac{4}{3} \pi (8 \times 10^{-16})^3 = 2.1 \times 10^{-45} \text{ m}^3$$

while the volume of the nucleus is

$$V_N = \frac{4}{3} \pi (3 \times 10^{-15})^3 = 1.1 \times 10^{-43} \text{ m}^3$$

and the atom has a volume of

$$V_A = \frac{4}{3} \pi (3 \times 10^{-10})^3 = 1.1 \times 10^{-28} \text{ m}^3$$

All of these are very small volumes, of course, but we can compare them. For example, we can ask what fraction of the volume of a typical nucleus would be taken up by a single proton. From the above numbers,

$$\frac{V_P}{V_N} = \frac{2.1 \times 10^{-45}}{1.1 \times 10^{-43}} = 1.9 \times 10^{-2} = 0.019 = 1.9\%$$

Thus, a single proton is roughly 2 percent of the volume of our typical nucleus. Since this nucleus has 10–20 proton-sized objects in it, we would say that these objects take up 20–40 percent of the nuclear volume. The nucleus, in other words, is definitely *not* mostly empty space.

Quite a different conclusion results when we compare the size of the nucleus to the size of the atom. In this case,

$$\frac{V_N}{V_A} = \frac{1.1 \times 10^{-43}}{1.1 \times 10^{-28}} = 10^{-15}$$

This means that, expressed as a percentage, the nucleus occupies 0.00000000000001% of the volume of the atom. The rest, except for the very small electrons, is empty space.

We can make this point in another way. Suppose the nucleus were blown up in size until it was a foot across; that is, the size of a large bowling ball or a small watermelon. How big would the atom be? Since our typical atom has a radius 10^5 times larger than the nucleus, the expanded atom would have to have a diameter of 10^5 feet. Since there are 5,280 feet in a mile, this would correspond to

$$\frac{10^5}{5,280} = 19 \text{ mi}$$

Hence, if the nucleus of our atom were as large as a bowling ball, the rest of the atom would consist of perhaps ten pea-sized electrons scattered around a sphere 20 miles across with the bowling ball at the center. Imagine putting a bowling ball in the center of a city and then scattering ten peas around through the rest of the city and you will have some idea how empty an atom really is.

II

The Nucleus

. . . the centre cannot hold.
—William Butler Yeats, "The Second Coming"

WHAT KEEPS IT TOGETHER?

ONCE WE RECOGNIZE that most of the matter in an atom is packed into a nucleus, our attention is naturally drawn to the way that the nucleus is put together. From Chapter I we know that the nucleus has a positive electrical charge and that it contains protons. One of the basic laws of electricity tells us that while opposite electrical charges must attract each other, similar electrical charges are repelled. This means that in any nucleus other than hydrogen (i.e., in any nucleus containing more than one proton), there will be a force that tends to push the protons apart. If this repulsive force were not countered by some other force, the nucleus of every atom would fly apart. Since they do not, we can conclude that there must be some sort of force acting in the nucleus that tends to hold things together.

The nature of this force was (and to some extent still is) mysterious. In the first place, the magnitude of the force is totally unprecedented in nature. Although electrical repulsion between two positive charges is a well-known phenomenon, the fact that within the nucleus the charges are separated by only 10^{-13} centimeter or so introduces an entirely new scale into the discussion. We can get some idea of what this means by performing a little thought experiment. First, we calculate the force between two protons in a nucleus, and then scale everything up so that the nuclei are spheres a foot across (with their centers perhaps 18 inches apart). If the repulsive force were scaled up by an equivalent amount, how large would it be?

One way of gauging its magnitude would be to imagine imbedding the two scaled-up protons in a block of solid steel. Even if we made the block of the strongest alloy known, the electrical repulsion between the protons would rip them apart, tearing the steel block as if it were tissue paper. Whatever it is that holds the nucleus together would have to be many orders of magnitude stronger than steel or the nucleus could not exist.

The simple fact that the nucleus does exist leads us to the conclusion that there must be some process in nature that is capable of overcoming this repulsion between protons. The process must produce much stronger forces than anything we know about in the ordinary macroscopic world. Physicists have named this process the *strong interaction,* and refer to the force generated by this interaction as the *strong force.* The development of elementary particle physics has been an attempt to understand what the strong force is and how it is generated. For now, we just note that the existence of stable nuclei poses a major problem in our understanding of the atom.

WHAT IS IT MADE OF?

The next question that has to be asked about the nucleus, once we accept its existence, concerns its composition. We know that there must be protons in it, of course, because it has a positive electrical charge. But when you start to look at the different chemical elements and compare the mass of the atom with the number of protons the atom must have, an unexpected fact becomes evident.

The simplest atom—hydrogen—is no mystery. The nucleus of this atom is a single proton, and a single electron orbits around it. The single positive charge of the proton cancels the single negative charge of the electron so that the electrical charge of the entire atom is zero, as it is known to be. The next most complicated atom—helium—has two electrons circling its nucleus. If helium is to be electrically neutral, then it must have a nucleus that contains two positive charges. Thus, we would expect the helium nucleus to contain two protons and, since virtually all of the mass of an atom is in the nucleus, we would expect the helium atom to weigh twice as much as hydrogen. In point of fact it does not. The helium atom weighs *four* times as much as hydrogen, so the helium nucleus must weigh roughly as much as four protons, but have the electrical charge of only two.

It turns out that this situation is pretty much the same for all of the chemical elements beyond hydrogen. The mass of the nucleus is close to a multiple of the proton mass and can be written approximately as

$$M_{\text{nuc}} \cong A \times m_p$$

where A is called the mass number of the atom. For hydrogen, $A = 1$, for helium $A = 4$, and so on. The charge of the nucleus, however, is expressed by a different integer, so that if we denote the charge on the nucleus by Q and the charge on the proton by q, we get $Q = Z \times q$, where Z is called the atomic number of the atom. Z is equal to the number of positive charge units found in the nucleus and therefore it is also equal to the number of electrons found in orbits around that nucleus. The experimental fact is that for elements beyond hydrogen, $A \gtrsim 2Z$; that is, the mass number is always at least as much as twice the atomic number. In heavy elements, such as uranium, it can be as much as 2.6 times greater than Z. Regardless of the details of how A and Z behave in different chemical elements, the central message is clear. There is twice as much matter in the nucleus as is needed to explain the charge. What form does this extra matter take?

One of several hypotheses that were put forward was that there ought to be another, hitherto unknown particle that had about the same mass as the proton but which carried no electrical charge. This particle was called the *neutron*. Finding and identifying the neutron then became a problem for experimental physicists.

In 1932 the British physicist James Chadwick, working at Cambridge, England, was studying something called the radiation of beryllium. It had been discovered that when particles from naturally radioactive sources were allowed to strike a target made of a thin sheet of beryllium metal, a type of radiation came out of the metal. This radiation had no electrical charge. Using a device called a cloud chamber (which we will describe later), Chadwick saw that when the radiation of beryllium hit an atom of a gas such as nitrogen or helium, that atom would recoil, losing an electron in the process. Only particles with electrical charge can be detected in the cloud chamber. The radiation of beryllium would enter the device, but since it had no charge it could not be seen. It would strike an atom, causing the recoil and loss of an electron. The atom would then have a charge, and its track would appear as a dark line.

Chadwick performed this experiment on atoms of different types, and for each type he measured the maximum recoil velocity of the atom. For example, he had data on collisions with helium, nitrogen, and oxygen atoms, among others. From these data and the known atomic sizes of the recoiling atoms, some simple laws of physics could be used to deduce the mass of the radiation of beryllium. When he carried out this calculation, Chadwick found that the unknown radiation consisted of particles whose mass was essentially the same as that of the proton. (Later measurements would show that it was a few percent heavier.) Thus, the radiation of beryllium could only be the missing piece of the nucleus—the neutron. For this discovery, Chadwick was awarded the Nobel Prize in 1935.

With this missing piece discovered, the atomic picture of matter seemed to satisfy the criteria, discussed in Chapter I, necessary for a truly simple picture. Every kind of material is composed of atoms, and although there are many different sorts of atoms, they are all built out of three basic constituents, the proton, the neutron, and the electron. Of these three, only two, the proton and the neutron, are to be found in the nucleus. Because of this fact, they are often referred to collectively as *nucleons*. Of course, the question of how the nucleons could be held together inside the nucleus remained unsolved, but this was a relatively minor point compared to the enormous simplification that resulted from the development of the nuclear picture of the atom.

How the attempt to solve the puzzle of the strong interactions ultimately led to a breakdown of this simplicity is a story that will be told in later chapters. For now, there are some very interesting properties of the neutron that we should discuss.

THE INSTABILITY OF THE NEUTRON

As mentioned in the previous section, the mass of the neutron turns out to be somewhat larger than that of the proton. In point of fact, the mass of the neutron is greater than the combined masses of the proton and electron. This larger mass means that it is possible (at least in principle) to make a proton and an electron from the amount of matter in a neutron. This, in turn, leads to one of the most striking properties of the neutron, its instability.

If (figuratively speaking) one were to take a neutron and set it on the table, the neutron would not remain there very long. In some hundreds of seconds, it would disappear, and in its place would be a proton, an electron, and another type of particle we will discuss later. In the language of particle physics, a neutron outside of a nucleus "decays," and the products of this decay include a proton and an electron. Since the neutron is the first example of an unstable particle we have encountered, we should take a little time to discuss this process of decay. Later on, we shall see that virtually all elementary particles share the characteristic of instability.

If we watched a large number of free neutrons, we would notice that they do not all decay at the same time. Rather, we would see one decay, then another, and so on at irregular intervals until all had completed the transformation into the decay products. It would be something like watching popcorn being made. Not all of the kernels "pop" at the same time: Each one goes when it is ready.

This means that if we started with 1000 neutrons and made a

graph showing the number of neutrons left at any time from the start, we would get a result something like the curve pictured in Illustration 3. The number of neutrons would decline steadily until, a long time from the start, none would be left.

In such a situation, we cannot say when an individual neutron will decay. There is no way of telling whether a particular neutron in our sample will be one of the first to go or the last. We can, however, talk about an average decay time for all of the neutrons in the sample by looking at the time it takes for a fixed fraction of the original neutrons to decay. One common way of describing decay times is to talk about the *half-life* of the sample. This is the time it takes for one-half of the original number of particles to decay. From the graph, we see that if we started with 1000 neutrons at time zero we would have 500 after 636 seconds had elapsed. The half-life of a free neutron would therefore be 636 seconds.

Although the term half-life is associated with nuclear physics, the idea behind it is a rather common one in the sciences in general. It will arise, in fact, whenever the amount being removed from a sample depends on how much of the sample there is. To take one example, if there are two chemicals undergoing reactions in a solution, the amount of each chemical, if plotted as a function of time, will show a curve exactly like that shown in Illustration 3 for the neutrons.

Another common measure of the decay time of a particle is derived by looking at the time it takes for the sample to be reduced to $1/2.718 = 0.367 = 36.7$ percent of its original size. The number 2.718 is usually denoted by the letter e, and is known as the base of natural logarithms. It happens to be a number that arises naturally in the calculus and that appears regularly in laws describing natural phenomena. The time for the sample to become $1/e$ of its original size is called the *lifetime* of the sample, and if we denote it by the letter T, then the number of neutrons left at a time t after we start the counting will be

$$N = N_0 e^{-t/T}$$

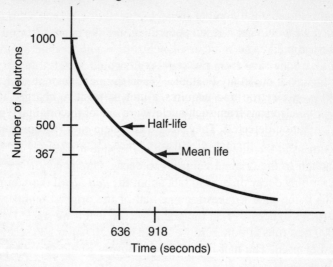

3. The number of neutrons left from a sample of 1000 as a function of time.

where N_0 is the number of particles in the original sample. The lifetime of the neutron is 918 seconds (about 15 minutes).

If a neutron sitting by itself decays in a matter of minutes, how can the atomic nucleus exist indefinitely? Why, for example, do all of the six neutrons in the carbon atom not decay?

The answer to this question lies in a rather subtle point about nuclear structure. When a free neutron decays, the proton that is a result of that decay can go just about anywhere it wants, since there are no other protons around. In a nucleus this is not true. In order for one of the neutrons in a carbon atom to decay, there has to be room in the nucleus for the proton that would result from the decay. In carbon, and in most other nuclei, however, all of the places that the decay proton could fit into are already filled by other protons. (The details of how this works depends on something called the Pauli exclusion principle, which we will discuss in detail in Chapter XII.) The net result of this fact is that neutrons in most nuclei cannot decay, simply because there is no place for the decay products to go. To use an everyday analogy, the proton "parking lot" in the nucleus is filled, so the neutron cannot "park" its decay proton. Thus, most common nuclei are stable.

There are, however, nuclei where the neutron can and does decay. Such nuclei are unstable. When this happens, the nucleus emits an electron and acquires a unit of positive charge. The emitted electrons from such nuclei were one of the original types of radiation detected. They were called beta rays (it took some time to identify them as electrons); hence, this sort of reaction is called beta decay. The electron emitted in such a process has a great deal of energy and therefore travels a long way from its parent atom.

After beta decay we are left with an atom that has one more proton in its nucleus than it did before the decay. This means that the new atom has a net positive charge (since what happens in the nucleus does not immediately affect the electrons in their orbits). Usually such an atom will pick up a stray electron from the surroundings, becoming electrically neutral in the process. The net result of all this is a new atom, one which has one more proton and one more electron than its predecessor. Thus, the beta decay of a nucleus is one way in which the alchemist's dream of transmutation of the elements can be realized.

THE NEUTRINO AND THE WEAK INTERACTION

There is one aspect of neutron decay that needs closer attention. When the decay occurs, two charged particles are created—a proton and an electron. Thus, the total electrical charge of the final decay products is zero, which is exactly the charge on the neutron. This means that in the beta decay of the neutron the electrical charge is the same before and after the decay. We say that the electrical charge is *conserved* in the reaction.

This is an important point, because there are quantities that are not conserved. The number of particles, for example, changes from one (the neutron) to more than one in the reaction. Particle number is a quantity that is not conserved. Neutron decay is just one example of a general law that holds in all known elementary particle interactions.

In every reaction involving elementary particles, the total electrical charge is the same before and after the reaction. This is called the Law of the Conservation of Electrical Charge.

Conservation laws play an extremely important role in physics, so it was natural to ask whether other well-known conservation laws hold in beta decay. For example, there are laws that tell us that the energy of a system has to be the same before and after every reaction,* and other laws that tell us the same thing about momentum.

If E_0 is the total energy of the neutron before it decays, and if the beta decay is really described by the reaction $n \rightarrow p + e$, then it follows that if energy is conserved, the final energies of the proton and electron must add up to E_0. From this, it follows that if we look at two decays in which the protons have the same energy, the energies of the two electrons will also have to be equal.

It turns out that they are not!

If we looked at a large number of decays in which the protons all have the same energy, which we can call E_p, and plotted the number of times we saw an electron of a given energy as a function of energy, we would get something like the graph shown in Illustration 4. The electron energies would have all values from some minimum value E_{min} (which would correspond to the electron hardly moving at all) to the maximum value allowed by energy conservation, $E_0 - E_p$. The number of electrons at this latter value—which is what the law of conservation of energy says all of the electron energy ought to be—is actually zero.

What to do? There are only two ways of approaching a problem like this. One is to give up the law of conservation of energy —something physicists were very reluctant to do because the law seems to apply everywhere else in nature. The other alternative is to assume that there is another particle involved in the interaction—a particle that for some reason is not detected but which carries away the missing energy. In 1934 the Italian physicist

---

* The reader who is unfamiliar with the concept of energy-mass equivalence will find it discussed in Chapter IV.

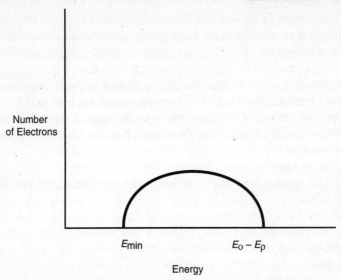

4. The number of electrons at each possible energy in beta decay.

Enrico Fermi (the man who later built the first nuclear reactor at the University of Chicago) put together the first successful theory of beta decay along these lines. Following some previous theoretical speculations, he supposed that beta decay is actually described by the reaction $n \to p + e + \nu$, with the Greek letter ν (nu) standing for the new hypothetical particle. This particle must be electrically neutral (otherwise, it would have been detected earlier); hence, it was given the name *neutrino* (little neutral one). If such a particle existed, then it could carry away the missing energy from a beta decay and all of the accepted conservation laws of physics would be maintained intact.

Searching for the neutrino to verify its existence, however, turns out to be difficult. Because it is not electrically charged, it does not affect the electrons of atoms as charged particles do, so detectors such as the Geiger counter will not show its presence. Unlike the neutron, it is unlikely to cause visible recoils when it comes near a nucleus. For one thing, like the photon, it is without mass. More important, it simply does not interact very

readily with other matter. One way of gauging this lack of interaction is to ask how thick a lead plate would have to be in order
for a neutrino to stand a good chance of interacting with at least
one lead atom. For the neutrino that results from Fermi's calculations, the answer comes out not in meters, not even in kilometers, but in *light-years!* In other words, if a tube of solid lead
stretched from Earth to the nearest star and we started a neutrino down that tube today, it would emerge at Alpha Centauri
more than 4 years from now without having disturbed a single
atom in the tube.

This property of the neutrino posed certain difficulties for
physicists who wanted to detect its presence. As we have seen,
the only way that any particle can be detected is for it to interact
with something and cause a change. For a charged particle, the
change may involve creating some free electrons. For a neutron,
it may involve creating a recoiling heavy ion. But no matter what
the change is, there must be some sort of interaction taking
place if the particle is to be observed.

If the neutrino is really as reluctant to interact with other
matter as our examples indicate, then the only way it can be
"seen" is for a large number of neutrinos to pass through a
detector, so that the small probability of interaction is counter-
acted by the large number of neutrinos. This, in fact, is how the
neutrino was first detected in 1956—almost 20 years after Fermi
first worked out the theory of beta decay. The source of the
"flood" of neutrinos was a nuclear reactor (the experimental
details will be given later). The neutrino is now fully accepted as
a particle, and, in fact, is routinely produced for experimental
purposes in many large accelerator laboratories around the
world.

Thus, the introduction of the neutrino saves some cherished
principles of physics while enlarging the roster of elementary
particles. Reactions such as beta decay, in which the neutrino
participates, typically occur on a very slow time scale. They are,
therefore, usually called *weak interactions* to distinguish them
from the strong nuclear interactions, which presumably occur in
much shorter times. The study of weak interactions has played

an important role in the development of our knowledge of elementary particles.

A REMARK ABOUT STABILITY

With the neutron we have entered a system where particles do not retain their identity, but decay into other particles after a time. *Every new particle we discuss from this point on will be unstable.* In fact, if we look at the massive particles, only the electron and the proton can exist by themselves for indefinite periods of time. Ultimately, therefore, every other particle must decay into some combination of photons, neutrinos, protons, and electrons.

Of course, like all statements of scientific fact, this statement is only as good as the observations that have been made to back it up. The statement that no one has yet seen a proton decay does not necessarily imply that no one ever will. All we can really talk about are those experiments that have been performed to search for the decaying proton and see what sorts of limits we can set on our knowledge.

One way of doing this is to ask how large the mean lifetime of the proton would have to be in order to be consistent with present experiments. Since there are a lot of protons around, a long mean lifetime (and the consequent small probability that a proton will decay while being watched) can be compensated for by watching a large number of protons. Right now, the accepted limit on the proton stability is this: The mean lifetime of the proton must be greater than 10^{32} years. From our discussion on neutrons, we know that this means that if we started out today with 1000 protons, only 367 of these would be left 10^{32} years from now.

To get some idea of the time scale that would have to be involved in the decay of the proton, recall that the earth itself is about 5 billion years old (5×10^9), while the age of the universe is reckoned to be a bit more than 10^{10} years. This means that even if we started with 1000 protons at the time of the formation of the universe 10^{10} years ago, not even one would have decayed by now! So when we say that the proton is stable what we mean is that its lifetime is at least many orders of magnitude greater than the lifetime of the universe, so that for all practical purposes the question of absolute stability is irrelevant.

By contrast, there are many naturally occurring radioactive substances with lifetimes of a billion years, the most familiar being uranium. Lifetimes of this length can easily be measured with modern techniques. Furthermore, unlike the proton and electron, decays of these atoms are seen in nature.

For reference, the lifetime of the electron is known to be greater than 10^{21} years, and, like the proton, no decay of the electron has ever been seen.

PROVISIONAL BOX SCORE

With the discovery of the neutron, the Bohr-Rutherford picture of the atom seemed fairly complete. There were two elementary particles—the nucleons—which formed the nucleus of the atom. Around the nucleus a third elementary particle—the electron—orbited. These three particles taken together can be thought of as the fundamental building blocks of matter that the Greeks had hinted at and whose discovery we described as one of the basic goals of science. As we pointed out in Chapter I, a fourth particle—the photon—can be thought of as the constituent of radiation. As such, it is not actually a building block in the sense that a brick is a building block of a cathedral, but it is certainly a particle that is present in nature and that must be included in our list of what is elementary.

The unusual properties of the neutron—specifically its decay —led to the introduction (and eventual verification) of a fifth particle, the neutrino. Like the photon, it is without mass, and is not a building block in the structural sense of the word. Nevertheless, it is intimately tied both to the weak interaction and to the decay process of the neutron (which *is* a building block), so it must be added to the roster. The list of elementary particles is now up to five.

While the experimental investigations that led to the five particles were being carried out, the 1920s also saw the development of a theoretical framework to describe the physics of the atom. The new science was called quantum mechanics (in analogy to the study of the motion of macroscopic objects, which is called classical mechanics). The concepts that arise from this new theory are now pretty well understood and accepted in the scientific world, but they are sufficiently unusual for us to devote Chapter III to them.

P stable

e stable

eak interactions: N → 15 min Life → P + e + neutrino

Photon

R → Lifetime?

Strong Force: holds nucleus of more than 1 Proton together, despite its natural repulsion.

III

A New Physics
for a New World

No phenomenon is a true phenomenon until it is an observed phenomenon.
—JOHN A. WHEELER, theoretical physicist

A man has to see, and not just look.
—LOUIS L'AMOUR, *The Quick and the Dead*

THE PHYSICS OF THE ATOM

SUPPOSE you were a Martian who liked to study languages and that after a long and difficult struggle you had managed to master German and French. Suppose that you had worked so long and so hard on these languages that you had come to believe that they were the only two languages that were spoken on Earth. Finally, suppose that because of your skill in language you were chosen to be a member of the first Martian landing expedition to Earth, and that your rocket ship put you down in the middle of North America. What would happen?

The first thing you would discover, of course, would be that the natives with whom you came into contact spoke a language that was very puzzling. Some of the words in the language (such as *hand*) would be exactly the same as German, and some of the other words (such as *cinema*) would be exactly the same as French. If you insisted that French and German were the only possible languages, you would have to conclude that the language you were hearing was a paradox, since it sometimes behaved like one and sometimes like the other. Philosophers back on Mars might even propound learnedly on the problem of "French-German duality" and ask whether the Martian mind was capable of understanding the true meaning of Earth's languages.

In this example, it is easy to see where the problem arises— our Martian friend has made the wrong starting assumption. There is no reason why a language has to be either German or French. Once she recognizes this fact, she will quickly see that English is a separate language with its own rules and vocabulary, but which happens to have similarities with both French and German.

In many ways, the map of the physical world is similar to a language map of the earth. Just as the languages spoken in different parts of the earth are different, so too are the laws that govern different aspects of physics. In Chapter II we saw how much smaller the atom and its constituents are than anything we encounter in everyday life. It should therefore come as no surprise that the behavior of subatomic particles is different from the behavior of particles in the larger world. In fact, assuming that circumstances in the atom have to be as they are in large objects with which we are familiar is something like the mistake our Martian linguist made when she assumed that every language spoken on Earth had to be one that she knew.

When physicists in the 1920s began to make a serious study of elementary particles, they were familiar with two sorts of things from their work with large-scale objects. They called these things by the names *particles* and *waves*. Particles are similar to baseballs; they are located at a specific region in space, they can move from one region to another, and they can be described in terms of their position and their velocity at any moment. Waves are also familiar to us (think of surf coming into a beach). They move from place to place, like particles, but they are not located at a specific point. The crest of a wave may move along with a given velocity, but the wave itself is spread over the extended region between successive crests. To describe a wave we therefore have to specify its velocity and the distance between crests, or wavelengths. As you might expect from the differences in their natures, particles and waves in the large-scale world behave differently from each other.

With this sort of background, it is not surprising that when physicists began to investigate elementary particles they did not ask "What are they?" but "Are they particles or waves?" In this sense, they were like the Martian linguist who assumed that everything she encountered had to be like the things she already knew. And just as the Martian soon found something that did not fit into her conceptual scheme, so too did the first particle physicists. They found that electrons, which normally exhibit the kind of localization associated with particles, could also behave like waves under certain conditions. Similarly, light, which normally displays the behavior of waves, would start to look like particles. In the end, nothing in the subatomic world looked exactly like a particle or exactly like a wave, and this state of things was profoundly disturbing to the physicists. They coined the term *wave-particle duality* to express this feature of the objects they were studying, and philosophers picked up this term and used it to "prove" that there are inherent limits to what the scientific method can uncover.

But in terms of our Martian analogy, we recognize that wave-particle duality does not arise because of anything paradoxical about the behavior of elementary particles, but simply from the fact that we have asked the wrong question. If we had asked "How does an elementary particle behave?" instead of asking "Does it behave like a particle or a wave?", we would have been able to give a perfectly sensible answer. An elementary particle is not a particle in the sense that a bullet is, and it is not a wave like the surf. It exhibits some properties that we normally associate with each of these kinds of things, but it is an entirely new kind of phenomenon. In this sense, it is like English in our Martian analog—it has aspects in common with familiar things, but it is neither of these things. It is something new.

Of course, it is easy for us, looking back to the 1920s, to see how pointless all the debate about the wave-particle duality was. We should also realize that taking this attitude is a sort of Monday morning quarterbacking that detracts nothing from the achievement of the men who first unraveled the laws of the sub-atomic world. And lest we start feeling too cocky, remember that 50 years from now someone may be pointing out how pointless some of *our* great debates are.

Once we have understood that the electron is not a particle and not a wave, we should ask "What is it?" A full answer to this question would take a long discussion of quantum mechanics, which we lack the space to present here, so I shall summarize some of the results. When physicists say that something is a wave, what they really mean is that the thing's behavior can be predicted according to a particular equation called, appropriately enough, the wave equation. For a wave on water, for example, this equation will predict all of the relevant properties of the wave—how fast it will move, how high above the normal surface the water will be at any particular time, and the shape of the wave. As the wave moves by a point in the water, the water rises, finally reaching the full height of the wave. This is called the amplitude of the wave. It is one number that characterizes the wave.

Another quantity that characterizes the wave is its shape. One way of indicating the shape of the wave is to say how high the water is at a given time for each point along the wave. The height of the wave at an arbitrary point is called the displacement, and it follows that the maximum value of the displacement is the amplitude of the wave.

The wave equation is a relation written in the language of differential calculus. It tells us the connection between the displacement of the wave, the time, and the properties of the medium on which the wave moves. Since there is no reason why a wave on water should have the same properties as a similarly shaped wave in alcohol, it is the properties of the medium that ultimately determine the speed of the wave. But the end result of the mathematical analysis of the waves is that if you give a physicist the shape of a wave at one point in time, he or she can then predict where that wave will be (and what its shape will be) at any future time.

It was therefore only natural that when evidence started to accumulate, which indicated that the electron had some properties normally associated with waves, someone would try to describe the electron by the wave equation. In 1926 Erwin Schrödinger, an Austrian theoretical physicist, wrote down such an equation, calling the displacement of the electron "wave" the *wave function* and denoting it by the Greek letter ψ (psi). Even though he did not know what the wave function signified in terms of the properties of the electron, Schrödinger found that he was able to use his equation to solve many of the outstanding scientific problems of the time—the structure of the hydrogen atom, for example. Thus, the statement that the electron exhibits properties normally associated with waves was transformed by Schrödinger into a precise mathematical description of the electron "wave."

But what is the wave function? Some physicists at the time wanted to interpret the displacement of the electron wave as being real in the sense that the "true" electron was spread out and the displacement measured how much of the electron was at each point in space. It remained for Niels Bohr in Copenhagen to provide what is now the accepted interpretation of the wave function. He reasoned that there was too much evidence for the particlelike properties of the electron to allow it to be smeared out in a classical wave. The electron, he said, should still be thought of as a localized object, but the displacement of Schrödinger's electron wave at a particular point is related mathematically to the *probability* that a measurement would show the electron to be located at that point. In this interpretation the Schrödinger equation predicts the properties of a *probability wave,* and with it we can predict the probability that an electron will be at a certain point if we know the wave function.

As the probability wave moves past a particular point, the probability of finding the electron at that point will change. We can calculate how much it will be at any instant by using Schrödinger's equation to find the wave function, and then getting the probability from the relation $P = \psi^2$.

A wave that is bunched up and confined to a specific region of space, like a tidal wave, is called a wave packet. The description of most elementary particles is in terms of this sort of wave. In this picture, it is clear that the particle has both particle and wavelike properties, since the wave packet is confined to a relatively small region of space (a property normally associated with particles), but the packet itself is a wave. In addition, the picture allows for the fact that an actual measurement will show a pointlike particle somewhere in the region of the packet.

But isn't this whole interpretation just saying that the electron is really a pointlike classical particle, and isn't the probability interpretation just a statement about the fact that we simply do not have good enough instruments to detect the electron's actual location? It is tempting to think that such an objection might restore a familiar view of the subatomic world, but there are some problems with making measurements on the subatomic level that lead to even stranger results than the idea of a probability wave. It is to these problems that we now turn.

THE UNCERTAINTY PRINCIPLE

Physics is a science that, in the end, rests on our ability to observe and to measure the world around us. The plain old nuts and bolts of measuring a quantity accurately have often made crucial differences in the development of science. For example, in 1572 the Danish astronomer Tycho Brahe, using newly designed instruments that were more accurate than any that had been used up to that time, showed that a nova seen in the sky was actually as far away as the other stars, thereby exploding the Aristotelean idea that the heavens, being perfect, could not change. The concept of measurement also occupied a large place in the thinking of the people who developed the quantum mechanics.

We must make one distinction very clear—the distinction between the measurement itself and the accuracy with which the measurement is made. For example, suppose you were going to measure your height. You could do it by standing up against a wall, having someone mark off your height on the wall, and then measuring the height of the mark on the wall. If you carried out this procedure, you would get a number—6 feet, for example—that would represent the result of the measurement.

How accurate would this result be? One way of finding out would be to do the measurement again. If you did this, you might be slouching a little more the second time, or your helper might make a slightly different estimate of the spot on the wall opposite the top of your head. There are many things in the measuring procedure that might be different the second time around, so you would expect the result you get to be a little different, too. You might, for example, find your height to be 6 feet ⅛ inch. A third try might give 5 feet 11¾ inches, and a fourth try 6 feet again. If you made a graph by plotting the number of times you measured a certain height against the height measured, you would get something like what is pictured in the left portion of Illustration 5.

A bar graph of this type is called a histogram, and this particular histogram shows two things. First, it shows that the measurements all cluster around 6 feet, and, second, that 6 feet is the average of all of the measurements. Thus, you would report that your height was 6 feet even.

But the histogram also shows that there is a scatter in the measurements; that they vary from one trial to the next. The

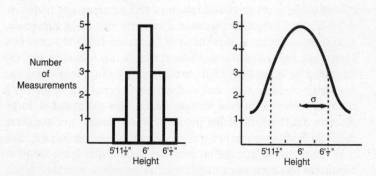

5. Results of measurement of height discussed in the text. The histogram on the left turns into the curve on the right when many readings are taken.

spread in these readings gives us some idea of how accurately the measurements were made. In the histogram in the illustration, for example, there are values of the height obtained that are $\frac{1}{2}$ inch above and $\frac{1}{2}$ inch below the average, with most readings being within $\frac{1}{4}$ inch of the average. From this you can conclude that your height is not 5 feet 11 inches or 6 feet 1 inch, that it probably is not 5 feet $11\frac{1}{2}$ inches or 6 feet $\frac{1}{2}$ inch, but that it might be 5 feet $11\frac{7}{8}$ inches or 6 feet $\frac{1}{8}$ inch. The spread in the measurements shown on the histogram indicates something about the certainty with which we can claim our average measurement is correct or, equivalently, what the uncertainty of the measurement is.

We can give a more precise definition of uncertainty by imagining our taking a very large number of height measurements. In that case, the histogram would develop into the continuous curve shown on the right in Illustration 5. This curve is called a normal or Gaussian distribution of measurement results. The quantity we have labeled by the Greek letter σ (sigma) on the graph is called the standard deviation. It is defined as that distance from the center of the curve within which 68 percent of the measurements fall. Hence, if the graph represents 1000 height measurements, then 680 measurements will be between $+\sigma$ and $-\sigma$. In this example, $\sigma \cong \frac{1}{4}$ inch.

The standard deviation of a large number of readings can therefore be used as an indication of the accuracy of the experiment and, by the same token, as a measure of the uncertainty of our knowledge of the true value of the quantity being measured. In the example of measuring height that we have been using, the existence of uncertainty is due to a number of factors that can vary from one measurement to the next—your posture, the judgment of your helper, how well you can read a ruler, and so forth. There would be little difficulty, however, in designing the experiment so that the uncertainty would be smaller than $\frac{1}{4}$ inch. You could, for example, take all measurement while lying down to eliminate the posture problem, and use accurate levels to locate the top of your head. There is no reason, in principle, why you could not measure your height to $\frac{1}{100}$ of an inch or even $\frac{1}{1000}$ of an inch. In fact, in the ordinary world of classical physics, there is no reason why the uncertainty of this measurement could not be zero. The only limitation has to do with how expensive and accurate your instruments are.

The second important point about the measurement of your height is that you will be the same person after the measurement as you were before. What I mean is this: If you wanted to know your weight in addition to your height, you could, in principle, know it with zero uncertainty as well. Furthermore, the fact that you had measured your height would not affect the measurement of weight. In the language we have been developing, in the classical world it is possible to measure both height and weight with zero uncertainty.

In the world of the atom the situation is different in one important respect. While it is still possible, in principle, to measure a quantity with zero uncertainty, the process of carrying out the measurement may make it impossible even in principle to measure another quantity to the same degree of accuracy. It is as if measuring your height to high accuracy meant that you could measure your weight only to a very low accuracy or vice versa. Let us see how this rather surprising state of affairs comes about.

When we talk about observing or measuring something in the macroscopic world, we always have in mind (at least implicitly) some method by which we can cause something to interact with the measured object. In the height measurement, for example, your helper either looked at the top of your head to see where to mark the wall, in which case he was seeing light that had interacted with you, or he laid a marker directly on your head. The entire discussion of height rested implicitly on the assumption that the interaction with the measured object did not change it in any important way—that the act of observation itself could be neglected. And, of course, when we talk about macroscopic objects, this is a perfectly reasonable assumption. No one would seriously suppose that looking at someone's head changes his height. The idea is too ridiculous to take seriously.

Why?

If pressed for an answer, you would probably say something like this: The photons that make up the light are so small and have so little energy that there is no possible way in which they could affect a large object, such as a human being. This is precisely the way that classical physicists have always justified the assumption that the observer does not affect the observed, and the assumption is reasonable when what is being observed is huge compared to what is interacting with it.

This argument will not work for elementary particles! In order to "look" at an electron, we must cause it to interact with a photon or some other particle. In either case, we can no longer argue that the interaction can be ignored, since the object being measured and the probe we use upon it are both of the same size and energy. This simple fact is the physical basis for one of the most celebrated laws of quantum mechanics, the uncertainty principle. Discovered in 1927 by Werner Heisenberg, this law states that there are situations in the subatomic world where it is not possible, even in principle, to know the values of two different quantities relating to an elementary particle because the act of measuring the first interferes with our ability to measure the second. Before giving a precise definition of the uncertainty principle, we can illustrate how it might work in our height-weight analogy.

Suppose we wanted to know someone's height and weight, but for some reason the only way we could measure height was by the following bizarre technique: The person to be measured would lie down, and then someone would drop small weights near him from above. When the weights fell straight down past the person, we would know we were in the region above his head, and when they were deflected we would know that we were below his head. By looking at the boundary between these two situations, we would have a measure of his height. (Although this probably would not be a good way to measure height, it is, in fact, quite close to the way in which the sizes of atoms and nuclei are measured.)

In the process of dropping weights, some weights would stay on the subject and some would slide off; therefore, when we finished we would not know how many weights had stuck. If we now weighed the subject, we could surely measure a weight, but this would be the man's weight plus the weight added by the measurement of height. In this example we would therefore say that measuring the man's height caused an uncertainty in the measurement of his weight, and that the processes of height and weight measurement interfered with each other. I leave it to you to think of a way of measuring weight that would interfere with subsequent measurements of height.

The uncertainty principle is stated as follows: If we call Δx the uncertainty in the position of an object and Δp the uncertainty in its momentum (for our purposes, momentum is simply the mass of an object multiplied by its velocity), then in any attempt to measure these two quantities, the product of the uncertainties is given by $\Delta x \cdot \Delta p > h,$ where the symbol $>$ means greater than, and h is Planck's constant. In units where mass is measured in grams and length in centimeters, h has the value $h = 6.62 \times 10^{-27}$.

There are a number of things to note about this principle. In the first place, Planck's constant is a very small number; thus, when we are talking about macroscopic objects, the limits that the uncertainty principle places on our measurements are very small indeed. For example, if we have a 300-gram object, such as an apple, and we can determine its position to within one millionth (10^{-6}) centimeter, then the uncertainty principle states that the error in determining the velocity must be greater than

$$\Delta p > \frac{h}{10^{-6}} = 6.62 \times 10^{-21}$$

$$\Delta V > \frac{6.62 \times 10^{-21}}{300} = 2.2 \times 10^{-23} \text{ cm/sec}$$

Saying that the uncertainty in the measurement of velocity must be larger than this number makes very little difference in the macroscopic world, since any real measurement would produce uncertainty of the ordinary type which would certainly be many orders of magnitude larger. Hence, the smallness of Planck's constant provides a kind of justification for ignoring the interaction of observer and observed in the macroscopic world.

If the object we were considering were a proton, however, and the uncertainty in position were 10^{-8} centimeter (you will recall that this is about the size of the atom), then the uncertainty in velocity would be

$$\Delta V > \frac{h}{m \Delta x} = \frac{6.62 \times 10^{-27}}{1.7 \times 10^{-32}} = 3.9 \times 10^5 \text{ cm/sec}$$

where we have used the result, from Chapter II, that the proton's mass is 1.7×10^{-24} gram. In this case, the uncertainty in the velocity is quite large, probably as large as the proton's velocity itself. Thus, on the quantum level, the uncertainty principle begins to be quite important.

Perhaps the best way to discuss what the uncertainty principle implies is to list some of the things that it does not imply.

1. *It does not imply that the particle's position cannot be measured exactly.*

You often run across the statement that the uncertainty principle proves that it is impossible to measure the position or momentum of a particle exactly, and from this statement innumerable incorrect conclusions are drawn. A quick look at the equation that defines the uncertainty principle shows that this is simply not the case. Measuring the position of a particle exactly would imply that the uncertainty in the position, Δx, would be zero. There is nothing in the uncertainty principle that says that Δx cannot be zero. All it says is that if $\Delta x = 0$, then Δp, the uncertainty in the momentum, must be infinite. If this were the case, then it would still be possible for the product of Δx and Δp to be greater than h. What this possibility implies is that if we decide that we want to know the position of a particle exactly, the measurement that we have to make to get this information will so alter the particle that we will be unable to get any information at all about its momentum. The resulting lack of a precise answer means that the momentum could be any number at all, which is just another way of saying that the uncertainty in the momentum is infinite.

2. *It does not imply that the particle "really" has a position and momentum, but that we just cannot measure them both.*

The uncertainty principle does such violence to our common sense feelings about what the world should be like that there is a strong temptation to think of it as a kind of technical limitation on what we can measure, rather than as a statement about a fundamental difference between the macroscopic and microscopic worlds. One therefore wants to think of the particle as if it "really" had a position and momentum (just as a billiard ball does), and to dismiss the uncertainty principle as a sort of pettifogging technical detail.

However, you have to ask what meaning you can give to a statement such as "The particle really has a precise position and momentum at any time, but we just cannot measure them both." There is no experiment one can imagine that could test this statement, because, as we have seen, measuring one variable puts a limit on how accurately we can determine the other. I would suggest that making a statement that cannot be verified by experiment, even in principle, is not a particularly useful way to do science. Whether or not the particle really has precisely defined values of position and momentum can make no difference in anything we can know about the subatomic world, so this sort of conjecture is in a class of statements that are not even right or wrong—they are just meaningless.

3. *If we could find some probe whose energy was very small compared to elementary particles, then the uncertainty principle would go away.*

This idea is known as the hidden variable theory of quantum mechanics. From our discussion of the physical basis of the uncertainty principle it seems fairly clear that if a new regime of sub-subatomic particles were discovered which could interact with the electron (for example) in the same way that light interacts with macroscopic bodies, then measurements made with these new particles would not disturb the electron and we could get around the uncertainty principle.

This is true, but people who work out the consequences of hidden variable ideas have been able to prove that if such particles existed, there are certain experiments in which ordinary quantum mechanics and hidden variable theories would predict different results. The experiments that have been done bear out quantum mechanics; so, while it is always dangerous to say that something in physics is impossible, it seems to me that it is very unlikely that anything resembling a hidden variable theory will eventually undo the uncertainty principle.

VIRTUAL PARTICLES
AND THE STRONG INTERACTION

In the previous section we talked about the uncertainty principle in terms of two variables—position and momentum—because these are quantities that are familiar to us all. It turns out that the uncertainty principle applies to a number of other pairs of variables as well. In the development of the ideas of elementary particles, the most important of these variables are energy and time. If we denote by ΔE the uncertainty in our knowledge of the energy of a quantum system and by Δt our uncertainty about the time at which it has a given energy, then reasoning that is precisely like that leading to the position-momentum relation leads to the equation $\Delta E. \ \Delta t > h$, where, again, h is Planck's constant. All of the remarks about the implications of the uncertainty principle made in the last section apply to this relation as well.

To interpret this type of uncertainty relation, let us think for a minute about what is involved in measuring the energy of a system. As in any other measurement, we can only observe the system by using some sort of probe. In order to find the system's energy, this observation will take a certain amount of time (at the very least, it will involve the time it takes for the probe to interact with the system). We can interpret Δt, the time uncertainty, as being the time it takes to make the observation, since obviously we are totally unable to answer questions about the energy in any lesser time scale.

Suppose, for example, that we were measuring a system in which it took our instruments 1 minute to take a measurement of the energy. We could then report this energy as our result, but if someone asked us whether our reported energy was the actual energy of the system at 10 seconds or at 50 seconds, we could not say. All we could say would be that during that minute, the average energy was as we reported it. It is even possible that the actual energy of the system changed during the minute, so that sometimes it was higher than the average and sometimes lower. In this case, the fact that the time determination was uncertain would lead to uncertainties in the energy as well. In classical macroscopic systems, of course, both the time and the energy uncertainties can be reduced simultaneously to zero (at least in principle).

In the quantum world, however, that reduction is not possible. If the probe's interaction is instantaneous (so that you know the time exactly and $\Delta t = 0$), the interaction is not long enough to give any indication of the energy, so $\Delta E = \infty$. If we observe for a long time in order to measure the energy exactly, then $\Delta t = \infty$ and $\Delta E = 0$. As with momentum and position, you can measure one or the other exactly, or both with some uncertainty, but you cannot get them both exactly.

The energy-time uncertainty relation leads to a very interesting concept in physics—the concept of the virtual particle. In Chapter IV we shall discuss the implications of the famous relation between energy and mass first written down by Einstein, $E = mc^2$. For our discussion here, we only have to understand that the energy associated with the mass of a particle can be included in the uncertainty relation as well as any other energy. In particular, if a particle of mass M is sitting by itself and we measure its energy in a time Δt, there is an uncertainty in the mass of the system of the amount

$$\Delta M > \frac{h}{c^2 \Delta t}$$

If Δt is small enough, it is even possible that the uncertainty in the mass may be large enough so that during the time Δt we cannot tell whether there is a single particle of mass M or a set of particles of total mass $M + \Delta M$ sitting at a particular point in space. In other words, we could have a sequence such as the one pictured in Illustration 6, on page 49, where a single particle becomes, for an instant, a pair of particles. No measurement we can make will tell us that this is or is not happening. We say that the original particle "fluctuates" into two particles in this process, and we call the extra object a *virtual particle*.

To get some idea of the times we are talking about for this

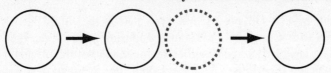

6. A particle (*left*) fluctuates to itself plus a virtual particle (*center*) and back to itself again.

fluctuation process, we can ask how small Δt would have to be in order for a single proton to fluctuate into a proton and an additional particle that has the same mass as the proton. For this sort of fluctuation, the mass uncertainty is

$$\Delta M = M_p = 1.7 \times 10^{-24} \text{ g}$$

and, recalling that the speed of light is $c = 3 \times 10^{10}$ centimeters/second, we find

$$\Delta t > \frac{6.6 \times 10^{-27}}{9 \times 10^{20} \times 1.7 \times 10^{-24}} = 4.3 \times 10^{-24} \text{ sec}$$

The speed at which the virtual proton can travel is limited by the velocity of light, and it might be interesting to know how far it could travel at that speed in the time Δt. We find

$$d = c\Delta t = 3 \times 10^{10} \times 4.3 \times 10^{-24} = 1.3 \times 10^{-13} \text{ cm}$$

—which is about the size of the proton itself.

This calculation has a very interesting interpretation. One way of visualizing the appearance of a virtual particle is to think of it as "sneaking out" while no one is looking. So long as it gets back "home" before a time Δt has elapsed, the uncertainty principle guarantees that no one will know the difference. In a sense, the principle plays the same role in this process as the clock played in the Cinderella story—so long as the particle gets home from the "ball" before the time runs out, it will not turn into the subatomic equivalent of a pumpkin.

The distance that the virtual particle can travel in its alloted time is also an interesting quantity. Suppose that we had two ordinary particles sitting a distance d_v apart. We could then have a situation in which one particle fluctuates into itself plus a virtual particle, and, provided that the mass of the virtual particle is such that it can travel a distance d_v during its life-span, the virtual particle could be absorbed by the second of the original two particles. (See Illus. 7.) This is called the exchange of a virtual particle, and, again, the uncertainty principle tells us that such processes can go on without our being able to detect any violation of the law of conservation of energy.

From our example, we know that a virtual particle whose mass is the same as that of the proton can travel about 10^{-14} centimeter, even if it moves at the speed of light. If, on the other hand, the virtual particle has a mass that is a fraction of that of the proton, it can travel several times this distance. This means that a virtual particle of such a mass can be exchanged between protons that are several fermis apart, and from Chapter II, we know that several fermis is about the size of an average nucleus.

In 1934, the Japanese physicist Hideki Yukawa, working at the Osaka Imperial University, wrote a landmark paper that showed that if two protons exchanged virtual particles, the result of the exchange would be an attractive force between the protons. Furthermore, he showed that if the virtual particles had a mass something like $1/9$ of the proton mass (the exact fraction was unimportant), then the result would be a force strong enough to overcome the electromagnetic repulsion between the protons and would, in fact, tend to hold them together. In other words, the exchange of such a virtual particle could generate the strong force needed to hold the nucleus together.

To understand how an exchange of particles could result in a force, consider what would happen during the time the virtual particle is in transit if it exerted an attractive force on the two

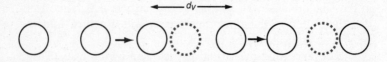

7. How two particles can exchange a virtual particle and still not violate energy conservation.

nucleons. During the brief moment of existence of the virtual particle, each nucleon would be pulled toward it. The net result would be that each nucleon would be attracted toward the other, prompting an outside observer who could not see the virtual particle exchange to say that there was an attractive force between the nucleons. In Yukawa's hypothesis, the new particle had precisely the properties needed to produce such an effect.

Of course, no particles with mass less than the proton and more than the electron were known at that time, but Yukawa suggested that it might be a good idea to look for what he called the *U* quantum. His hypothetical particle was later termed the *meson* (intermediate one) because of its mass. Although the strange story of the search for this particle will be told in a later chapter, let us spend a minute on the implications of this new idea of the strong force.

In the first place, the nucleus is now seen not as a static collection of protons and neutrons locked into place like some sort of Tinker Toy construction, but as a dynamic system in which mesons are whizzing around from particle to particle. They are continuously being created at one spot and absorbed at another, and this process is what holds the nucleus together. In a sense, the mesons form the "nuclear glue" that we saw was needed in order to understand nuclear structure.

Since both protons and neutrons can emit and absorb mesons, it is easy to see the role that the neutrons play in the nucleus. They have no charge, so they do not contribute to the electromagnetic forces that are pushing the nucleus apart; but they can contribute to the forces binding it together. Thus, they play the dual role of diluting the repulsive forces and enhancing the attractive ones. It should therefore come as no surprise to learn that most stable nuclei have as many, or more, neutrons as protons.

Perhaps more important than the explanation of the strong interaction given by Yukawa is the idea that we can think of a force as being due to the exchange of particles. Although we have discussed this idea only for the strong interaction, it should not be too hard to accept that modern physicists think of *all* forces as being due, ultimately, to the exchange of particles. The electrical force, for example, is thought to arise because of the exchange of virtual photons; the gravitational from the exchange of particles called gravitons (no gravitons have been detected as of this writing). So the principles of quantum mechanics as developed by Yukawa have led us to a whole new way of looking at forces in nature—a way that is intimately tied to the structure of matter and to elementary particles.

Finally, lest the reader leave this discussion with the impression that the whole thing is some sort of elaborate shell game, I should point out that modern nuclear physics experiments provide strong (albeit indirect) evidence for the existence of virtual particles. When high-energy electrons are allowed to collide with nuclei, the results of the collisions seem to indicate that although the electrons usually encounter nucleons, one of them occasionally happens to be in the right place at the right time to hit one of the virtual particles during its transit. In this sense, the virtual particles are really there.

IV

Energy, Matter, and Antimatter

Things equal to the same thing are equal to each other.
—EUCLID, *Elements*

MASS AND ENERGY

ALTHOUGH the results of the theory of relativity are many, one of them has become so familiar that it could almost be classified as a piece of folklore: It is the famous relation between energy and mass, $E = mc^2$. We have already seen that this equation has far-reaching consequences in the world of elementary particles, leading ultimately to the idea of virtual particles and the concept of force as a result of the exchange of such objects. In this chapter we will see that this equation also leads to another consequence for elementary particles—the existence of antimatter. But before we go on to that subject, we should take a moment to discuss the most common question that people have about the Einstein equation: "Why is it the speed of light figures in this equation and not something else, such as the speed of sound?"

Actually, this question goes right to the heart of the theory of relativity, which is where the Einstein equation comes from. The basic postulate of the theory says that the laws of physics have to appear to be identical to every observer, whether that observer is moving or not. Thus, relativity places a heavy emphasis on the laws of physics, and anything that is built into these laws will occupy a special place in the theory.

The laws that govern the fields of electricity and magnetism are called Maxwell's equations, after James Clerk Maxwell, the British physicist who first wrote them down in 1873. It turns out that these equations predict that there will be waves that can move through the vacuum, and that the speed of these waves is related to measurable quantities, such as the force between electrical charges and the forces exerted by magnets on each other. What led Maxwell to identify these new waves with light was the fact that when he took the measured forces and put the appropriate numbers into his theoretical value of the wave velocity, it came out to be 3×10^{10} centimeters/second, or 186,000 miles/second, which is precisely the measured velocity of light. In this sense, the speed of light plays a special role in physics because it is built into the laws of electricity and magnetism, whereas no other speed plays a similar role in any other branch of physics. Basically, this is the fact that leads to the c^2 term in the mass-energy relationship.

I must make one other point before we leave this topic. It is customary to refer to c as the *speed of light*, which, of course, it is. A more correct terminology, however, would be *speed of electromagnetic radiation*. Visible light is only one type of an enormous variety of such waves, and all of them move with velocity c. This family includes infrared and ultraviolet light, radio waves, X rays, microwaves, and gamma rays. All of them can be thought of as consisting of photons, although the wavelength of the photon corresponding to each type is different. All move at the same speed, and visible light has no special place other than as an example of the entire class of electromagnetic radiation.

Energy is defined as the ability to do work and work is defined to be the product of a force times the distance through which that force acts. Thus, energy must ultimately be related to the ability to produce a force that can act through a distance. For example, if you lift a weight, your muscles supply a force that overcomes the force of gravity exerted on the object you are lifting and that acts through the distance through which the object is lifted. You have done work (force × distance) on the object, and consequently the object has thereby also acquired the ability to do work. For example, if the object fell, it could exert a force on an object in its path. Thus, when we lift an object, we do work on it, but once it is lifted it can do work on something else. It therefore has energy. This energy is associated with the object's position—the higher it is, the more energy it has; this is called *potential energy*.

In this example, we have talked about the energy an object has because it has been lifted in the presence of a gravitational field. There are other kinds of potential energy associated with other kinds of forces. For example, if the electron in the atom is moved farther away from the nucleus, we have to do work to overcome the attractive electrical force between the nucleus and the electron. The electron has acquired potential energy in this process, just as the object acquired energy when it was lifted against the force of gravity. Consequently, moving electrons around near a nucleus, or rearranging electrons in a set of atoms, will change the energy of the system. Since this type of rearrangement takes place in chemical reactions, this type of energy is sometimes referred to as *chemical potential energy* to distinguish it from the gravitational variety. When you burn gasoline in your car, for example, the energy to run the car comes, ultimately, from the chemical potential energy liberated when long hydrocarbon molecules are broken up into smaller ones, with an accompanying rearrangement of electrons.

There can also be energy associated with motion, and this is called *kinetic energy*. For example, you have to apply a force with your hand over a distance to get a bowling ball rolling. When the ball hits the pins, some of them are knocked into the air (acquiring some potential energy in the process) and some just move away from the ball. Clearly, you imparted energy to the ball with your hand just as surely as you impart energy to something that you lift against the gravitational field.

A somewhat more subtle form of kinetic energy is evidenced by the process of lighting a fire by rubbing two sticks together. On first glance, it would seem that the kinetic energy of the sticks was dissipated in the rubbing process, but if you could look closely at the atoms in the sticks, you would find that as the temperature of the wood increases, the atoms move faster and faster. Thus, heat energy can be thought of as energy of motion at the atomic level.

By the end of the nineteenth century, physicists knew of two general types of energy—kinetic and potential—and of many subclasses of each. In addition, they knew that, while it was often possible to change the form of the energy in a system, the energy in an isolated system had to remain constant in time. For example, we discussed how the chemical energy in gasoline can be converted to the kinetic energy of an automobile. This is a process in which energy changes form, but the energy of the car will always be less than (or at best, equal to) the change in potential energy in the gasoline. This conclusion, which tells us that energy can neither be created nor destroyed, is called the law of conservation of energy, and it is one of the foundation stones of classical physics.

The theory of relativity does not change this principle, but it does add a new category for energy. In addition to potential and kinetic energy, Einstein's equation tells us that mass is also a form of energy. And just as energy can be converted from potential to kinetic and back again, so too can energy be converted from familiar forms to matter and back again.

Does this mean that it is actually possible to create matter where none existed before? Einstein's equation tells us that the answer to this question must be yes! In fact, much of the rest of this book will be devoted to a study of what happened when this process of creating matter began being performed routinely in laboratories all over the world.

From the fact that the velocity of light is a very large number, it follows that it takes an enormous amount of energy to make even very modest amounts of mass. The unit of energy in the system where mass is measured in grams and length in centimeters is called the erg. A 100-watt light bulb uses about 1 billion ergs each second when it is operating, so the erg is not a large energy unit. Each day Americans use about 2.3×10^{24} ergs for all of their energy requirements. This includes home heating and lighting, transportation, electricity, and industry. According to Einstein's formula, how much mass would we need to convert to energy in order to supply this energy? From the equation, the mass is

$$m = \frac{2.3 \times 10^{24}}{9 \times 10^{20}} = 2.5 \text{ kg} \approx 5 \text{ lb}$$

In other words, if we could convert matter to energy with 100 percent efficiency, 5 pounds per day would supply all of the energy needs of the United States! It was the realization that quite modest amounts of matter could produce enormous amounts of energy that was one of the main motivations for developing nuclear reactors.

Conversely, if our interest is in making new kinds of particles —in studying matter that has been created artificially—then the Einstein equation tells us that it will require large amounts of energy to do so. We will see that the usual method of gaining such energy is to allow a very fast particle to collide with one at rest, and to convert the kinetic energy of the fast particle into the mass needed to make the new particles in the process. This method, of course, requires a supply of particles of very high energy—particles whose velocity is greater than 90 percent of the speed of light. In the early 1930s, when the first important results of elementary particle experiments were being gathered, the only source of particles that had energies as high as this were cosmic rays.

COSMIC RAY EXPERIMENTS

Deep inside most stars the temperatures and pressures set off nuclear reactions that, in turn, produce energy that percolates through the body of the star and prevents the star from collapsing. When this energy reaches the surface of the star, it is radiated into space. Some of this radiation is in the form of visible light, and this is what we see when we look at the star. Other portions, not visible, are in the form of radio waves, X rays, and other types of radiation that have only lately been detected and measured by astronomers. And, almost as an afterthought, the surface of the star emits streams of elementary particles, primarily protons, that move off into space. If the star happens to be our sun, these particles form what is called the solar wind—a tenuous stream of particles blowing by the earth. If the star happens to be far away, these particles may travel for millions of years without encountering any solid body. Eventually, however, some infinitesimal fraction of the particles from the sun and other stars in our galaxy will strike the upper atmosphere of the earth; when they do, we call them cosmic rays. Although the emission of cosmic rays is not a particularly important part of a star's energy budget, these particles had an enormous impact on our understanding of elementary particles.

The great majority of the cosmic rays that come to the earth are protons, although, as one might expect because of their origin as debris from star processes, there are also scatterings of all sorts of other particles. For example, entire nuclei of uranium atoms have been seen in Skylab experiments. Those cosmic rays that come from our own sun are mostly of quite modest energies; in fact, comparable to the energies of particles emitted by radioactive nuclei here on Earth. A few cosmic rays, however, have been accelerated somewhere in the Galaxy to extremely high energies by a process that we do not really understand. This small minority of high-energy cosmic rays are the ones that are of interest to us. As we shall see in later chapters, it is possible for the energies of these protons to be even higher than those that can be produced in the largest machines that have been built or are planned today.

When one of these high-energy protons enters the earth's atmosphere, it will descend until it collides with a nucleus. On the average, a cosmic ray will penetrate about a half mile into the atmosphere before such a collision occurs, although individual protons may travel shorter or longer distances than this. As a result of this collision, some of the cosmic ray's enormous store of kinetic energy will be converted into mass, and a spray of particles will be produced. Some of these secondary particles will also have a high kinetic energy, so that when they, in turn, interact with nuclei still deeper in the atmosphere, they, too, will produce sprays of secondary particles. In this way a process called a cascade develops, in which the secondary products of collisions at one level become primary projectiles and initiate collisions on the next level down.

At each level of the cascade some particles collide with atmospheric nuclei and produce more particles, while others continue to move toward the earth's surface. When they strike the surface, they are termed a cosmic ray shower. Depending on the energy of the incoming cosmic ray, these showers can contain millions of particles and can extend over several square miles of the earth's surface. As you are sitting and reading this book, particles from processes like this cascade are passing through your body at the rate of several a minute.

From this discussion, it is clear that cosmic ray interactions in the atmosphere provide a sort of natural laboratory in which the interactions of elementary particles can be studied and, more particularly, in which the process of matter creation can be observed and analyzed. Before we go on to describe some of the important experiments, two of the general principles about the particle creation process should be examined.

If we took all the energy (kinetic plus mass) of all the particles created in the cascade and added to it the increased kinetic energy of molecules in the air due to the cascade, the result would be equal to the energy of the incoming cosmic ray. In other words, even though particle creation is occurring, the law of conservation of energy holds in the cascade. Furthermore, if we took any single collision in the cascade and added up the energies before and after the collision, they too would balance. In other words, the processes in the cascade may convert energy from its kinetic form to mass, but they do not violate any known law of energy conservation.

In a similar way, the electrical charge is conserved in the cascade as a whole and at each step in the cascade. For example, if the initial particle is a positively charged proton, then if there are a million negatively charged particles created during the cascade, there will have to be a million and one positively charged particles created, so that the net total charge of the cascade is always +1. But, with the exception of the constraints imposed by the conservation of energy and charge (and a few other conservation laws to be discussed later), anything can happen in the cascade, which means that the cosmic ray "laboratory" can provide us with a look at a wide variety of the processes that can occur when elementary particles collide with each other. This variety is one good reason why so much of our knowledge of the properties of elementary particles came originally from cosmic ray experiments.

THE DISCOVERY OF THE POSITRON: A TYPICAL COSMIC RAY EXPERIMENT

To illustrate the richness of the information that was gleaned by physicists in the 1930s and 1940s from cosmic ray experiments, consider one of the most important—the discovery of the first piece of antimatter by Carl D. Anderson in 1932 at the California Institute of Technology. The experiment was designed to detect and identify those portions of cosmic ray showers that reached ground level. To this end, it was necessary to do three things: (1) find out when a particle had passed through the experimental apparatus; (2) find the charge of the particle; and (3) find the mass of the particle.

"Seeing" an elementary particle is not an easy thing to do, since they are so small. If we build the right apparatus, however, we can use some natural processes to tell us what path a particle has followed. In the 1930s, the right apparatus was the Wilson cloud chamber. A container with a moveable piston for one wall is filled with air that is saturated with something similar to alcohol vapor. A particle passing through this chamber will, because of its electrical charge, leave behind it a trail of atoms from which electrons have been torn. These atoms are called ions. If the piston is suddenly lowered just after the particle has left the chamber, the alcohol in the air will start to condense into drops, much as water comes out of air as dew when the temperature falls. It turns out that the ions that have been left in the wake of the particle can act as nuclei around which drops of alcohol will form, so the net result of this operation is that there will be a trail of visible drops in the otherwise hazy and diffuse cloud of condensate formed in the chamber. This trail will be located along the path that the elementary particle traversed. Although this process does not allow us to see where the particle *is,* it does allow us to see where it *was,* which for our purposes is just as good.

The presence of a track of droplets in a cloud chamber shows that a particle has been in the chamber and has been detected. In addition, it turns out that the number of drops (which is related to the number of ions created by the particle passing through the chamber) gives us some information about the velocity and mass of the particle. To determine these two variables exactly and to ascertain the electrical charge of the particle, one new feature has to be added to the experiment.

Suppose that the cloud chamber were placed between the poles of a large magnet, so that the entire chamber was inside a magnetic field. The laws of electromagnetism tell us that a particle entering the chamber will not go through in a straight line, but will follow a curved path. Hence, the droplet track will not be straight, but will be part of the arc of a circle. The amount of curvature of the track will tell us the momentum of the particle, and the direction of curvature will tell us the charge of the particle.

Thus, a cloud chamber located between the poles of a magnet is a good device for identifying particles. From the number of droplets and the amount of curvature of the track we can determine the mass and velocity of the particle, and from the direction of curvature we can find the electrical charge. We shall see later that the twin measurements of the ability of a particle to ionize nearby atoms and its reaction to a magnetic field are still the main criteria for the identification of particles in the laboratory.

In 1932 Carl Anderson used an apparatus that he and Robert Millikan had built at the California Institute of Technology to look at cosmic ray showers. The device consisted of a set of powerful magnets to bend the paths of particles and a cloud chamber so placed that those particles coming from a vertical or near-vertical direction would pass through the longest dimension of the chamber. In a series of papers in *Physical Review,* Anderson proved conclusively that a large number of the particles passing through his apparatus had a mass about equal to that of the electron, only they had a positive electrical charge. He called this new particle the *positron*—a contraction of positive electron —and this name has stuck. His suggestion that in view of this discovery we ought to refer to the electron as a "negatron," fortunately, was not widely adopted.

In the usual manner of experimentalists reporting a new finding, Anderson's original papers did not spend much time on speculation about the nature of the positron, but were devoted to establishing its existence beyond question. It was only later that the full import of the discovery was realized. For the first time, there was conclusive laboratory evidence that an entirely new type of matter existed—a type we now call antimatter.

ANTIMATTER

In normal matter, electrons always have negative electrical charges and the nucleus of the atom always has a positive charge. The positron, therefore, is not a particle that "hides" in a normal atom until it is shaken loose in a cosmic ray collision. Hence, it must be one of the particles created in the cascade by the conversion of energy to matter. This process is now well understood and is seen daily in modern laboratories.

As an example, consider a high-energy photon impinging on an atom in the atmosphere. After the photon strikes the nucleus of the atom, some debris (pieces of the nucleus) will result, perhaps with some miscellaneous particles created from the energy of the photon, and, most important, an electron and a positron. We denote the former by the symbol e^- and the latter by e^+. Often a photon collides with a nucleus and only the electron and positron come out. In either case, we refer to the occurrence as pair creation and speak of the electron-positron pair. It is important to realize that in this process the electron and positron are both created from energy, and that neither is part of the nuclear debris.

Energy is conserved in this process; that is, the kinetic energy of the incoming photon is converted to the mass and kinetic energies of the pair. Charge is also conserved, since every time a positively charged positron is created a negatively charged electron is created with it. This is a specific example of the general principle we discussed in the previous section—the principle that every process at the level of elementary particles is subject to the restrictions imposed by the conservation laws.

If we now allow the positron that we created to encounter an electron, a startling event takes place. The electron and positron disappear, and in their place we find some high-energy photons. This process is called *annihilation*, and it represents the inverse of creation. When a particle and its antiparticle meet, their total energy (including the energy in their masses) is converted into photons, and the original particles cease to exist.

In general, after a positron is created, it will wander until it encounters an electron with which it can annihilate. This electron could be the one with which it was created, but is much more likely to be a particle met more or less at random.

The creation and annihilation of electron-positron pairs provides a vivid and compelling verification of the principle of the equivalence of mass and energy. Since 1932, it has been found that for every particle in nature there is an antiparticle. (We will be discussing some of these discoveries later.) Actually, the existence of a particle like the positron was predicted in 1930 by the British theoretical physicist Paul A. M. Dirac. His reasoning provides an interesting way of visualizing the pair creation and annihilation processes.

Dirac did not start by thinking about antimatter, but instead was trying to resolve a rather difficult theoretical problem that had arisen when physicists had tried to combine ordinary quantum mechanics with the principles of relativity in order to get a quantum mechanical description of high-energy particles. They found that the equations predicted that particles, such as the electron, could exist in states where the energy was actually negative.

This, of course, is nonsense. If electrons had negative energy states into which they could fall, the tendency of every system in nature to reach the lowest possible level of energy would cause the electrons to start falling into these negative states. The electron's situation would be analogous to that of a stone poised on the side of a hill of infinite length. Once it started rolling, it would keep on going, falling to lower and lower energy levels as time went by. If negative energy states for electrons did exist, and if there was nothing to prevent electrons from falling into them, all of the electrons in the universe would fall into states at −∞, and the radiation released when they did so would fill the universe. Consequently, physicists tended to ignore the possibility of negative energy states, or to think of them as some sort of quirk in the mathematics that would be resolved later. Dirac chose, instead, to take the possibility of negative energy states very seriously. Suppose, he said, that the states are really there, but that they are completely filled up with electrons already? In other words, if we supposed that the possible energy states of the electron are like the ones shown in Illustration 8, then the electron could have an energy mc^2 (this would correspond to an electron with no kinetic energy) or it could have some higher energy (corresponding to the electron having both mass and kinetic energy). The predicted negative states would then extend from $-mc^2$ on down to −∞.

If these states are there and if they are always filled with electrons (a condition referred to as a "filled negative-energy sea"), two consequences follow. First, electrons cannot fall into these negative states and descend to −∞ for the simple reason that there is no room for them to do so—any state they could descend into is already filled. Second, the entity we normally call the vacuum would be a filled negative-energy sea with no particles in the positive energy states.

In terms of our stone-on-the-hill analogy, Dirac's suggestion was that the stone could not start rolling down the hill because the lower slopes were already completely covered with stones.

Illustration 8 shows that pair creation works like this: A photon comes along and propels an electron from the negative-energy sea to a positive energy state. The final result will be a positive energy electron and the absence of a negative-energy negatively charged particle. If you think about the double negatives for a while, you can probably convince yourself that the absence of a negative-energy negative charge is the same as the presence of a positive-energy positive charge, and that is what we have been calling a positron. Thus, in Dirac's picture, the positron is thought of as the effect of the empty state that results when an electron is moved to a state of positive energy.

In the same way, pair annihilation occurs when a positive energy electron encounters a hole in the negative-energy sea (which, remember, is a positron) and falls into that hole. From a state of having two "particles," we wind up with a filled negative-energy sea again, with some photons moving off.

One way of visualizing the Dirac theory of the positron is to imagine a plot of level ground. If you take a shovel and dig a hole, there will be two things left—a pile of dirt and a hole (which is the absence of the pile of dirt). The former would be the electron and the latter the positron. Annihilation, in this analogy, would correspond to putting the dirt back in the hole. When this is done, you are back to level ground again.

Having understood Dirac's picture, it is important to emphasize that you do *not* see the absence of a negative energy electron in the laboratory. What Anderson saw (and what you could see in a modern laboratory) is a positively charged particle with positive energy and the same mass as the electron. The Dirac picture is just a way of interpreting this observation in terms of positive and negative energies and, at the same time, resolving a rather difficult theoretical problem.

There is another interesting thought about antimatter we

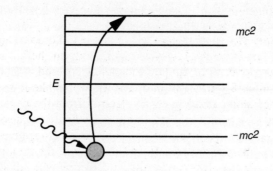

8. How pair creation works.

should discuss. We have mentioned the fact that there is a kind of symmetry in nature between matter and antimatter in the sense that there can exist an antiparticle for every known particle. On the other hand, there is a manifest asymmetry in the world we know in that everything is made of normal matter, and antimatter is a rare and exotic species.

Antimatter seems to enter the laws of nature on an equal footing with ordinary matter, yet is very rare in the universe. Indeed, all of the antimatter ever created by all the accelerators on earth wouldn't weigh as much as a mote of dust. Why should the universe favor ordinary matter so?

For a while, a number of theoretical physicists, most notably Hannes Alfvén, argued that there were equal amounts of matter and antimatter in the universe, but that the two had become segregated. The earth, they argued, exists in a little bubble of ordinary matter, but somewhere else there is a compensating "anti-earth," where antimatter dominated. In this scheme, matter and antimatter in the universe resembled material in a poorly mixed salad.

At first glance, it would seem to be impossible to disprove such an idea. After all, light emitted from anti-atoms (that is, atoms in which positrons orbit nuclei made of antiprotons and antineutrons) would be indistinguishable from light emitted from ordinary matter. If we just look at light from a distant star or galaxy, there is no way to tell if it is made from matter or antimatter.

The phenomenon of annihilation, however, does provide us with a way of seeing whether other parts of the universe are like us or not. For example, we know that Venus, Mars, and the Moon are made from ordinary matter because we have sent space probes to them. Had they been made from antimatter, the probes would have disappeared in a titanic explosion when they landed (or entered the planetary atmospheres).

We also know that the sun is constantly sending out a stream of ordinary matter in the form of the solar wind. If the other planets were made from antimatter, particles in the solar wind would annihilate in their atmospheres. These annihilations would produce X rays that would be easy to detect. We see no such X rays; hence we can conclude that the entire solar system is ordinary matter—there are no "anti-Martians" out there.

Many of the cosmic rays that strike the earth come from other stars in our own and neighboring galaxies. And while these cosmic rays do contain some antimatter, they have no more than you would expect from chance productions by collisions of particles in deep space. This is one piece of evidence that antimatter is rare in the Milky Way and its neighborhood.

Another piece of evidence comes from noting that if two regions in a galaxy (or two neighboring galaxies) were made of matter and antimatter respectively, there would have to be a region where the thin solar winds of the two regions overlapped. Annihilation would then cause this region to be a diffuse source of X rays. When we examine the sky with X ray telescopes, we find no such regions. Consequently, astronomers are now in general agreement that the entire visible universe is made from ordinary matter.

And this, of course, brings us back to the original question: Why isn't there any antimatter? In fact, one of the great triumphs of the Grand Unified Theories discussed in Chapter XIV has been to explain the observed absence of antimatter. Thus, instead of being a puzzle, this absence becomes one of the strongest pieces of evidence for the current theories of elementary particle interactions.

V

The Discovery of Mesons and Other Strange Things

INTERLOCUTOR: *Here's the shaggy dog you advertised as lost in the paper yesterday.*
MR. BONES: *Good Lord, it wasn't that shaggy.*
　—"The Shaggy Dog Story," Anonymous

THE DISCOVERY OF A MESON

THE TYPE of cloud chamber in which the positron was discovered proved to be a tremendously useful piece of equipment. As so often happens when a new detection method is developed to explore new fields, discoveries come thick and fast. The discoveries of Galileo after he built his first telescope are historical examples of this generality. Within months of the time when the telescope was first turned toward the heavens, he had seen the moons of Jupiter, mountains on the moon, and sunspots. In a similar way, when scientists first began looking at cosmic ray showers with an instrument capable of measuring both ionization and the amount of bending in a magnetic field, entirely new and unexpected results began to show up. These results were obtained during the 1930s and 1940s, and they led to the creation of an entirely new area of research—the field of elementary particle physics.

In the early 1930s, cloud chamber experiments of the type that led to the discovery of the positron were being performed in many laboratories around the world. By 1934 it had become apparent that something was wrong with the way these experiments were being interpreted. Particles were seen whose ability to create ions in the cloud chamber did not seem to correspond with the behavior of electrons, positrons, or protons. These new particles, which went by the name of "penetrating rays" (because they could penetrate through the atmosphere to the apparatus at sea level), were characterized by either positive or negative electrical charges. When the new quantum mechanics was used to calculate the progression of electrons through a cosmic ray shower, it showed that very few would be expected to reach sea level with energies like those seen. On the other hand, the energies that the particles lost through ionization in cloud chambers were too low for them to be protons. This meant either that penetrating radiation was not made up of known particles or that the quantum theory gave incorrect answers when applied to particles of very high energy. The resolution of this quandary became one of the central issues in cosmic ray physics in the late 1930s.

The fact that a problem such as this existed illustrates an important point about the way scientific research is done, a point that is often ignored when we look back at important achievements. There is a temptation to pick out the significant results and put together a chain of reasoning which, *in retrospect,* seems very precise and logical. In real life, however, the scientist is confronted with evidence that is ambiguous. In this case, for example, the discrepancy could arise because the theory was wrong (after all, it was only in 1936 that the quantum mechanical theory for electron showers was worked out). On the other hand, the technical details of the analysis of cloud chamber pictures made it very difficult to determine the mass of particles going through at high speed. Such particles would be only slightly bent by the magnets, and the normal experimental errors could easily mask significant results. Only when all these ambiguities had been eliminated could the conclusion be drawn that particles were being seen that had a mass different from both the proton and the electron.

By 1938 enough experimental evidence had accumulated to convince physicists that the explanation of penetrating radiation did not lie in an error in quantum mechanics. In an effort to give a final resolution to the problem, Seth Neddermeyer and Carl Anderson, working with the California Institute of Technology cloud chamber, tried a new mode of operation. They inserted a Geiger counter into the chamber, and then arranged to expand it and photograph the drops only when a particular reading was seen in the counter, a reading that indicated that a very slowly moving particle was in the chamber. In this way, they hoped to obtain a series of photographs of the tracks of penetrating rays, which would be easy to analyze. Luck was with them and their program ultimately proved successful. In a letter to *Physical Review* on June 6, 1938, they reported a single event in which a penetrating ray of positive charge was slowed down enough by passing through the Geiger counter so that it came to a complete stop in the cloud chamber. It turns out that this was the best possible experimental situation for determining the mass of a particle, and when Neddermeyer and Anderson analyzed the photograph, they reported that a new particle had been discovered, with a mass about 240 times that of the electron. (The modern value for this number is 210.) They named their new discovery the "mesotron," from the Greek root "meso" (which means middle). This was later shortened to *meson,* a term that is both more convenient and more correct linguistically. This particle was customarily denoted by the Greek letter μ (mu); it is therefore now called the mu-meson, or muon.

There were actually two mu-mesons, one with a positive charge and one with a negative, and the masses of the two were identical. Were these, then, the particles that Yukawa had predicted would be responsible for the strong force? Again, while it is tempting to think that science proceeds in this straightforward way, in point of fact there had been many predictions of intermediate mass particles. By 1938, Yukawa's was only one, and the suggestion, coming as it did from a research center far away, was not even the most prominent one in the minds of American scientists. Thus, the question that had to be answered was whether this meson behaved as one would expect if it were really the "nuclear glue" we previously mentioned.

The first thing that became obvious was that the muon was not stable in the sense that the electron and proton are. Like the neutron, it decays in a certain amount of time into other particles. The lifetime of the muon is measured to be about 10^{-6} second, and it decays by the process $\mu \rightarrow e +$ two neutrinos. Of course, the two neutrinos cannot be seen in the cloud chamber, since they are uncharged and therefore do not create ions.

When a mu-meson enters a block of material, it will encounter the atoms that make up that material. If the mu-meson is indeed the particle responsible for holding the nucleus together, we would expect that when it came near a nucleus it would interact strongly with it. We know that a typical time scale for strong interactions is approximately 10^{-24} second, and since this is much shorter than the lifetime of the muon, we would expect that most of the muons that are slowed down and stopped in a block of material would wind up interacting with a nucleus long before they had a chance to decay. This, in turn, means that if we look at what comes out of the other side of such a block of material, we should not see the characteristic electrons that result from muon decay, but rather the kind of nuclear debris that is associated with nuclear reactions themselves.

Actually, there is a slight quibble to be made about this argument. When the theory of the process by which mu-mesons are captured in atoms was worked out, it appeared that the positively charged muon would have to be repelled from the nucleus by the ordinary electrical force, so that it would decay normally without entering a nucleus. The argument given above, however, held for the muon with negative charge, since it would be attracted toward the nucleus by the electrical force. This quibble lost its importance, however, when experiments showed that decay electrons were seen for *both* kinds of muons when they entered a block of material. Both kinds were somehow staying alive for the full 10^{-6} second required for them to decay.

Thus, by 1947 physicists were faced with a real dilemma. The mesons were supposed to be the particles that bound the nucleus together, yet when a meson came near a nucleus it showed no inclination to interact strongly with it. If the meson *inside* the nucleus interacted strongly enough with protons and neutrons to overcome the electrostatic repulsion, how could it interact so weakly when it was *outside* the nucleus? Physicist I. I. Rabi of Columbia expressed the sentiments of the physics community about the mu-meson very well with the query, "Who ordered this?" The predicted meson had been found all right, but it turned out to be the wrong one!

THE PI-MESON: ONE DILEMMA RESOLVED, ANOTHER CREATED

We have seen that one of the major experimental problems that arises in the study of elementary particles is finding ways of detecting the presence of the particles. The cloud chamber solved this problem by using ions created by the particle as condensation centers for alcohol drops. During the period immediately after World War II, a similar technique using photographic emulsions came into widespread use in cosmic ray studies.

In ordinary photographic film, light striking the emulsion causes a chemical reaction which, when the film is developed, leads to grains of silver being deposited on the negative. These grains are opaque, so that after development the film will be dark where light was present and lighter where it was not. In this way, the characteristic reversed negative of a photograph is formed.

A very similar process can be used to detect the presence of a charged particle. When such a particle passes through a specially prepared photographic emulsion, it causes reactions which, upon development, lead to silver grains being deposited along the path that the particle followed. Someone examining the emulsion with a microscope will then be able to see where the particle had been by following the trail of these grains, which show up as dark spots on the lighter background.

This new technique had several advantages over the cloud

chamber. A stack of emulsion plates could be left on a mountaintop for months at a time, and every particle that passed through them would be recorded. In this way, the emulsion could "see" many more particles than could a cloud chamber that was operated sporadically. In addition, the small size of the silver grains made it possible to see particles in the emulsion even if they only traveled 10^{-6} centimeter, something that could not be done when the detection depended on the formation of droplets. Finally, and most important, the photographic emulsion was much denser than the air-alcohol mixture used in a cloud chamber, so that particles produced by collisions in the emulsion were likely to encounter another nucleus and interact with it before they could leave. In this way, the emulsion could be used as a target in which new particles were created, and, at the same time, as a detector that measured the way the new particles interacted with nuclei.

In 1948 a group of physicists headed by Cecil F. Powell at the University of Bristol in England began publishing the results of their examination of emulsions that had been exposed at the height of 10,000 feet on the Pic du Midi in the French Alps and on other high mountains around the world. They saw tracks of the mu-mesons, of course, but, in addition, they saw collisions of energetic particles with nuclei that produced another kind of meson that was heavier than the muon. Once this heavier meson was produced, one of two things would happen. The new meson could decay in about 10^{-8} second into a muon and an uncharged particle (which was assumed to be a neutrino), or it would interact with another nucleus in the emulsion. When it did the latter, the evidence of the tracks in the emulsion showed that the new meson interacted strongly, breaking up the nucleus and creating a spray of debris. In other words, this particular meson seemed to be the one that Yukawa had been talking about. It not only had a mass intermediate between that of the proton and the electron, but it interacted strongly when it came near a nucleus, something that the mu-meson did not seem to do.

The new particle was christened with the Greek letter π (pi) and called the pi-meson, or pion. Because of the possibility that the new pi-meson played a fundamental role in binding the nucleus together, it quickly became an object of intense study throughout the physics community. In Chapter VI we will talk about the development of particle accelerators—machines that can take protons (or electrons) and accelerate them to energies comparable to those of cosmic rays. In the early 1950s machines were becoming available that were capable of creating pions for use in physics research, so that detailed studies of their properties could be done. It turned out that the meson come in three varieties—there are pions with a positive electrical charge, pions with a negative electrical charge, and pions that are electrically neutral. These are denoted, respectively, by the symbols π^+, π^-, and π°. The mass of the charged meson is 273 times that of the electrons, and each of the charged mesons decays via the reaction $\pi \rightarrow \mu + \nu$, with a lifetime of about 10^{-8} second. The π° has a mass 265 times that of the electron and decays via the reaction $\pi^\circ \rightarrow$ two photons in about 10^{-16} second. It is now believed that most of the strong force in the nucleus is generated by the exchange of mesons, as we discussed in Chapter IV. The pi-meson is, therefore, an extremely important addition to the ranks of known particles.

Why was it not discovered sooner? Given the importance of the meson in the theory of the nuclear force, why did it take a decade to unravel the π-μ problem?

Part of the answer to this question is historical—the decade between the discovery of the muon and the identification of the pion with the Yukawa meson spans World War II, a period when most physicists were focusing their attention on more pressing questions. But perhaps more important are the properties of the pi-meson itself. When such a meson is created by a cosmic ray collision high in the atmosphere, one of two things can happen: It can decay before it hits the ground or it can interact with a nucleus in the atmosphere. Typically, a pi-meson of moderate energy will travel only a few meters or tens of meters before it decays, and even if the energy were high enough to bring the meson to sea level before decay, it would be able to travel only a few hundred meters through the atmosphere before interacting with a nucleus. In either case, the original meson will not reach ground level and therefore would not be seen in a ground-based cloud chamber experiment. And since most pi-mesons in a cosmic ray shower are created at high altitudes, before the energy of the particles is degraded below the level needed for particle production, pi-mesons are never seen at sea level. This fact explains, incidentally, why the Bristol group found evidence for the pion when they exposed emulsions on a mountaintop rather than at sea level. It also explains why, in the acknowledgments of one of the original pi-meson papers, the authors thank the leader of a mountaineering expedition for carrying some plates to the top of Mount Kilimanjaro in Tanzania—a height of 19,000 feet.

But even a cloud chamber on a mountaintop would not stand a very good chance of actually detecting and identifying a pi-meson. As we have discussed, the material in a cloud chamber is not very dense, so that a meson passing through the chamber has a relatively low probability of encountering a nucleus and interacting. It would, therefore, be very difficult to tell the difference between a fast muon and a fast pion by looking at droplets. We have already seen how critical it is for a particle to be brought to rest in a cloud chamber in order for it to be identified, and how the first identification of the mu-meson depended on a lucky event in a ground-level experiment. Given the difficulties attendant on operating cloud chambers on top of a mountain peak and the resulting paucity of high-altitude cloud chamber data, it is not too surprising that there was no equivalent lucky event for the pi-meson.

With emulsions, however, the situation was different. They are quite dense, so that pions entering the emulsion are likely to be stopped. In addition, it turns out that it is possible to make much more accurate determinations of particle mass by counting silver grains through a microscope than by analyzing droplets.

By 1948, therefore, the riddle of the mesons had been solved. Not one, but two groups of particles with a mass between that of the electron and the proton had been found. The pions are the particles predicted by Yukawa, and all three of them are routinely exchanged within the nucleus to generate the strong inter-action. In a sense, once Rutherford had discovered the nucleus, the existence of such a particle was inevitable. Thus, the number of elementary particles is increased by one (it is customary to refer to all of the pion family as a single particle). In return for this complication in our picture of the universe, we gain an understanding of the strong interaction.

The case of the mu-meson is not so clear. It adds another elementary particle to our fast-growing collection, but it is not obvious just what role it plays. It is not essential to our under-standing of the nucleus, and in many ways it seems as if nature, having created the electron, went ahead and repeated the process for a particle 200 times heavier. One of the major unsolved mysteries of particle physics remains the question of why the muon should exist at all or, in the words of Nobel Laureate Richard Feynman, "Why does the muon weigh?"

Today, we understand that the muon is one of a small group of particles that are "elementary" in the sense that they constitute the basic building blocks of matter. We will return to the question of the muon in Chapter XIV, when we talk about the Standard Model.

STRANGER STILL

At the same time that the evidence for the pi-meson was ac-cumulating on photographic plates on mountaintops, two re-searchers at Manchester University in England began reporting some very unusual events in their cloud chamber photographs. We have seen that one of the drawbacks of the cloud chamber is the fact that the air-alcohol mixture in the chamber has such a low density. To get around this difficulty, experimenters began inserting plates of heavy materials, such as lead, into the cham-ber to slow the particles. Pictures similar to Illustration 9, on page 77, would then be seen. A particle would enter the cham-ber from the top and go into the lead plate. It would strike a lead nucleus, and the debris of this collision would then be seen emerging from the other side of the plate.

In December 1947, however, a rather unusual event was reported. It is shown in the right-hand portion of the illustration. The usual particle-above-debris-below pattern was seen, but in addition a set of V-shaped tracks seemed to appear from nowhere on the far side of the plate. The only possible interpretation of this event was that an uncharged particle was created in the lead plate. Such a particle would not create ions and would not be revealed by the droplets in the chamber. At the point labeled V this uncharged particle then decayed into two charged particles, which were visible in the chamber. Later work showed that these two charged particles were, in fact, a proton and a negative pi-meson. In modern terminology, the uncharged particle is called the Λ (lambda) particle, and is customarily written Λ° to emphasize its lack of electrical charge. Thus, the reaction sketched in Illustration 9 would be $\Lambda^\circ \rightarrow p + \pi^-$.

There are a number of extraordinary things about this event. In the first place, the fact that the lambda can decay into a proton plus something else clearly implies that it must have a mass greater than that of the proton. No one had expected or anticipated that such a particle could exist. Even more surprising, the fact that the lambda seemed to travel several centimeters in the chamber before it decayed indicated that it had a very

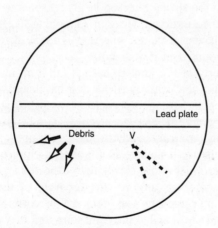

9. Discovery of the lambda particle.

long lifetime. We can get a rough estimate by noting that the time it takes light to travel 3 centimeters is

$$t \approx \frac{3}{3 \times 10^{10}} = 10^{-10} \text{ sec}$$

so that the lifetime of the particle would have to be of this order of magnitude in order for its tracks to be seen in the cloud chamber. In fact, the lifetime of the lambda is now known to be 2.6×10^{-10} second. While this may seem like a very short time on the human scale of things, it is very long when compared to the 10^{-24} second that we saw was "natural" for particles participating in the strong interaction.

This fact poses a serious problem for our understanding of the nature of the lambda. Physicists had been able to accept the relatively long lifetimes of the pi- and mu-mesons because they decayed by a process something like the weak decay of the neutron. The fact that the end products of pion and muon decay are particles that do not participate in the strong interaction made it easy to accept the fact that, on the nuclear scale, they lived a long time. With the lambda, however, this argument cannot be made. The lambda is created in a strong interaction (the disintegration of a nucleus) and it decays into particles that manifestly participate in the strong interaction. By reasonable analogy, the lambda *ought* to undergo decay in something like 10^{-24} second rather than having a decay time that is more characteristic of weak interactions. But it does not.

This property of the lambda earned for it an unusual name. Together with a few other particles that have similar properties, it was dubbed a "strange" particle. As we shall see later, the quantity that physicists call "strangeness" can be given a quantitative meaning (contrary to what you might suppose), but for the moment let us use it merely as a label for those particles that seem to decay much more slowly than expected.

At about the same time as the discovery of the lambda, a group of strange mesons was detected in cosmic ray experiments. They are now called the K-mesons, and they come in two pairs. The first pair contains a positively charged meson called the K^+ and a neutral member called the $K°$. The second pair contains a negatively charged member called the K^- and a neutral member that is the antiparticle of the $K°$ and is called the $\overline{K}°$. All of these particles have a mass about 1000 times that of the electron.

With the discovery of strange particles in cosmic ray experiments, the whole nature of elementary particle physics changed. From the study of the few known particles and a search for a few predicted ones, scientists turned to the much more general question of how many particles could actually be produced and what their properties were. To carry out this sort of study, the kind of cosmic ray experiment we have been discussing is really not suitable. Having to wait and hope to see a favorable event may be a good way to map the general features of a new field, but when exhaustive studies need to be made, it becomes necessary to have a source of energetic particles that can be controlled. With such a source, it should be possible to create the new particles at will so they can be studied in much more detail than is possible with cosmic rays.

Fortunately, during the 1930s machines capable of producing such energetic beams were being developed. They are called accelerators and bear such exotic sounding names as cyclotron, synchrotron, and linac. We will talk about the development of these machines and the discoveries made with them in Chapter VI.

A QUESTION OF UNITS

Up to this point, we have talked about the masses of the elementary particles either in terms of grams or in relation to the mass of the electron. Once a serious study of these particles starts, however, it is pretty clear that neither of these two sets of units is particularly useful. For scientists who used them all the time, writing all of those 10^{-34}'s soon became tiresome, and there seemed to be little point in referring everything to the mass of the electron in processes in which that particle was not involved. The system of units now used in discussing particles is based on the equivalence of mass and energy, and has as its basic unit the electron volt, written eV.

The electron volt is defined as the amount of energy gained by a particle whose charge is equal to that of the electron when it moves through a 1 volt potential difference. For example, a single electron that moves from one pole of an ordinary car battery to another would acquire 12 electron volts of energy. Ten electrons following the same route would acquire a total of 120 electron volts of energy, as would a single electron moving across a larger battery rated at 120 volts. The electron gains energy in such a process because it takes work to move it from one side of the battery to the other against the electrical force. In the system where mass is measured in grams and length in centimeters, 1 electron volt = 1.6×10^{-12} erg.

Since energy and mass are related by the Einstein equation, we can talk about the energy equivalent of the mass of the electron. In terms of electron volts, we have for the electron

$$m_e c^2 = 0.51 \times 10^6 \text{ eV}$$

while for the proton we have

$$m_p c^2 = 939 \times 10^6 \text{ eV}$$

From these two numbers, we see that the masses of elementary particles turn out to be relatively large numbers when expressed in terms of electron volts. Consequently, an abbreviation scheme is normally used, as indicated in the following table:

ABBREVIATION	FULL NAME	VALUE
eV	electron volt	1
keV	kiloelectron volt	10^3 eV
MeV	million electron volt	10^6 eV
GeV	gigaelectron volt	10^9 eV
TeV	teraelectron volt	10^{12} eV

The prefix *giga,* to designate 10^9, arises because the term "billions" is given a different meaning on the two sides of the Atlantic. To an American, a billion is a thousand million (10^9). To a European, however, it is a million million (10^{12}). For a while, the unit BeV (billion electron volts) was in wide usage, but it caused enough confusion to generate the set of prefixes in the table above. These are now used universally by international agreement.

In terms of energy units, the masses of all of the particles we have discussed so far are given in the following table:

PARTICLE	MASS (for particles) OR ENERGY (for photons)
Electron	0.511 MeV
Muon	105.7 MeV
$\pi^{\pm}$	139.6 MeV
π°	135.0 MeV
K	493.7 MeV
Proton	938.3 MeV
Neutron	939.6 MeV
Λ°	1,115.6 MeV
Photons (visible light)	~10 eV
X rays	~100 keV–10 MeV

VI

The Coming
of the Accelerators

Silently we went round and round . . .
—Oscar Wilde, "The Ballad of Reading Gaol"

FROM NATURAL TO MAN-MADE

ALL THE EXPERIMENTS we have described up to now have involved the use of natural sources of energetic particles to trigger the events being studied. Rutherford, for example, got his alpha particle projectiles from pieces of naturally radioactive material, and Anderson discovered the positron as an end product of the collisions of cosmic rays. There are obvious advantages to this kind of experiment. They are relatively cheap, since you only have to build equipment to detect interactions: the projectiles are free. On the other hand, as we pointed out in Chapter V, there are serious limitations as well. If you depend on the natural supply of particles for an experiment, then you have no choice but to wait until the particles you want happen along.

In addition, the cosmic ray experiments revealed that there were many previously unsuspected particles in nature, and that all these particles are unstable. If we want to see what these particles are like, we must have some way of producing them in large enough quantities to study. From the fact that many of the cosmic ray discoveries involved an element of luck it was clear that any systematic work on the elementary particles would also have to wait until "lucky" events could be produced routinely in the laboratory. This, in turn, would depend on the ability of physicists to produce large quantities of energetic particles. With these "artificial cosmic rays," experiments such as the ones we have been describing could be carried out under controlled conditions.

To take a normal particle and give it high energy requires that the particle be accelerated. Machines that do this job are called accelerators (terms such as *atom smasher* never really appealed to physicists). While cosmic rays can supply us with high-energy protons, accelerators can be (and are) used to accelerate any charged particle, from protons and electrons to the nuclei of heavy atoms. By and large, however, it has been machines that accelerate protons and electrons—the most abundant and stable of the elementary particles—that have occupied the forefront of modern research.

There are two general classes of accelerators. In one kind, particles are accelerated while they travel down a long straight tube. This is a linear accelerator. In the other type, the particle is made to move in a circular path by applying a magnetic field and then boosting its energy each time it comes past a given spot on the circle. This sort of machine is called a cyclotron or a synchrotron, depending on how the magnetic field is applied. Within each of these general categories are variations and adaptations which, in the final analysis, are limited only by the ingenuity of the designers. Both protons and electrons can be (and are) accelerated in linear and circular machines.

Historically, a conference held in Bagnères, France, in July 1953 is thought of as the point at which the main research work on elementary particles shifted from cosmic rays to accelerators. Before such an event could occur, though, there had to be a long period of development during which these machines were transformed from experimental ventures into reliable tools that could be used in daily research.

E. O. LAWRENCE AND THE CYCLOTRON

On a California evening in 1929, Ernest O. Lawrence, at the time an assistant professor of physics at Berkeley, sat in the university library catching up on the technical journals. In the German journal *Arkiv für Electrotechnik,* he came across an article devoted to a scheme for producing accelerated particles. This started him thinking of another way of accomplishing the same task, and he jotted down a few notes.

Basically, his idea went like this. It had already been seen that a charged particle will be deflected by a magnetic field; in fact, if the field covers a large enough region of space, the particle will move in a circle. If the magnetic field has a strength denoted by *B,* then it will exert a force equal to *Bqv* on a particle of charge *q* moving with velocity *v.* In order to keep the particle moving in a circle of radius *R,* the magnetic force has to balance the centrifugal force. The formula for the latter is mv^2/R, where *m* is the mass of the particle. This means that for a particle in a magnetic field we must have

$$\frac{mv^2}{R} = Bqv$$

so that the radius of the circle in which the particle moves is just

$$R = \frac{mv}{Bq}$$

If we imagine a large magnet with one pole above the page and one below it, then to each velocity that a given type of particle can have, there is one radius for its motion. For example, a particle moving with velocity v_1 would move in a circle of radius R_1, a particle of velocity v_2 would move in a circle of radius R_2, and so on. The faster the particle moves, the larger its radius.

These circular paths have come to be called *cyclotron orbits.* Electrons in the same magnetic field will move in the opposite (clockwise) direction through circles of much smaller radius from protons. The fact that the electron radius is smaller is simply a consequence of the fact that the orbit depends on the mass of the particle, and the electron has a smaller mass than the proton.

The next question is how long it takes for a particle to get around a cyclotron orbit. Suppose the particle is moving with a speed v and has to travel a distance $2\pi R$, which is the circumference of the circle that defines its orbit. This means that

$$2\pi R = vt$$

If we substitute for v from equation $mv^2/R = Bqv$, this equation becomes

$$2\pi R = \frac{Bq}{m} R \cdot t$$

This equation contains the basic insight that allowed Lawrence to build the first cyclotron. Notice that the radius of the orbit, R, cancels out on both sides of this equation—

$$2\pi \cancel{R} = \frac{Bq}{m} \cancel{R} \cdot t$$

so that the time it takes to complete an orbit depends only on the magnetic field and the charge and mass of the particle. It does *not* depend on how fast the particle is moving. The reason for this lies in the fact that faster moving particles go in larger orbits, so the increased speed of the particle is exactly canceled by the larger distance it has to travel. The time remains exactly the same.

The significance of the cancellation is this: If we installed an accelerating device at a particular spot in the magnetic field and waited until a particular particle in orbit came around to that line, we could accelerate the particle. In fact, if we timed the pushes just right, we could arrange to give the particle a boost each time it came around. In this way, we could imagine giving the particle a great deal of energy in small doses, just as one can get a child's swing going to great heights by a series of small, properly timed pushes. The only problem would be that as the particle went faster, it would move to an orbit of greater radius. And that is where the cancellation becomes important, because the pushes that are timed to a particle in the first orbit will also be properly timed for the second or any other orbit. This means that if we start with a particle in orbit and time our accelerations to give it energy, the accelerations will continue to be properly timed as the particle gains energy and moves to higher orbits. In this way, it should be possible to accelerate the particle to a very high energy by supplying a series of small appropriately timed voltages, rather than a single large one.

According to people at Berkeley at the time, when Lawrence realized the implications of this fact, he raced around the laboratory like a modern-day Archimedes, except that instead of shouting "Eureka," he kept stopping people and telling them that "*R* cancels *R! R* cancels *R!*"

The machine that Lawrence and his co-workers eventually developed is known as the cyclotron. The name itself started as something of a laboratory joke at Berkeley. The machine consisted of two semicircular, D-shaped magnets (called dees) arranged as shown in Illustration 10. There was a gap between these magnets across which a voltage could be applied. This voltage changes sign periodically, much as ordinary household voltage changes sign 60 times each second.

A proton that arrives at the right-hand gap when the far side of the gap is at a negative voltage with respect to the near side will be attracted toward the far side and pulled across the gap,

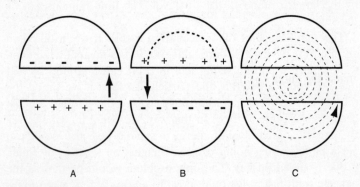

A   B   C

10. The principle of the cyclotron.

acquiring energy in the process (part A). While the proton is moving through the magnetic field of the upper dee, the voltage across the gap is changing so that when the proton arrives at the left-hand gap it is again attracted by the negative voltage of the far side (part B). Once more it is accelerated, and the process repeats.

A particle injected near the center of the magnets will move in a spiral path like the one shown in part C. Each time the particle crosses a gap, it is accelerated and moved to a higher orbit until it is extracted at the outer edge of the machine. The end result of this process is a beam of energetic particles that can be used in place of cosmic rays in experiments.

The development of the cyclotron in the 1930s makes a fascinating story. Lawrence commandeered an old shack on the Berkeley campus for his laboratory and went around knocking on the doors of private foundations to raise money for the project. Meanwhile, starting with a pillbox-shaped machine 4 inches in diameter, he and a former student, Stanley Livingston, worked day and night to overcome the technical difficulties involved (some of which were truly prodigious). An improved version of the 4-inch machine yielded protons with energies of 80 keV in 1931, and in February 1932 an 11-inch version of the machine achieved Lawrence's original goal of 1 million volts (1 MeV). Livingston's recollection of the event is vivid: "I wrote the figure on the board. Lawrence came in late. . . . He saw the board, looked at the microammeter to check the resonance current, and literally danced around the room."*

The figure of 1 million volts was chosen as a goal for a number of reasons. It is a nice round number, and as such it undoubtedly made the fund raising easier. It was also generally accepted at the time that energies of this magnitude were needed to penetrate the nucleus, so Lawrence felt that an energy of at least an MeV would have to be achieved before any significant experiments could be done. People in the laboratory were so confident that this was the case that they had wired the Geiger counter,

---

* As quoted in Nuel Pharr Davis, *Lawrence and Oppenheimer* (New York: Simon and Schuster, 1968).

which monitored the radiation from the beam target, into the same circuit as the machine, so that it operated only when the beam was on.

Imagine their surprise, therefore, when they learned that a group of English scientists, using an accelerator of the old-fashioned design, had produced a collision in which the target nucleus had been broken apart, producing new chemical elements in the process. Realizing that if a beam of a few hundred keV could produce an artificial transmutation of elements, then their cyclotron would have been doing the same thing, Lawrence's group went into the laboratory and rewired the circuits so that the Geiger counter would stay on after the machine was turned off. Sure enough, the counter kept clicking, indicating that the cyclotron beam had produced new chemical elements that were now decaying like any other radioactive material. There must have been some pretty sick faces in the laboratory on that day. It is bad enough to be beaten to an important discovery, but to be beaten because you did not leave a machine on . . . In the words of one of the participants at that scene, "We felt like kicking each other's butts."*

But if this achievement eluded Lawrence's new radiation laboratory, other honors came thick and fast. In 1933, at the age of thirty-two, he was elected to the National Academy of Sciences as its youngest member (and first South Dakotan). In 1939 he received the Nobel Prize. The cyclotron quickly developed into the principal instrument for the study of nuclear physics, and it also became one of the prime sources for exotic radioactive materials, which were used in medicine as diagnostic tracers and for the treatment of cancer. In 1940 Edwin M. McMillan identified the elements neptunium and plutonium in targets that had been irradiated by the cyclotron beam, an achievement for which he later received the Nobel Prize. These elements do not occur in nature, but are the first such artificial chemical elements ever produced—the first of a dozen or so that are now known, and most of which were created at Berkeley.

With his brother John (an M.D.), Lawrence was quick to ex-

* Nuel Pharr Davis, ibid.

ploit the medical uses to which his machine could be put. In addition to the production of radium and other radioactive elements for cancer therapy, experiments were undertaken to assess the usefulness of the cyclotron beam in the direct treatment of tumors. This is still a field in which beams of accelerated particles are used, and there are treatment facilities in the United States where beams of protons, neutrons, and even pi-mesons are used in cancer treatment.

This aspect of the cyclotron must have acquired a very special significance for the Lawrence brothers when their mother was diagnosed as having terminal cancer in 1938. In what must be one of the most dramatic and little-known episodes in the history of physics, the brothers brought her to the experimental treatment facilities in Berkeley and treated her with large amounts of radium. She may even have been the first person to have been treated with the neutron beam from the cyclotron, although there is no official evidence of it. Whatever the treatment was, it was successful, and she lived a vigorous life until the age of 83.

In many ways, Lawrence was the prototype of a figure that has become rather common in modern physics—the big-time operator. Most of the experiments that we have discussed previously have been rather modest in scope. They required relatively little in the way of financial support and could be run by a research team consisting of a scientist and a few students. Developing the cyclotron could not be done in this way. It required a major laboratory with engineers, technicians, and scientists from many different areas of specialization. It is probably the first example of what we now call team science. The man who can run such an operation needs to have abilities beyond those we normally associate with the scientist. He has to be able to "shake the money tree," coordinate the work of many different individuals, and, in his spare time, think up good experiments for his team to perform. Today, when even quite ordinary experiments at modern accelerators require dozens of workers and millions of dollars, these skills are even more in demand.

THE SYNCHROTRON

As Lawrence saw things, there was no limit to the amount of energy that a particle could be given in a cyclotron. Just make the machine bigger, he said, and we can go on accelerating forever. Just before World War II, he even managed to obtain funds for a machine that would achieve 100 MeV.

There are, however, some fundamental limits on cyclotron operation discovered through the study of the theory of relativity, one of which is that as particles start to move at speeds approximating that of light, they become more massive. Since the velocity of the particle in a cyclotron depends on the mass, this places limits on the energy of the particles in the beam. The theoretical limit is about 25 MeV, well below Lawrence's goal of 100 MeV. In fact, the highest energy cyclotron in use today produces a beam of about 22 MeV protons.

How important is this limit? Twenty MeV is an energy that is large enough to do almost any experiment involving the nucleus of the atom. But if we want to study elementary particles, it is a rather small energy. For example, to produce a pi-meson in a collision between the beam and the target, we need an absolute minimum of 140 MeV, the energy equivalent of the pion mass. Clearly, this sort of experiment cannot be done with a 20 MeV machine.

In fact, if we want to talk about accelerators that could be used to produce and study elementary particles, it is probably a good guess that we would need energies in the range of a few GeV, rather than in the range of hundreds of MeV. It follows that we would need a machine that works on an entirely new principle. This new principle was realized in the synchrotron, the basic prototype of the present generation of accelerators.

The basic limitation of the cyclotron is that it was designed to contain ever faster moving particles with a single, constant magnetic field. This difficulty was overcome in the synchrotron by increasing the magnetic field as the particle becomes more energetic. A typical synchrotron apparatus consists of a series of magnets in the shape of a hollow ring and one (or more) places where forces can be applied to accelerate the particle. Suppose that the field in the magnets is adjusted so that a given particle is moving in a circle whose radius is exactly that of the ring itself. If this particle were not accelerated, it would simply continue to move around the ring. If it is accelerated, then the same law that applied in our discussion of the cyclotron says that the particle will move to an orbit of larger radius. If nothing were done to counteract this tendency, the particle would soon strike the wall of the machine and be lost.

Suppose, however, that we arrange things so that as soon as the particle passes the acceleration point, the field in the magnets is increased. There will then be two competing effects: The acceleration will tend to make the particle move to a larger radius, but the increased magnetic field will tend to move it to a smaller one. If we adjust the acceleration and the magnets just right, we can make these two effects cancel, so that the particle will keep moving around the ring, even though it is more energetic. Thus, by constantly stepping up the magnetic field as the particle's speed increases (a process that accelerator people call *ramping*), we can keep increasing the energy of the particle in small increments until it reaches the energy we need for a particular experiment. The only limit on the synchrotron is the size of the ring and the expense of building large machines.

Unlike the cyclotron, which can deliver a continuous beam of accelerated particles, the synchrotron runs through the ramping cycle for a bunch of particles and then starts the process again for the next bunch. It consequently delivers its accelerated particles in short bursts, rather than continuously, but this is a small price to pay for the higher energies it can attain.

The first synchrotron to break into the GeV range was built at Brookhaven National Laboratory on Long Island, New York. In 1953 it delivered a beam of 3 GeV protons from a ring 60 feet in diameter. The largest proton synchrotron at this time is located at the Fermi National Accelerator Laboratory (Fermilab, FNAL, or, to old-timers, NAL) near Chicago (we will discuss it in more detail later). It delivers 1 TeV protons from a pair of rings over a mile in diameter. Thus, we have come quite a way from the first 4-inch cyclotron that Lawrence put together in his laboratory.

The synchrotron is the machine of choice for delivering protons at very high energies. For electrons, though, it has a fundamental limitation. In Chapter I we saw that any accelerated electrical charge will emit photons. Electrons moving around in a ring are being accelerated and will therefore produce radiation. This radiation results from the forces associated with the magnetic field (i.e., with the forces that keep the electron moving in a circle), and not primarily from the modest boost the electron receives each time it comes around the ring. We thus have a situation in which energy is being added to the electrons by the machine and being lost by radiation.

Because they are rather light, electrons radiate much more than heavier particles, such as protons. You can see a bluish glow in electron machines, which results from this so-called synchrotron radiation. Thus, for electrons, the limit at which as much energy is lost through radiation as is added by the acceleration device occurs at a fairly low energy. Because of this problem, getting high energy from circular electron accelerators has always been more challenging than from proton machines. At any given time, the energy available from proton machines has always been 5–10 GeV greater than that from electron machines.

To get past this limit, use is made of devices in which electrons are accelerated in straight lines. These are called linear accelerators, or linacs. They, too, were developed in the 1930s and have played an important role in particle physics. The cross section of a typical linear accelerator is a long hollow tube divided at intervals by rings that form separate compartments. Both the tube and the rings are made of a conducting material, such as copper. Each compartment has an independent power supply that can create an electric field. The power supplies are operated in such a way that the resulting electromagnetic wave appears to travel down the tube from one compartment to the next. The electrons "ride" this wave in much the same way that a surfer rides a water wave. As the electrons speed up, the velocity of the wave increases accordingly, so that they stay at the point on the wave where they get maximum acceleration.

The center for research on linear accelerators was at Stanford University, near San Francisco. Starting just after World War II, a series of linear electron accelerators (named, appropriately enough, the Mark I, Mark II, and Mark III) eventually produced electron beams of 1.2 GeV. Robert Hofstadter used these machines for detailed studies of the shapes of nuclei and of the proton, for which he was awarded the Nobel Prize in 1961. On May 21, 1966, the first electron beam was brought through the ultimate electron machine—a 2-mile-long accelerator operated by the Stanford Linear Accelerator Center (SLAC). This machine produces 20 GeV electrons and has been the source of electrons used in several of the important discoveries we will describe later.

SECONDARY BEAMS

The acceleration of electrons and protons remains the center of concern in modern high-energy machines, but over the years a number of other extremely important and useful functions have been developed for these machines. The most interesting of these functions is the secondary beams.

Suppose that the high-energy proton beam from an accelerator is allowed to strike a target. The target could, in principle, be any material, but it is usually a block of metal, such as copper. When the protons collide with the nuclei of the target, all kinds of particles are produced. These secondary particles emerge from the target in a narrow cone. By running the particles through a suitable arrangement of magnets and slits, we can arrange it so that only the positive pi-mesons (for example) of a certain desired energy come out into the experimental area.* By this process, we can use the primary proton beam to produce a secondary beam of pions and use these pions in our experiments. In this way, it is possible to carry out detailed studies of the interactions of these secondary particles with matter.

In modern accelerators, beams of pi- and K-mesons, neutrons,

---

* This is known as magnetic selection, a technique to be discussed later in more detail.

high-energy photons, muons, and neutrinos are routinely available, along with beams of other particles that we have not yet mentioned, such as antiprotons.

If we start with a beam of pi-mesons and wait for a while, they will start to decay into mu-mesons. The mixed beam of pions and muons that results can be "cleaned up" by running it through an appropriate magnet, so that a beam of pure mu-mesons results. This beam is then ready for use in experiments. Over the last decade, as accelerator technology has improved, it has become possible to go one step further. If the mu-meson beam is allowed to go on for a while, the mu-mesons will decay into electrons and neutrinos. If *this* mixed beam is run through a large block of material (typically, steel plates or hundreds of yards of dirt), all the particles will be removed from the beam by interactions, leaving only the neutrinos. The result is a beam of neutrinos. As we shall see later, this beam at the Fermilab has been used to make some important discoveries about the nature of weak interactions.

One more point of interest about secondary beams: The lifetime of the pi-meson is 2.5×10^{-8} second, so if it were traveling at the speed of light you might expect it to go a few centimeters or so before it decayed. Since secondary beams must be meters (and even hundreds of meters) long, do we not face a fundamental contradiction here?

Actually, there is a contradiction unless we apply one of the corollaries of the theory of relativity. The theory tells us that a moving clock appears to run slower than a stationary one (or, more precisely, that a moving clock will appear to be running slower than a stationary one to an observer who is also stationary). The relation between the times measured by the two clocks

$$T_s = \frac{T_v}{\sqrt{1 - v^2/c^2}}$$

where T_s is the time on the stationary clock, T_v the time the moving observer sees on the moving clock, and v the velocity of the moving clock.

If we imagine an observer sitting on a meson, then as far as he is concerned the velocity of the meson is zero, and the meson will decay in 2.5×10^{-8} second, as expected. But if the meson is moving at a high velocity with respect to the laboratory, someone in the laboratory will see the clock that moves along with the meson going much more slowly than his own. Thus, the meson will travel farther than the expected few centimeters in the laboratory.

For example, if the meson is moving at 99.999 percent of the speed of light, then an interval of 2.5×10^{-8} second on the clock moving with the meson translates into

$$T_s = \frac{2.5 \times 10^{-8}}{\sqrt{1 - (0.99999)^2}} = 5.5 \times 10^{-6} \text{ sec}$$

for a clock in the laboratory. In this time, the meson will travel a distance $D \approx 3 \times 10^8 \times 5.1 \times 10^{-6} = 1,530$ meters in the laboratory —ample space for the construction of the secondary beam. Hence, the fact that secondary beams exist at all can be taken as evidence for the theory of relativity!

FERMI NATIONAL ACCELERATOR LAB: A TYPICAL
MODERN MACHINE

To get a sense of what a modern accelerator facility is like, you could drive about 50 miles east of Chicago to a facility near the town of Batavia. The Fermi National Accelerator Laboratory, or Fermilab, the nation's premiere high-energy physics installation. Located on some 7000 acres of flat Illinois prairie, the lab is interesting for all sorts of reasons as well as for the accelerator operating there.

For example, when construction on the lab first started in the late 1960s, the site included a small post–World War II subdivision known as Weston. Once the houses had been acquired by the state of Illinois, the town had to be legally disincorporated. Imagine the legal team's surprise when they found that there was no provision in Illinois law for this procedure. Thus, Fermilab may be the only scientific center that required a special act by a state legislature before construction could start.

In the early days, the old ranch houses served as offices and laboratories. Several were joined, for example, to make the science library and meeting room, and I can remember the slightly disjointed feeling of being assigned an office in what had obviously been a bedroom a few years earlier.

Today, of course, there is a modern (and striking) headquarters building designed by the lab's first director, Robert Wilson. It consists of two multistory upraised piers, with the space between them covered with glass. Inside, there is a tall atrium that is one of the most pleasant indoor environments I've ever seen. Between the old town and headquarters is another unique feature of Fermilab—a ring of two-story white buildings around a circular drive. These are the farmhouses that used to stand on the land occupied by the lab. They were moved to a central site, refurbished, and are now used as living space for scientists who have to spend weeks or months on site as their experiments run.

Much of the remaining land has been restored to tall grass prairie—the ecosystem that used to occupy this area before the land was tilled. Thus, one unanticipated effect of mankind's drive to understand the universe has been the restoration of some of Illinois's original prairie habitat.

But, of course, the real business of Fermilab is accelerating protons. It is the highest energy accelerator now operating in the world. In fact, as in most modern machines, the final energies are too high to be achieved all at once in a single machine. Instead, in a process analogous to shifting gears in a car, the protons go through several stages before they come out of the machine and are used for experiments.

At Fermilab, protons begin by being boosted to 200 MeV in a linear accelerator ("first gear"). They then enter a small booster synchrotron that takes them up to 8 GeV ("second gear"). At this point, they are ready to go into the main ring, which is *three miles* around. The original machine, which was capable of accelerating protons to 500 GeV, is still in place in the underground tunnel. It consists of 1000 ordinary magnets that keep the protons in the ring. Today, this machine is just the "third gear" at Fermilab, accelerating protons to about 150 GeV before injecting them into the final stage of acceleration.

This final stage takes place in a ring of magnets that were installed below the original ones in the tunnel in the early 1980s.

When the protons have reached 150 GeV, they are directed a few feet downward into the "Tevatron," where they can be accelerated up to 900 GeV.

One important point about the Tevatron is that the magnets in it represent a major technological breakthrough. They are superconducting—the Tevatron was the first major installation of such magnets in the world. A superconductor is a material that, if maintained at temperatures near absolute zero, conducts electricity without loss of energy. If you make a magnet whose windings are superconducting and start a current going, then the current will keep going and the magnet will keep working even if you disconnect it from the power source.

Today superconducting magnets constitute a billion-dollar-a-year business. If you've ever had an MRI scan, for example, a superconducting magnet was part of the apparatus. It was the demand for superconducting wire for Tevatron magnets that stimulated American manufacturers to develop the capability to play an important role in the worldwide superconductor industry. Whenever people ask me about practical payoffs from high-energy physics, I always point to this example (there are others, of course).

But the most important aspect of Fermilab is not the production of an external proton beam, but its use as a kind of machine we have not yet discussed—a collider.

COLLIDERS

You know that a head-on collision between two cars going 30 miles per hour is likely to cause a great deal more damage than a 30-mile-per-hour collision with a fixed object like a wall. In the language of physicists, we say that there is more energy available in the head-on collision, and we see the effects of that extra energy in the extra damage.

In just the same way, the use of accelerators as we have been describing them, in the so-called "fixed target" mode, produces only a limited amount of energy that can be used to produce new particles or explore the properties of a known one. Starting in the late 1960s, physicists began to think about building machines that made full use of the energy available in head-on collisions between elementary particles. Today, the highest energy machines available, as well as those that exist only in the minds of their designers, operate in this way.

The way a collider is built depends on the particles involved. Suppose, for example, that the accelerator produces high energy protons. When they come out of the main ring, instead of being sent into an experimental area, they are sent into one of two separate structures called storage rings. Like the main ring, the storage rings have magnets to keep the particles moving in a circular track. Unlike the main rings, the storage rings do not accelerate the particles—they just store them.

Typically, in this sort of system there will be two storage rings, one built over the other to save space and construction costs. The protons will circulate clockwise in one, counterclockwise in the other. Then, when the experimenters are ready, the two beams are brought together to produce a head-on collision.

There is a particularly elegant version of the collider apparatus when we wish to produce head-on collisions between a particle and its antiparticle. In this case, only one storage ring is needed, because the opposite electrical charges of the two particles ensures that they will circulate in opposite directions within the same ring. In a sense, a particle-antiparticle collider represents the ultimate in energy production in accelerators. We have the energy not only of the head-on collision, but of the mass of the two particles as well (if they annihilate).

Let's look at Fermilab as an example of how a modern collider works. A bunch of protons from the main ring are allowed to hit a block of copper. In the resulting collisions, about 10 billion antiprotons are created. These particles are brought to a storage ring (which is actually shaped like a triangle). In a triumph of modern fast electronics, the condition of the stored antiprotons is monitored on one leg of the triangle, then, while the antiprotons are going the "long way around," signals are sent to magnets on another leg, so that when the antiprotons arrive at that point, any stability problems in the beam can be corrected. Over a period of several hours, physicists "build their stack" of antiprotons.

When there are enough antiprotons on hand, they are injected into the main Fermilab ring. There, they circulate in the opposite direction from the protons. The two countercirculating beams are accelerated simultaneously to 900 GeV, at which point they are allowed to collide.

The whole operation of creating, storing, injecting, and accelerating the antiprotons is called a "shot" and, believe it or not, is overseen by a scientist called a "shotmaster."

VII

The Proliferation of Elementary Particles

Things were not slow in becoming curious.
—THOMAS PYNCHON, *The Crying of Lot 49*

THE DISCOVERY OF THE ANTIPROTON:
A TYPICAL ACCELERATOR EXPERIMENT

BY THE EARLY 1950s, a number of lines of research came together in a way that made the next big push in elementary particle work possible. The cosmic ray data had shown that there were many more particles than had been expected. One textbook went so far as to title a chapter "Particles We Might Do Without." At the same time, it was pretty clear that cosmic ray experiments had reached the limit of the results they could produce, and the advent of accelerators in the GeV range came at just the right time to keep things going. One of the major experiments that was performed resulted in the discovery, at Berkeley, of the antiproton.

We have already encountered the idea of antiparticles in connection with the positron. That a particle should exist with the same mass as the proton but with a negative electrical charge was accepted on faith by most theorists. There were even a few cosmic ray events that could, with some stretch of the imagination, be interpreted as evidence for such a particle. Getting hard laboratory evidence, however, was another matter. With a proton beam, the only way an antiproton could be produced was through the reaction $pp \rightarrow ppp\bar{p}$. The symbol $\bar{p}$ is to represent the antiparticle. (We will use the convention of the bar over the particle symbol to denote antiparticles from this point on.) The reason that the production of an antiproton has to be accompanied by the production of an extra proton arises from conservation laws that we will discuss later.

How much energy is needed to create the antiparticle? The first impulse is to say that since two extra particles of mass 938 MeV have to be produced, we need a beam with kinetic energy $2 \times 938 = 1,876$ MeV. It turns out that energy and momentum conservation require that the four particles in the final state of the reaction cannot be sitting still: They must have some minimum amount of kinetic energy. When this is taken into account, the kinetic energy of the proton from the accelerator has to be about 5.6 GeV. This requirement was very much in the minds of the men who designed the bevatron at Berkeley. In fact, that machine has been characterized as the accelerator "designed to produce the antiproton."

Once the required energy was achieved, the problem that faced the experimenters was how to tell which of the many negative particles produced in collisions were antiprotons, and, most important, how to distinguish them from negative pi-mesons. There were several tools available, which should be described separately before we can grasp how they were put together in an experiment.

We know that when particles of a given mass and velocity move into a magnetic field, they move in a circle whose radius is given by

$$R = \frac{mv}{Bq} = \frac{P}{Bq}$$

where we have replaced the quantity mv in the second equality by the letter P. This quantity is known as the momentum of the particle.

From this equation it follows that if two particles of different momentum enter a magnetic field, each will start to move in a circle with a different radius. For example, in Illustration 11 we show two particles of momentum P_1 and P_2. The orbits of these particles will be as shown, where the radii R_1 and R_2 are given by the equation above.

Suppose we put a narrow slit at point A, as shown. A particle of momentum P_1 will pass through the slit, but a particle of momentum P_2 (or any momentum other than P_1) will hit the solid material. As a result, on the other side of the slit we would see only particles of momentum P_1. In this way, we can say that the magnetic-field-plus-slit arrangement "selects" momentum P_1 and rejects all others. This apparatus is called a magnetic spectrometer, or momentum analyzer.

Imagine, then, that particles produced in a collision are run into such a magnet. If we are looking for negative charges, then positively charged particles will curve in the opposite direction to the orbits shown in this illustration, and will come nowhere near the slit. Furthermore, only those negatively charged particles with a preselected momentum will pass through the slit. Thus, this simple operation will immediately reduce the unwanted particles by a large factor.

However, it is possible for a negative pion and an antiproton to have the same momentum even if they have different masses. All that is required is that their mv be the same. Another technique must be used that will give an independent determination

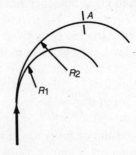

11. The magnetic spectrometer.

of the particle velocity after the momentum selection has been made. The simplest way to do this is to allow the beam of particles to pass through two thin layers of scintillating material a known distance D apart. A particle will cause a flash of light from each of these as it passes. By measuring the time between these flashes, the velocity of the particle can be determined. This is called a time-of-flight measurement, and is quite common in experimental physics. Obviously, its success depends critically on the experimenter's ability to build fast, accurate electronic timers.

A second way of determining the velocity of fast particles was developed by the Russian physicist Pavel A. Čerenkov in 1934. When a fast particle enters a material such as a gas or liquid, it can emit a burst of light. This light, called Čerenkov radiation, is somewhat analogous to the sonic boom emitted by supersonic aircraft. From our point of view, the important aspect of this radiation is that the angle at which it is emitted depends on the velocity of the particle. From the angle of the radiation we can determine how fast the particle is moving through the medium.

With this background in experimental technique, we are in a position to understand the details of the antiproton search. The apparatus is sketched in Illustration 12, on page 104. The proton beam from the accelerator strikes the target, thereby creating a flood of particles and, perhaps, an antiproton or two. The secondary particles are run through a magnet to select the proper momentum, and are then brought out through the concrete shielding to a scintillation counter S_1, another magnet, and a second scintillation counter S_2. From time-of-flight measurements, the velocity of the particle is now known. The beam continues to two Čerenkov counters, the first of which is set to register if a particle faster than an antiproton goes through it, and the second of which is set to register if a particle having the correct speed for an antiproton is there.

Why are the Čerenkov counters there at all? Is the time of flight not a sufficient criterion for distinguishing between antiprotons and mesons?

In principle, one velocity measurement should suffice to iden-

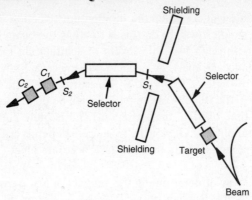

12. Apparatus used in the antiproton search.

tify the antiproton. But in an experiment of this type it is not at all unusual for the beam coming from the selector to contain tens of thousands of pions for each antiproton. It might be possible, therefore, for one pion to go through the first scintillator and for a different pion to go through the second in just such a way as to produce a spurious time of flight that is exactly equal to that of the antiproton. Such "accidental coincidences" are the bane of the experimental physicist's life. The Čerenkov counters are one way of dealing with them.

To identify a particle as an antiproton, three requirements must be met: The time of flight must be correct; the second Čerenkov counter must register; and the first Čerenkov counter must not register. (In technical terms, we say that we "veto" with this counter.) Only if all three of these conditions are satisfied can we be sure we have seen the particle we are looking for and not some accidental collection of pions.

This sort of careful "overkill" on the identification of particles is typical of accelerator experiments. In the antiproton experiment a total of fifty particles were seen in months of work, so we could think of the search as an analog of the needle in the haystack. For their discovery of the antiproton, Owen Chamberlain and Emilio Segrè of Berkeley were awarded a Nobel Prize in 1959.

THE PION-NUCLEON RESONANCE

In Chapter VI we saw how the intrinsic limits of the cyclotron made it impossible to produce pi-mesons. Nonetheless, the artificial production of pions remained an important goal in physics. They are, after all, the lightest of the strongly interacting particles, and therefore the easiest to produce. In addition, they are the primary particle involved in the nuclear force. Therefore, whether the purpose is to study the pions themselves or to use them to study other interactions, producing them becomes a very important goal.

In the late 1940s and early 1950s, before the synchrotrons became widely available, a machine called the synchrocyclotron was used for this purpose. This machine was a cross between the cyclotron and the synchrotron: Like the former, it guided particles with a constant magnetic field; like the latter, it accelerated particles in bunches. It overcame the cyclotron limit by lowering the rate at which the voltage across the gaps was changed as the particles moved toward larger radii. In this way, the slowing down of the particles because of increased mass could be taken into account. In a typical machine (like the one installed at the University of Chicago), the diameter of the magnets might be 170 inches (almost 15 feet) and the frequency might drop by 60 percent as the particles spiraled out. With such a machine, proton energies in the range of 400–500 MeV could be attained—enough to produce pi-mesons.

One of the first things measured when pion beams became available was the way in which pions interacted with protons. This information could be obtained by letting the pion beam strike a target composed of hydrogen and seeing what happened when the pion hit the hydrogen nucleus. One of the simplest questions to ask was this: "If a pi-meson comes near a proton, what are the chances that the two will interact with each other?"

If two pions in a beam come near a proton target, one may not be deflected at all, so we would say that it did not interact with the proton. From an experimenter's point of view, we could recognize this situation by noting that this pion remained in the beam after it had passed the target. A different pion, on the other hand, may interact with the proton. This interaction could be through the electrical force if the pion is charged, or it could be through the strong interaction. In either case, the result is the same. The pion is scattered out of the beam.

Physicists usually express interaction probabilities in terms of something called a *cross section.* If you imagine holding a circular disk in the pion beam, it is clear that the disk will also scatter mesons out of the beam. The cross-sectional area of the disk that will scatter out just as many pions as does the proton is called the *pion-proton cross section.* It can be measured by allowing a beam with a known number of pions to enter a target that has a known number of protons in it. By counting the number of pions left in the beam after it has traversed the target, we can deduce the interaction probability, and, hence, the cross section.

Starting in 1952 with a group under the direction of Enrico Fermi at Chicago, physicists began collecting data on the scattering of charged pions from hydrogen. When they plotted their results, a graph similar to the one shown in Illustration 13 emerged. There was a large peak in the cross section at a pion energy of about 200 MeV—a peak that was about 100 MeV wide.

This sort of bump in a cross section means that when the pion has exactly the right energy with respect to the proton, it is much more likely to interact than if it has some other energy. One way of thinking about these situations is to imagine that at this precise energy, the pion and the proton can "lock together" for a short time, whereas at other energies they just bounce off one another. If we think of things this way, there is no reason why we should not think of the locked-together state as a particle. If we do so, we can represent the interaction of a pion and a proton near the peak of the cross section with a diagram similar to the one on page 107 (Illus. 14). The two particles come together and fuse into another particle which, after a short while, separates into the original pion and proton again. The intermediate particle in this diagram is called a *resonance,* and in modern terminology is denoted by the Greek letter Δ (delta). Since the particle in the diagram must have two positive charges, it is written Δ^{++}.

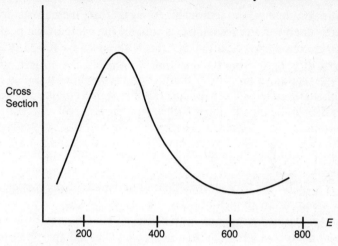

13. The cross section for the scattering of a positive pi-meson from a proton, showing the delta peak.

How long would such a particle live? One way of estimating the lifetime is to use the uncertainty principle we introduced in Chapter III. We can see from Illustration 14 that the Δ must have an energy equal to the sum of the pion and proton energies. But what is the uncertainty in the energies required to produce the particle? Clearly, if the pion is at the energy corresponding to the peak of the bump in the cross section, the resonance will be formed. But what if the pion energy is a little lower or higher, so that it corresponds to a point one-third of the way down the peak? Or one-half or one-fifth? Will these energies correspond to Δ^{++} production as well? In fact, if we follow

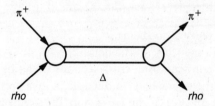

14. Interaction of a pion and a proton near the cross section's peak.

this principle, it seems reasonable to say that ΔE, the uncertainty in the energy of the resonance, is roughly the width of the peak in the cross section. For the Δ^{++} this is about $\Delta E \approx 100$ MeV = 1.6×10^{-4} erg, so from the uncertainty principle, the uncertainty in the time must be

$$\Delta t = \frac{h}{\Delta E} = \frac{6.6 \times 10^{-27}}{1.6 \times 10^{-4}} \sim 4 \times 10^{-23} \text{ sec}$$

This, in turn, can be taken as a reasonable estimate of the lifetime of the Δ^{++}.

This lifetime is very short compared to the life of any particles discussed so far. In fact, physicists in the 1950s were very reluctant to consider resonances as particles at all. One author went so far as to compare them to automobiles that fell apart before they left the factory. The reason for this attitude rests, I think, on the fact that it is so difficult to detect particles with short lifetimes. The strange particles live long enough to travel several inches in a cloud chamber, so their tracks can be seen in the usual way just by looking at the droplets. It requires a stretch of the imagination to extend the term *particle* to something that can be seen only indirectly and that never leaves a visible track in a detection device.

But the idea that something is a particle only if it is easy to detect seems a little artificial. The delta, after all, lives long enough to travel from one side of a nucleus to the other; therefore, its lifetime is quite respectable compared to the characteristic time of the strong interactions. Because many particles with lifetimes comparable to the delta have been discovered, and because these particles seem to play an important role in the strong interactions, physicists have become accustomed to applying the term particle to them.

To complete the story on the delta, careful study of the different pion-nucleon scattering cross sections shows that they are a family of particles, similar to the pions. They come in four charge states: Δ^{++}, Δ^+, Δ°, and Δ^-, where the superscripts refer to two positive charges, one positive charge, neutral, and one negative charge, respectively. The mass of the family is about 1,236 MeV.

The idea that a bump in a scattering cross section can be interpreted as evidence for a short-lived particle immediately suggests that there might be resonances in systems other than those involving the pion and the nucleon. We could ask, for example, whether there might be a resonance in the cross section for the scattering of one pion from another, or in the scattering of a pion from a $\Lambda°$. Unfortunately, although it is possible to make beams of pi-mesons, none of the other particles we have studied live long enough to be made into targets. Consequently, the direct identification of resonances in such systems cannot be accomplished in the same way as it was for the delta.

If you think of the mechanism by which a resonance is formed, however, you will realize that it is not really necessary to have a conventional beam-plus-target arrangement to make one. All that is necessary is that the particles that are to form it be near each other for a period of time that is characteristic of the strong interaction. This can happen in a target-beam experiment, of course, but it can also happen when the two resonating particles are produced in the same reaction. For example, it is possible to start with a beam of pions and have a reaction such as $\pi^-P \rightarrow \pi^+\pi^-n$, in which two pions are present in the final state. These two pions will interact with each other just as surely as they would if one were a target and the other a beam. Consequently, we can ask about whether they interact in such a way as to form a resonance or not.

If the particles form a resonance, then the reaction described above can be represented schematically as in Illustration 15. Here, again, the double line represents the resonance. In order to see whether the data support this kind of picture, we can look for bumps again—not bumps in a cross section, as in the case of the delta, but bumps in something called a phase space diagram.

We start this procedure by imagining that we are riding on the resonance in the previous diagram. After the resonance decays

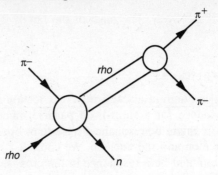

15. The process by which a rho-meson is produced and decays.

we will see the two pions moving off. Each of them will have some kinetic energy that we could, in principle, measure. We could then make a plot of the number of pion pairs produced with a certain total energy as a function of that energy. The result of such a plot could be either of the graphs in Illustration 16. In the left-hand graph the pion pairs are distributed evenly over the allowed range of energies. Such a graph would be interpreted as evidence that the two pions were produced independently of each other and therefore as evidence for the absence of a resonance. The right-hand graph, however, shows an excess of pion pairs at a given energy. This bump is what corresponds to the peak in a cross section. It tells us that there is some sort of interaction between the pions that causes them to be produced preferentially at this particular energy, which, in turn, means that there must be a resonance between the two particles.

The idea behind this method of analysis is that *if* we could make a pi-meson target and bombard it with a pi-meson beam, we would see a bump in the cross section corresponding to a graph similar to the bump in the cross section that led us to the delta. A bump in the phase space diagram, in other words, leads us to other resonances.

In 1961 a group of scientists working at Brookhaven carried out the same kind of analysis that we have been describing. There was a definite peak in the phase space diagram at around 760 MeV, with a width of a little over 100 MeV. This new parti-

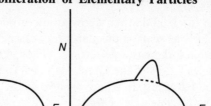

16. The number of particle pairs of a given energy for no resonance (*left*) and with a resonance (*right*).

cle lives long enough to participate in the strong interactions, but not long enough to leave a track in a cloud chamber, which is also true of the delta. The new particle is called the ρ (rho) meson, and like the pion, it comes in three charge varieties—that is, positive, neutral, and negative.

LIFE WITH THE STRANGE SET

The 1950s also proved a productive time in the realm of strange particles. You will recall that these were particles that decayed rather slowly so that they were visible in cloud chambers, either directly or through their decay products, as a ∨ particle. About the same time that the Brookhaven accelerator was coming on-line, the last in a series of cosmic ray discoveries were made. Not one, but two, new strange particles were discovered, each heavier than a proton.

In 1953 a particle of mass of about 1,190 MeV was seen. It was denoted by the Greek letter Σ (sigma). We now know that there are, in fact, three particles in the Σ "family," one with positive charge, one with negative, and one neutral. The charged particles decay into a nucleon and a pion in about 10^{-10} second, while the neutral Σ decays somewhat more quickly into a lambda and a photon. Like the lambda, this particle was totally unexpected and seemed to play no known role in physics.

Then, in 1954, the last of the "lucky" cosmic ray events was recorded. This time it was a particle of mass about 1,320 MeV, which decayed in 10^{-10} second into a lambda and a pion. Because the lambda then decayed into a nucleon and a pion, this particle was seen as the start of a "cascade" of decays, and was christened the *cascade* particle. Following the custom of giving particles letters of the Greek alphabet, this particle was denoted by the letter Ξ (xi). There are two xi particles, one neutral and one with a negative charge.

One thing to note about the cascade is that not only does it decay slowly itself, but its decay products include a particle, the lambda, which also decays slowly. Therefore, whatever it is that makes the lambda and the sigma "strange," should appear in a double helping in the cascade. We might even label it *doubly strange*.

The appearance of all these strange particles set the theoretical community to wondering, and by 1957 some interesting ideas started floating around. If the nucleon has a resonance (the delta), why should the strange particles not have resonances as well? In 1961 a group at Berkeley looked at the reaction $k^-p \to \Lambda°\pi^+\pi^-$ and did a phase space analysis of the lambda-pion state. Sure enough, a bump showed up with a mass of about 1,385 MeV and a width of about 40 MeV. This particular strange resonance decayed into a lambda and a pion in about 10^{-22} second, the characteristic time scale for strong interactions. From this we conclude that not every decay involving a strange particle has to be slow. This is a puzzling fact, but it is a fact. This particular strange resonance is now called the $\Sigma(1385)$. (Nomenclature will be discussed in Chapter IX.)

THE BUBBLE CHAMBER

Experiments of the type used to discover the antiproton are known generically as *counter experiment*. Provided that we have a pretty good idea of what we are searching for, they provide a very precise way of measuring particle properties. In the cosmic ray experiments, however, we saw that some of the most important results are those that are not expected. The ability to visualize the particle events provided by the cloud chamber was very important in the early work in particle physics.

In order for an interaction to be seen in the chamber, however, it is necessary that the incoming particle collide with a nucleus while it is passing through. When the energies of the incoming particles get into the GeV range, such a collision becomes less and less likely. For a cloud chamber to be effective in a modern accelerator beam, for example, it might have to be 100 yards across. A shorter chamber would simply not provide enough interactions to study the incoming particles. Clearly, construction of a cloud chamber on these dimensions is out of the question.

Consequently, visualization for high-energy particles was accomplished by a device somewhat similar to the cloud chamber, but without its disadvantages. It is called the bubble chamber.

Like the cloud chamber, the bubble chamber depends on the ionization of atoms in the wake of a fast particle in order to provide the centers around which the track will develop. Unlike the cloud chamber, the bubble chamber is filled with liquid held near its boiling point and under pressure. Instead of using the ions as centers of condensation, the ions serve as centers around which boiling occurs in the fluid. You can see a very similar phenomenon when you open a bottle of carbonated beverage. Bubbles start to rise to the top of the bottle, but if you look closely you will notice that the bubbles come in streams from definite spots on the inner surface of the bottle. These are spots where some local roughness provides a nucleus around which the bubbles can form. The folklore in the physics community has it that Donald A. Glaser, the man who developed the bubble chamber at the University of Michigan and who received the Nobel Prize in 1960, got his inspiration when he opened a bottle of beer in an Ann Arbor saloon. (The folklore does not include the brand name).

The bubble chamber operates as follows: A particle passes through the fluid (which is ready to boil if a nucleus for bubbles appears), leaving a trail of ions in its wake. A piston on top of the chamber is then pulled out quickly, lowering the pressure on the fluid, which starts to form bubbles around the ions. The bubbles are photographed, the ions are "swept" out of the chamber by a magnetic field, the piston is lowered, and the whole cycle can be repeated. Typically, the chamber will be operated each time a pulse of particles from an accelerator comes through, and the resetting goes on while the next pulse is being accelerated.

The size of a bubble chamber is normally measured in meters. Modern low-temperature technology has led to the construction of bubble chambers in which the working fluid is liquified hydrogen. In chambers of this type, the particles in the incident beam interact with hydrogen nuclei (i.e., protons), and these interactions are photographed as described above. In this way, the fluid serves both as the experimental target and the means by which the charged particles are detected. In such chambers, it is possible to run experiments of the "Let's put the beam in and see what happens" variety, and also to look for very rare situations where only a handful of events of the desired type might be expected. In addition, using bubble chambers makes it possible to record data on all the final states in a given reaction, so that the search for resonances can be carried out in each of the states separately after the experiment is finished. For example, in an experiment where a pion beam enters a hydrogen bubble chamber we could study the delta by looking at the reactions where only a pion and a nucleon are produced in the final state, and the rho-meson by looking at other reactions in which two pions (and a nucleon) are in the final state.

Although the bubble chamber was an important instrument in the history of particle physics, it has several important drawbacks. It produces film, rather than an electrical signal, and this, in turn, means that human beings have to examine the experimental output. Also, the bubble chamber records what happens after each collision, whether or not it produced a reaction of interest to the experimenters. Thus, a lot of the picture isn't particularly useful. Because of these shortcomings, bubble chambers are no longer in use at major installations—the last one at Fermilab, for example, was shut down in the early 1980s.

THE PROLIFERATION ORGY

If you are confused by this time, imagine how physicists must have felt in the early 1960s. It seemed that every time they turned around someone was discovering another particle. Far from solving the problem of the strong interactions, the advent of high-energy accelerators and sophisticated detection equipment seemed to lead to more and more confusion. As accelerators in the 20–30 GeV range came on-line at Brookhaven and at the European Center for Nuclear Research (CERN) in Geneva, the discovery of new particles (primarily of the resonance variety) became an everyday occurrence. In fact, bump hunting became something of a cottage industry for physicists throughout the 1960s.

Perhaps a nontechnical example will serve to illustrate this point. In 1963, the Finnish physicist Matt Roos compiled the first comprehensive table of elementary particles and resonances. Published in the *Review of Modern Physics,* Roos's article contained two tables and ran about five pages. There were seventeen "particles" and twenty-four "resonances." This article appeared periodically after that, expanding and being updated each time it appeared. In 1972 the distinction between particles and resonances was finally dropped. The last edition of the article, in 1976, ran 245 pages (with a thirty-page supplement) and took up an entire edition of the journal. Literally hundreds of particles are now known and cataloged in these tables.

Obviously, when particles start appearing in such profusion, the search for order among them takes on a very high priority. This will be the topic of Chapter VIII. Before we turn to it, there is one question that we ought to face.

We started the search for elementary particles in the hope of finding simplicity in nature. For a while it seemed we had succeeded, but the developments we have just described appeared to say otherwise. If there are more elementary particles than there are chemical elements, then the world has not been made any more simple than it was before. Obviously, the particles that have been discovered cannot be "elementary" in the intended sense of the word.

There is, in fact, a strong temptation to say that most of the very short-lived particles are not really elementary at all. We have already seen how reluctant physicists were to apply the label particle to them. But if we are going to exclude some of the newly discovered particles from the class elementary, we must have some criterion for doing so.

What will it be?

Stability? If we excluded all unstable particles, we would have to throw out the neutron and the pion, both of which are necessary to explain nuclear structure.

Lifetime? If we kept only long-lived particles, we would still have the puzzle of strangeness. In addition, we know that resonances like the rho and delta play a much more important role in keeping the nucleus together than do relatively obscure but longer lived particles, such as the cascade. Finally, saying that a lifetime is "long" or "short" implies that there is some standard of comparison against which a lifetime can be judged. For particles, the only reasonable standard is the characteristic time of the strong interaction. By this standard, *every* particle ever discovered is "long-lived," since all of them have lifetimes that would allow them to participate in the interaction.

Hence, there is no logically consistent way of calling some particles elementary and other, nonelementary. This situation was called nuclear democracy by Geoffrey F. Chew of Berkeley. Basically, as far as the strong interactions are concerned, all particles are equally important. We simply have to accept the fact that there are a lot of them and see where it leads us.

VIII

The Search for Order amid Chaos

A place for everything and everything in its place.
  —Anonymous

INTRODUCTION

THERE ARE two different senses in which one can talk of imposing order on a large group of seemingly unrelated objects, such as the elementary particles. For the sake of discussion, let us call them classification and reduction. The difference between the two can be illustrated by an architectural example.

The buildings in a large city would seem to form a class of unordered objects. No two are exactly the same, and a school might be found next to an apartment building, a high-rise office block next to a residential area, and so forth. If we were to impose an order on these buildings by classification, we would start by trying to find groups of buildings with common characteristics. When we had found such a group, we would give it a name and then lump under the one name all the buildings that shared this characteristic. In this way, the large number of individual buildings would be replaced by a few categories of types of buildings, and most people would agree that the situation was more orderly.

The classification of buildings could also be based on use—residential, commercial, and industrial (these are, in fact, the categories used by the Census Bureau). We could classify them by type of construction—wood frame versus steel and concrete. We could classify them by height, by year of construction, by value, or by any other criterion that we considered useful. I can recall being involved in a study of solar-energy potential in a city where we were trying to classify homes by the direction in which the roof pointed.

One important point about building classification schemes such as this is that an individual building may very well belong in several categories. Think of a wood-frame one-story residential building with a roof facing south that was worth $50,000. Which of the different attributes of the house was important would depend on your ultimate purpose. The price would be important if you were buying it, but the orientation of the roof would be more important if you were thinking about installing a solar collector.

If we took the same city and attempted to impose order by reduction, we would proceed in a different way. We would start by taking some buildings apart (either literally or figuratively) to see what they were made of. We would then start putting together a list of the kinds of materials we found—lumber, bricks, roofing shingles, glass, and so on. This list would probably not be long. We would then say that the materials in the list constituted the elements of buildings in the city, and that every structure was made of these elements arranged in different ways. The long and complex list of individual buildings would be replaced by a short list of basic building materials, and this, too, would result in a more ordered way of thinking about the city.

To use an analogy from science, someone who wants to find order among the collection of chemical elements can either group them according to their chemical properties (as Dmitri I. Mendeleev did when he put together the periodic table of the elements) or the researcher can find the basic constituents of the atoms (as Rutherford did in evolving the modern atomic theory). These are simply complementary ways of approaching the same problem.

Both classification and reduction have been used to bring the proliferation of elementary particles under control. In this chapter we will take up the various ways of categorizing these particles that have proved useful, and in Chapter IX we will discuss the idea that the particles we have seen are made up of a small number of basic building blocks.

CLASSIFICATION BY INTERACTION :
LEPTONS AND HADRONS

We have so far listed three different types of interactions that can affect an elementary particle. All charged particles are affected by the electromagnetic force, and hence can be said to participate in the electromagnetic interaction. Most of the particles we have studied are either created or decay via the strong interaction, and a few seem to be involved only in the weak interaction. We can use this difference in the type of interaction to introduce one way of classifying particles.

The electron, the muon, and the neutrino do not seem to be part of the strong interaction at all. These three particles are called *leptons* (weakly interacting ones). This is a rather small group of particles, but a very important one. As we shall see in Chapter XIV, leptons comprise one of the fundamental building blocks of matter in the universe.

All the other particles we have discussed, with the exception of the photon, are involved in one way or another with the strong interactions. They are called *hadrons* (from the Greek root hadrys, or strong). The proliferation of particles that we discussed in Chapter VII is entirely in this category. Consequently, most of the effort that physicists have made to try to sort out and categorize elementary particles has involved the hadrons.

Finally, there are particles like photons that are involved in mediating forces between elementary particles. If hadrons and leptons are the bricks from which the structure of the universe is made, then particles like the photon are the mortar that holds everything together. We will discuss them in more detail later.

CLASSIFICATION BY DECAY PRODUCT:
MESONS AND BARYONS

If we watch any hadron long enough, we will eventually see it decay into some collection of the stable particles—the proton, electron, photon, and neutrino. A possible decay scheme for the negative cascade particle is shown to illustrate this point:

$$\Xi^- \rightarrow \Lambda^\circ \qquad\qquad + \pi^-$$
$$ \searrow \mu^- + \nu$$
$$ \searrow e^- + \nu + \bar{\nu}$$
$$ \rightarrow n + \pi^\circ$$
$$ \searrow \gamma + \gamma$$
$$ \rightarrow p + e + \bar{\nu}$$

Some of these decays will be fast and some will be slow, but the end products are the stable particles.

There are two possible ways in which decay chains of this type can come out. There may be only leptons and photons in the final collection, or, as in the example above, there may be a proton as well. The presence or absence of a proton therefore becomes a criterion that we can use for classification.

Particles, such as the cascade, in which a proton does appear in the end product of the decays, are called *baryons* (heavy ones). The proton itself is included in this class, as are the lambda, the sigma, and the delta.

Particles whose final collection of decay particles is made up entirely of leptons and photons are called *mesons*. This definition of the term meson now supersedes the original one, in which the meson was thought of as a particle intermediate in mass between the proton and electron. The pi- and *K*-mesons obviously satisfy both definitions of the word. For example, the decay scheme for a π^+ is shown below:

$$\pi^+ \rightarrow \mu^+ + \nu$$
$$ \searrow e^+ + \nu + \bar{\nu}$$

On the other hand, with the new definition it becomes possible to talk about mesons that are more massive than the proton. Many particles of this type have been discovered. For example, there is a particle of the resonance variety called the A_2 meson with a mass of 1,310 MeV, which decays by a series of fast and slow decays as follows:

$$A^+{}_2 \rightarrow \rho^\circ \qquad + \pi^+$$
$$\qquad\qquad \hookrightarrow \mu^+ + \nu$$
$$\qquad\qquad\qquad \hookrightarrow e^+ + \nu + \bar{\nu}$$
$$\qquad \hookrightarrow \pi^+ + \pi^-$$
$$\qquad\qquad \hookrightarrow \mu^- + \bar{\nu}$$
$$\qquad\qquad\qquad \hookrightarrow e^- + \nu + \bar{\nu}$$
$$\qquad\qquad \hookrightarrow \mu^+ + \nu$$
$$\qquad\qquad\qquad \hookrightarrow e^+ + \nu + \bar{\nu}$$

Even though the A_2 is heavier than the proton, there is no baryon among its decay products. By our new definition it is a meson, something it would not be under the old scheme of things.

The concept of baryon and meson classifications is given a slightly more quantitative aspect by defining a quantity known as the *baryon number, B*. This is the number of protons that appear in the final state of a decay. For all of the baryons we have discussed above, $B = 1$, while for all of the mesons, $B = 0$. For antibaryons, $B = -1$.

CLASSIFICATION BY SPEED OF DECAY:
STRANGE VERSUS NONSTRANGE

In the last few chapters we have seen how a series of hadrons can be produced in the laboratory and in cosmic ray experiments. All of these are created in a time scale characteristic of the strong interactions, but some of them seem to take an awfully long time to decay. Thus, the speed of decay provides another way of distinguishing among particles. The strange particles seem to decay in times on the order of 10^{-10} second. The nonstrange particles (which we have been calling resonances up to this point) decay in 10^{-23} second or so.

In 1953 two physicists, Murray Gell-Mann (then at the University of Chicago) and Kazuhiko Nishijima (at Osaka University in Japan) independently suggested a theory that seemed to provide a good way of thinking about this phenomenon. They reasoned that in the case of the two stable particles, the proton and electron, the infinite lifetimes could be thought of as being due to a conservation law. For the electron, there is no lighter negatively charged particle into which it can decay, so the law of conservation of charge tells us that the electron cannot decay. In the same way, the proton does not decay because of the conservation of baryon number, a law that we shall discuss later. Thus, long lifetimes seem to be associated with conserved quantities.

Could it be that "longish" lifetimes are associated with quantities that are almost (but not quite) conserved? Gell-Mann and Nishijima postulated that there was another quantity similar to electrical charge carried by every particle. Nishijima called this the η (eta) charge, while Gell-Mann called it S. For the nonstrange particles, S is zero. For the lambda, sigma, cascade, and K-meson families, however, S is not zero. For technical reasons, the S-charge of all of these particles is taken to be -1. In a process such as $\Lambda^\circ \to p + \pi^-$ we then have the S-charge changing from -1 on the left to 0 on the right. If the S-charge were conserved like an ordinary electrical charge, this decay would be absolutely forbidden and the lambda would be a stable particle. The fact that the lambda is not stable means that S-charge, whatever it is, is not conserved exactly. The fact that the lambda has a long lifetime, however, does mean that it is almost conserved.

The other systematics of the strange particles are also explained by this hypothesis. If we assign the cascade an S-charge of -2, for example, then the double slow decay chain

$$\Xi^- \to \Lambda^\circ + \pi^-$$
$$\qquad \hookrightarrow p + \pi^-$$

can be understood, since the S-charge changes by one unit in each step.

When a strange resonance decays quickly into a lambda and a pion, however, we say that both the resonance and the lambda have an S-charge of −1. Although the S-charge is nonzero before and after the decay, it does not change during the decay. This means that the reaction can proceed quickly, as indeed it does.

Since the nonstrange particles have $S = 0$ and the strange particles have $S \neq 0$, Gell-Mann called the new quantity *strangeness,* a name that has stuck. We say, therefore, that the $\Lambda°$ has strangeness −1, the cascade strangeness −2, and so on.

Physicists readily adopted this whimsical turn of phrase because it tended to give a rather serious and abstract subject a lighthearted side. Perhaps in the aftermath of the Manhattan Project it was felt that a little humor would do the profession some good. This lightness has been bought at a price, however. Although physicists mean something precise when they talk of strangeness, and can relate it to measurable quantities, such as the length of tracks in a bubble chamber, it is almost impossible to disentangle the precise concept from the everyday connotations of the word. When a middle-aged savant speaks learnedly of baryons, hadrons, and unified field theories, he can sound impressive. When he talks of strangeness (or, more recently, of such concepts as "color" and "flavor"), he sounds slightly frivolous. The names tend to trivialize a rather serious endeavor, and, in the present climate of research funding, may prove a positive embarrassment to the field.

This is all water under the bridge at this point, though, since both the concept and the term strangeness are firmly entrenched as a property of elementary particles.

CLASSIFICATION BY INTERNAL DYNAMICS : SPIN

Although it is an incorrect depiction, in this section we will think of elementary particles as small spheres of material rather than as smeared out wave functions. By so doing, we can see that there is one possible type of motion of the particle that we have not yet discussed—the possibility that the particle may rotate (or spin) around an axis. There are many examples of spinning spheres in the macroscopic world. Perhaps the most familiar is the earth itself, which turns on its axis once a day.

In describing such a rotational system, physicists like to define a quantity called the *angular momentum*. This is a quantity analogous to ordinary linear momentum (mass times velocity). Just as linear momentum is conserved and is related to the tendency of a moving object to keep moving unless acted on by a force, angular momentum is also conserved and expresses the tendency of a rotating object to keep on rotating unless a force acts to slow it down. The angular momentum of a spinning object is usually represented by an arrow; therefore, to define it completely we have to specify both the direction and length of the arrow.

In Illustration 17 we show two rotating spheres. They are identical in every respect except for the direction of rotation. We could express this difference by saying that one sphere is spinning clockwise and the other counterclockwise. It turns out, however, that it is more convenient to say that the angular momentum arrow representing the angular momentum for the two spheres points in a different direction for the two cases. By convention, the direction of this arrow is defined by something called the *right hand rule*. The rule says that if you wrap the fingers of your right hand in the direction of the rotation of the sphere, the angular momentum will point in the direction of the thumb of that hand. Thus, the angular momenta for the two spheres in the figure are as shown.

Since angular momentum is supposed to be related to the tendency of a body to keep rotating, we would expect the momentum to become larger as the mass and size of the body increase, and to grow smaller as the rate of spin is reduced. For a sphere of radius R and mass M, turning around every T seconds, the length of the arrow representing the angular momentum turns out to be

$$L = \frac{2}{5} MR^2 \cdot \frac{2\pi}{T}$$

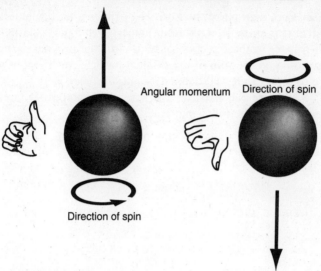

17. Angular momenta of two spheres.

For a classical macroscopic sphere, such as the earth, the period of rotation, T, can be any number at all. When we go to quantum mechanical objects, such as elementary particles, however, the situation is different. Just as in quantum mechanics an electron in an atom can only be at certain specified distances from the nucleus, the same laws state that a particle can spin only at certain specified rates. According to quantum mechanics, the angular momentum of a spinning particle can only have the values

$$L = J(J+1)\frac{h}{2\pi}$$

where J is called the spin quantum number of the particle. J can have only half- or whole-integer values—that is, it can be $\frac{1}{2}$, 1, $\frac{3}{2}$, 2, $\frac{5}{2}$, 3, and so forth. It cannot be $\frac{3}{4}$ or $\frac{2}{3}$ or any other value in between.

The spin of unstable elementary particles is usually deduced from looking at the directions in which the decay products are emitted when the particle disintegrates. The usual procedure is to construct a graph of the number of times a particular decay product (for example, a pi-meson) emerges at a particular angle with respect to the direction of motion of the original particle. Depending on the spin of the original particle, such graphs will have different (but well-defined) shapes.

The spins of the particles we have encountered so far are summarized in the following table:

Spin 0

pion

Spin $\frac{1}{2}$

electron, proton, neutron, muon, neutrino, Λ°, Σ, Ξ,

Spin 1

photon, ρ

Spin $\frac{3}{2}$

Δ

The highest spin particle found so far is called the *h*-meson. It has a mass of 2,040 MeV and a spin of four. There is, however, no theoretical limit to the spin—it can be as high as it wants. Consequently, the fact that very high spin particles have not been found is probably related more to the lack of interest in designing experiments to find them than to any intrinsic natural limitation.

There is still another important difference between the classical and quantum mechanical properties of particles with angular momentum. If we define a direction in space (for example, by putting the particles in a magnetic field that points in a fixed direction), there is no necessary connection between the spin of a classical object and this direction. The classically spinning particles can have their angular moment pointing in any direction at all. In the quantum mechanical case, however, the angular momentum can only point in certain specified directions in space.

The allowed directions for the spin of a quantum mechanical particle are determined in the following way: For spin ½ parti-

cles, there are only two—"up" and "down." For spin one p
cles, there are three—"up," "down," and "sideways." The n
ber becomes higher as the spin increases.

CLASSIFICATION BY ELECTRICAL CHARGE: ISOSPIN

We have seen that one of the most important properties of a
particle is its electrical charge. We have also seen that many of
the elementary particles seem to occur in families where the
members are identical in all respects—even to the point of having roughly equal masses—except in electrical charge. The pion,
coming in positive, neutral, and negative varieties, is a good example of this kind of occurrence. Furthermore, it appears that as
far as the strong interaction properties of a family of particles
are concerned, it makes very little difference which member of
the family is involved. The strong interactions, in other words,
do not seem to depend on the electrical properties of an individual particle.

In an attempt to provide a simple unified way of understanding these facts, theoretical physicists have drawn an analogy between the laws that govern spin (see the previous section) and
the electrical properties of particle families. Suppose, they say,
that the laws that govern ordinary spin also govern another kind
of quantity—a quantity related to the charge. To stress this
purely mathematical analogy, this new quantity is called *isotopic
spin*. Suppose further, they say, that there is some abstract,
mathematical space in which we can imagine the isotopic spin to
be an arrow, just as we represented ordinary spin as an arrow in
the last section. In this abstract isotopic-spin space the isotopic
spin can therefore only be oriented in certain directions, just as
the ordinary spin can only be oriented in certain directions in
ordinary space. This means, for example, that an isotopic spin of
one would correspond to a situation in the abstract space in
which the projections of this spin along the axis are +1, 0, and
−1.

Do we know of any family of particles in which the charge
comes in three states, positive, zero, and negative? The answer,
of course, is yes. The pi- and rho-mesons both qualify. We then

agree that instead of talking about these particles as coming in different charge states, we shall regard them as a single particle that can have three possible orientations of its isotopic spin vector. In this way of looking at things, the members of any family are now considered (at least as far as the strong interactions are concerned) the same particle. The differences between them are thought of purely as differences in the orientation of the isotopic spin, and we would no more say that this made them different types of particles than we would claim that two electrons were different kinds of particles just because their angular momenta pointed in different directions.

Because the number of different orientations of an isotopic spin vector is given by $N = 2I + 1$, where I is the isotopic spin, the number of members of a family can be used to determine the isotopic spin directly.

The connection between electrical charge and isospin projections, in general, is given by the equation $Q = I_z + B/2 + S/2$, where I_z is the projection of the isotopic spin, B is the baryon number, and S is the strangeness. For the rho- and pi-mesons, both B and S are zero; hence, nothing in the above argument is changed.

We can look at the proton and neutron as examples of how isotopic spin works. Here are two particles in the same family; therefore, we have $2 = 2I + 1$ or $I = \frac{1}{2}$. An isotopic spin of $\frac{1}{2}$ will have two projections, $+\frac{1}{2}$ and $-\frac{1}{2}$. For the nucleons, $B = 1$ and $S = 0$, so the two possible charges are $Q = \frac{1}{2} + \frac{1}{2} = 1$ and $Q = -\frac{1}{2} + \frac{1}{2} = 0$. Thus, the proton corresponds to the $+\frac{1}{2}$ orientation of the isotopic spin and the neutron to the $-\frac{1}{2}$ orientation.

For the delta, on the other hand, there are four charge states, so $I = \frac{3}{2}$. Since $B = 1$ and $S = 0$, the possible charge states are:

$$Q = \frac{3}{2} + \frac{1}{2} = 2$$

$$Q = \frac{1}{2} + \frac{1}{2} = 1$$

$$Q = -\frac{1}{2} + \frac{1}{2} = 0$$

$$Q = -\frac{3}{2} + \frac{1}{2} = -1$$

In the following table the results of this kind of calculation are presented for some of the particles we have studied:

PARTICLE	N	I	B	S	Q
π	3	1	0	0	+1, 0, −1
ρ	3	1	0	0	+1, 0, −1
p, n	2	$\frac{1}{2}$	1	0	+1, 0
Δ	4	$\frac{3}{2}$	1	0	+2, +1, −1, 0
Λ	1	0	1	−1	0
Σ	3	1	1	−1	+1, 0, −1
Ξ	2	$\frac{1}{2}$	1	−2	0, −1

One point that should be emphasized before we move on is that the families that are described by isotopic spin exist only among the hadrons; therefore, we do not speak of the isotopic spin of leptons. Furthermore, although we have talked about only one use of isotopic spin (its ability to simplify our picture of particle families), there are many predictions about reactions that can be made under the assumption that the laws governing ordinary spin also govern isotopic spin. Since these predictions are invariably correct, there is a good deal of experimental evidence to support the rather abstract presentation given here.

OTHER METHODS OF CLASSIFICATION

There are a few other properties of elementary particles which, while not of great importance for the process of classifying, become important in the discussion of conservation laws. We will present them briefly here.

Parity. In Chapter III we saw that every quantum mechanical particle can be described by a wave function that is ultimately related to the probability that a measurement would find the particle at a particular point in space. The parity of the particle has to do with what happens to the wave function when a particular kind of mathematical operation is performed. The operation is essentially the interchanging of right and left—something like looking at the wave function in a mirror. In technical terms, the value of the old wave function at a point x is replaced by the value of the wave function at the point $-x$. If the wave function is such that performing this operation (or, equivalently, looking at the wave function in a mirror) just reproduces the original wave function, the particle is said to have positive parity.

On the other hand, if this process produces a new wave function that is precisely the negative of the old one, the particle it describes is said to have negative parity.

Charge Conjugation. If we imagined taking a collection of particles and changing every one of them into its antiparticle, we would have performed the operation known as charge conjugation. In general, there is no particular relation between the wave function of a particle and that of its antiparticle, but for a few neutral particles, such as the $\pi°$, the operation will produce a wave function that is either plus or minus the original one. Thus, as with parity, we can speak of positive or negative charge conjugation for this limited class of particles.

Time Reversal. Time reversal is not, strictly speaking, a property of particles, but we include it here for completeness. If you watch an interaction involving elementary particles, you can imagine putting the whole process on film. The operation of time reversal corresponds to running the film backwards. This operation does not change a particle into plus or minus itself, but it obviously has an effect on what is seen. To take one example, a particle that is spinning so that its angular momentum is "up" will, upon time reversal, be seen to have spin "down."

A SUMMARY

To classify a new elementary particle we therefore have to ask first whether it is a hadron or a lepton. If it is the latter, there is not much else to do. If the former, we proceed to see if it is a baryon or meson, strange or nonstrange, and to ascertain its spin, isotopic spin, and, if applicable, its charge conjugation. This information, together with the mass of the particle, gives us all of the properties we are likely to need to know. In the following table we present this information for all the particles discussed so far.

SUMMARY OF PARTICLE PROPERTIES

	PARTICLE	MASS	J	S	I	P	B
Photon	γ	0	1	0			0
Leptons	e	0.51	$\frac{1}{2}$	0			0
	μ	105	$\frac{1}{2}$	0			0
	ν	0	$\frac{1}{2}$	0			0
Mesons	π	140	0	0	1	−	0
	K^-	494	0	−1	$\frac{1}{2}$	−	0
	K°	498	0	−1	$\frac{1}{2}$	−	0
	ρ	770	1	0	1	−	0
Baryons	p, n	938	$\frac{1}{2}$	0	$\frac{1}{2}$	+	1
	Λ	1,115	$\frac{1}{2}$	−1	0	+	1
	Σ	1,190	$\frac{1}{2}$	−1	1	+	1
	$\Xi^{\circ,-}$	1,318	$\frac{1}{2}$	−2	$\frac{1}{2}$	+	1
	Δ	1,232	$\frac{3}{2}$	0	$\frac{3}{2}$	+	1
	$\Sigma(1385)$	1,385	$\frac{3}{2}$	−1	1	+	1

IX

The Road
to the Quark Model

The quark model is to physics as the folk song is to music.
  —Anonymous

SOME SYSTEMATIC RELATIONSHIPS
AMONG PARTICLES

WHEN the first resonances were discovered, the general feeling among physicists was that they were somewhat anomalous and special. This feeling was reflected in the names given them. What we have called the delta, for example, was called the N*, the assumption being that this would be *the* pion nucleon resonance. Similarly, the strangeness −1 baryon resonance was called the Y*. But as the number of known resonances began to increase dramatically in the 1960s, this nomenclature became more and more unwieldy. Other resonances in the pion-nucleon system began to appear and were christened the N** and the N***. As Matt Roos said in his first review of elementary particles, "We expect the starred notation to become unpopular by the time a resonance is discovered which needs eight stars."

A look at the pion-nucleon resonances will serve to introduce the modern nomenclature for resonances and to illustrate some of the regularities that are seen in them. By 1976, the *Review of Particle Properties* listed an even dozen resonances, which have the following properties: $B = 1$, $S = 0$, and $I = \frac{1}{2}$. All these particles decay into a nucleon and a set of pi-mesons. Consequently, we would be justified in calling all of them by the name N*. In addition, the quantum numbers listed above are exactly those we have described for the nucleon itself. The other properties of these particles (such as mass, parity, and spin) vary as we go through the list. The convention that has been adopted is to denote them all by the letter N and to add in parentheses the mass measured in MeV. For the nucleonlike resonances, the list includes:

N(1470)	N(1780)
N(1520)	N(1810)
N(1535)	N(2190)
N(1670)	N(2220)
N(1688)	N(2650)
N(1700)	N(3030)

Since 1976, more particles have been added to this list.

The decay of all these particles takes place primarily through the creation of a nucleon and a single pion. However, all of them can decay by other processes, such as

$$N(\ ) \rightarrow \Delta\pi$$

or

$$N(\ ) \rightarrow N\rho$$

In fact, the decay is very similar to the process by which light is emitted from an atom (see Chapter I). When an electron is in a high orbit in an atom, it can move into lower orbits in a number of ways. It can make one big jump (emitting a high-energy photon), or it can make a series of smaller jumps through intermediate orbits (emitting a series of lower energy photons).

In the same way, it appears that these nucleon resonances can decay to a nucleon plus pions through a single transition, or they can decay in a series of steps. This similarity with the electron is the first hint we have that under the avalanche of elementary particles we might still be able to find an underlying simplicity. Maybe calling all the resonances in the above list separate particles is as pointless as insisting that two atoms are fundamentally different because their electrons are in different orbits.

The nucleon is not the only low-mass baryon that has a "family." We could compile a list just like the one above for non-strange baryons with isotopic spin 3/2. This would be the delta family, and the lowest mass member of the family, the $\Delta(1236)$, is the familiar resonance we have been calling the Δ. Other families are listed in the table below. One member of the sigma family, the $\Sigma(1385)$, was the first strange resonance, whose properties we discussed in Chapter VII.

B	S	I	NAME OF FAMILY	NUMBER KNOWN IN 1976
1	−1	0	Λ	10
1	−1	1	Σ	10
1	−2	$\frac{1}{2}$	Ξ	3

There are fewer known meson resonances than resonances associated with the baryons. This fact is related more to the problems involved in doing a complicated phase space analysis than to any fundamental differences between mesons and baryons. We shall return to this point later, but what evidence we have indicates that there are meson families similar to the baryon families we have just enumerated.

Thus, by using the classification schemes outlined in Chapter VIII we can apply an enormous conceptual simplification to the problem of hadron proliferation. If each new resonance is regarded as a new member of a family, having no more fundamental significance than a new electron orbit in an atom, we have already succeeded in bringing a large measure of order into a previously chaotic world. In this way of looking at things, the number of families is rather small, even though the number of particles is large.

THE EIGHTFOLD WAY

The first really successful scheme for showing the fundamental connection between particles in different families was developed independently in 1961 by Murray Gell-Mann at the California Institute of Technology and Yuval Ne'eman, who at the time was filling the roles of physicist at Imperial College, London, and military intelligence attaché at the Israeli embassy. This scheme bears the same logical relation to elementary particles as the periodic table does to chemical elements. Perhaps if we expand on this analogy a bit it will help us to understand what was done.

When the number of known chemical elements began to approach the hundred mark during the last century, it was recognized that some sort of order would have to be discovered among the proliferating elements. The Russian chemist Dmitri I. Mendeleev noticed that if the chemical elements were arranged in rows so that the atomic weight increased from left to right, and if the number of elements in the rows was adjusted properly, then elements in the same column would have similar chemical properties. Thus, a connection between atomic weight and chemical properties was established. In addition, the few "holes" in the table—places where elements should have been but weren't—led to the discoveries of scandium and germanium, two elements that were unknown up to that time.

In considering the periodic table, however, it is extremely important to realize that if someone had asked Mendeleev why the first row contained two elements while the next contained eight, he would not have been able to answer. The periodic table worked, but until the advent of quantum mechanics in the twentieth century no one understood why.

This same spirit of order-without-explanation led to the development of the periodic table of the elementary particles, which has been called the *eightfold way* (because it predicts that many hadrons will be grouped in sets of eight) and SU(3) (a technical mathematical term describing the properties of the groupings). Perhaps the best way to understand this table is to plunge right in and start ordering the particles.

Suppose we make a graph in which the vertical axis is the

quantity $B + S$ for any particle, and the horizontal axis is I_z, the projection of the isotopic spin. On such a graph, any particle will be represented by a point. Consider, for example, the Σ baryons. All of them have $B = 1$, $S = -1$, so that $B + S$ (a quantity sometimes called the hypercharge) is zero. At the same time, recalling that

$$I_z = Q - \frac{B + S}{2}$$

the three charge states of the Σ give $I_z = 1$, 0, and -1, respectively. Consequently, on such a graph, the Σ family appears as three points spaced out along the horizontal axis.

Referring to the table at the end of Chapter VIII, we see that the nucleon, the sigma, the lambda, and the cascade all have the same spin, baryon number, and parity, and have masses that are roughly equal. What if we plotted all of these particles on a graph such as that shown above? You can test your facility with the concepts we have introduced so far by verifying that you will get a set of points like the ones shown in Illustration 18. In this graph the sigma and the lambda are placed apart from each other so that they can be easily seen, but both are at the point $(B + S) = 0$, $I_z = 0$.

The eight particles in the graph (Illus. 18) are similar in many respects. It turns out that grouping them in this way tells us a great deal about the way they contribute to the strong interaction. For example, if we read across the rows in the graph, all the particles have the same strangeness. If we read along a diagonal from upper left to lower right, they all have the same electrical charge. The grouping shown is thus similar to the periodic table of the chemical elements—it tells us about the way the particles behave, but does not tell us why.

All the hadrons we have discussed so far can be clustered in groups of particles with equal spin, baryon number, parity, and (roughly) equal masses. The mathematical formalism tells us how many should be in each group. With a few exceptions, the lower mass particles we have been discussing fall into groups of eight, as the baryons pictured in Illustration 18 do. Such a grouping is called an octet, and the origin of the term eightfold way is now obvious.

During the early 1960s, it became clear that grouping the elementary particles as suggested by the eightfold way was an enormously fruitful way to think about them. The underlying mathematical formalism (a branch of modern mathematics called group theory) allowed theoretical physicists to relate the lifetimes of different resonances to each other and to predict the differences in the masses of particles within a given grouping. Being able to *predict* the lifetime of a resonance that previously had to be accepted from experiment meant that the amount of arbitrariness in the properties of the particles was being reduced. Instead of having to accept the new particles as they showed up in experiments, physicists could now begin to predict where these particles ought to be seen and to predict some of their properties. (We will see later how one of these predictions was obtained.) Like the periodic table, the eightfold way led to an

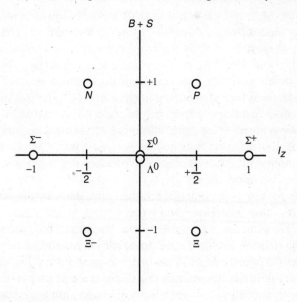

18. The baryon octet.

ordering of the elementary particles, an ordering that significantly reduced the previous chaos.

Yet there is still cause for an underlying uneasiness about the whole procedure. We saw that Mendeleev could not answer the question about why the periodic table should be arranged as it was. In the same way, we cannot say why particles should be grouped on a graph like the ones shown above. Simply stated, why should the graph have $B + S$ on one axis and I_z on the other? Why not some other combination of quantum numbers? In mathematical terms, why is it SU(3) and not some other symmetry that operates in the elementary particle world? Just as a full understanding of the chemical elements had to wait for the development of quantum mechanics, a deeper understanding of the eightfold way had to wait for a few years until a new simplification took place in our way of looking at the hadrons.

THE QUARK MODEL

In 1964, Murray Gell-Mann at the California Institute of Technology and George Zweig in Geneva, Switzerland, independently suggested a simple physical explanation for the success of the eightfold way. Just as the periodic table could be explained once it was accepted that the atom possessed a structure and that electrons moved around a nucleus in orbits, the tendency of elementary particles to form octets could be understood if they, too, were made up of some still more elementary constituents. Hence, the question that had to be asked was, "What sort of constituents would be necessary to produce the elementary particles as we know them?"

The answer turns out to be rather surprising. In the original picture—also based on group theory and SU(3)—there were three constituents making up all the hadrons. These are called *quarks* (from a line in James Joyce's *Finnegans Wake*—"Three quarks for Muster Mark"). In order to produce the known elementary particles, the quarks must have unusual properties. For example, they have to have electrical charges that are fractions of the charge on the electron and proton. This alone would make them unique, since every other known electrical charge is either equal to one of these or some whole number of them.

On the other hand, no matter how unusual these new particles might be, there is plenty of circumstantial evidence indicating that the particles we have been calling elementary up to this point are, like the atoms before them, made up of constituents. The existence of particle families certainly suggests this, as does the unexplained ordering of the particles brought about by the eightfold way. Thus, no matter how unusual the quarks might be —either in name or in properties—it is worth spending time to see what the evidence for them is.

In the table we present the relevant properties of the three original quarks (the names are those now in use, rather than those originally proposed). For each of these there is a corresponding antiparticle—the antiquark—with opposite charge, baryon number, and strangeness. The antiquarks are denoted by the symbols $\bar{u}$, $\bar{d}$, and $\bar{s}$.

NAME OF QUARK	SYMBOL	J	Q	S	B	I	I_z
Up	u	$\frac{1}{2}$	$\frac{2}{3}$	0	$\frac{1}{3}$	$\frac{1}{2}$	$\frac{1}{2}$
Down	d	$\frac{1}{2}$	$-\frac{1}{3}$	0	$\frac{1}{3}$	$\frac{1}{2}$	$-\frac{1}{2}$
Strange	s	$\frac{1}{2}$	$-\frac{1}{3}$	-1	$\frac{1}{3}$	$\frac{1}{2}$	0

The quark names are not so unusual as they may seem. The u and d quarks form a family of isotopic spin $\frac{1}{2}$, and "up" and "down" refer to the two projections of that spin. "Strange" refers to the fact that this quark has $S = -1$, while the others have $S = 0$.

How could we go about constructing the hadrons from these building blocks? Let us start with the baryons. These particles all have $B = 1$. Since all of the quarks have $B = \frac{1}{3}$, we can deduce an important principle of the quark model: *Baryons are made from three quarks.* The values of B for the three constituent quarks in the baryon will therefore add up to $B = 1$, as they should.

To find which quarks go with which baryons, we have to see which combination will give the correct spin, charge, and strangeness. Consider the delta baryon of charge +2 as an example. This is a particle with charge +2, spin $^3/_2$, and strangeness 0. We see immediately that there can be no s quarks in the Δ^{++}. A single s quark would give the particle $S = -1$, while a combination of s and $\bar{s}$ (which would give $S = 1 - 1 = 0$) would give $B = ^1/_3$ (remember that the antiquark has $B = -^1/_3$).

Is there any way we can combine the u and d quarks to make a Δ^{++}? There is only one way to do this and still have a charge +2, and that is to have three u quarks. The total charge of this combination will be $Q = ^2/_3 + ^2/_3 + ^2/_3 = 2$. Similarly, to get a total spin of $^3/_2$ from three objects of spin $^1/_2$, we have to have all three spins lined up, pointing in the same direction. Thus, in terms of quarks, the Δ^{++} must look something like Illustration 19a, on page 141. In the same way, the Δ^+ must also have three spins lined up, but must have two u quarks and a d quark to give total charge $Q = ^2/_3 + ^2/_3 - ^1/_3 = 1$. It will look something like Illustration 19b. The proton, on the other hand, must have spin $^1/_2$ and charge 1. Consequently, it must be made up of two u quarks and a d quark, but the spins must be arranged differently from the Δ^+. A possible arrangement for the proton is shown in Illustration 19c.

We can now see the essential point of the quark model. Up to this point we have had to regard the delta and the nucleon as two separate particles. All of the regularities we have discussed have not altered that essential fact. In terms of quarks, however, we see that there is a fundamental connection between these two particles. They are both made of the same kinds of quarks, but these quarks are arranged differently. The philosophical implications of this fact will be discussed later, but for the moment we note that it gives us a very simple picture of the process by which one particle changes into another.

Consider the fast decay $\Delta^+ \rightarrow p^+\pi^\circ$. In terms of particles, we have to think of this decay as something that can be seen in the laboratory but not understood. In terms of quarks, however, we can picture this event as the d quark flipping its spin from up to down, emitting a neutral pion in the process. Since this sort of phenomenon happens all the time in nuclei (the pion is simply

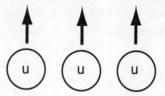

19a. Δ^{++} in terms of quarks.

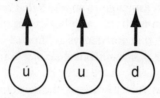

19b. Δ^+ in terms of quarks.

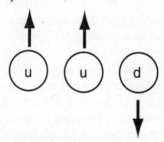

19c. Possible arrangement for the proton, in terms of quarks.

replaced by an X ray), physicists felt very much at home with it. Not only that, but once a simple picture like this is produced, it is possible to use the techniques of nuclear physics to calculate lifetimes of elementary particles, including lifetimes of particles in different octets. In the mid-1960s, many such calculations were made and, considering the extreme simplicity of the model, they were quite successful.

The quark model also gives us a way of understanding the fast-versus-slow decay systematics that first became apparent in cosmic ray experiments. To see how this works, let us consider the positively charged sigma (Σ^+). This particle has $S = 1$, so it must contain one s quark. In order to make up the charge of +1, the other two quarks must be u quarks. To get spin $1/2$, we must have something like what is shown in Illustration 20, on the left, below.

Now consider the decay $\Sigma^+ \rightarrow p + \pi^\circ$. In order for this decay to proceed, the s quark in the Σ^+ must change not its spin, but its identity. It must transform into a d quark in a process similar to that shown in Illustration 20.

It is not too unreasonable to suppose that it takes longer for a quark to transform itself into another quark than to flip its spin. The quark model rule for decays is thus: *A decay can proceed quickly only if the quarks involved do not have to change their identity.*

We can use this rule to discuss the decay of the $\Sigma(1385)$, a strange resonance that decayed quickly into a lambda and a pion. This particle has $S = -1$, $B = 1$, and spin $3/2$. If we consider the positively charged $\Sigma(1385)$, it must have a quark content like the Σ^+, except that the three quark spins must be aligned. It must, in other words, look similar to the representation on the right in Illustration 21. If we consider the decay $\Sigma^+(1385) \rightarrow \Sigma^+ + \pi^\circ$, then it must proceed as pictured. The similarity between this spin-flip process and the spin flip that allowed the delta to decay to a proton cannot be overemphasized. The processes are identical except for the type of quark that is flipping its spin. Both are fast because they involve only a spin flip.

With this rule, we can also understand the double slow decay

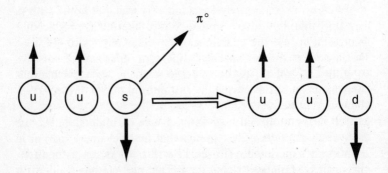

20. The s quark is a Σ^+ transforming into a d quark.

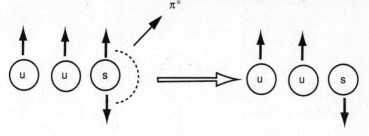

21. The decay process of $\Sigma^+(1385)$, in terms of quarks.

of the cascade. An $S = -2$ particle must have two s quarks in it. The cascade must therefore look similar to Illustration 22a. In order for this particle to decay, the two strange quarks must change to ordinary d quarks. This happens in two steps. First one of them changes while emitting a neutral pion, which leaves a particle similar to that represented in Illustration 22b. It has 0 charge, spin $\frac{1}{2}$, and strangeness -1. This, of course, is the Λ°. This new particle then decays slowly by converting the remaining s quark into a d quark. Thus, we see that double slow decays correspond to situations in which we have to change two s quarks into two d quarks to get to the final state, rather than changing just one s to one d.

One caveat to this decay rule should be made at this point. In the decay $\Delta^+ \rightarrow n + \pi^+$, a u quark flips its spin, emits a pion, and becomes a d quark. Doesn't this process involve a change in identity?

If you recall that the u and d quarks are members of the same isotopic spin family, you will see that the change in identity in this process is just another spin flip. The isotopic spin changes from up to down in isotopic spin space just as the ordinary spin changes from up to down in ordinary space. Consequently, changes between u and d quarks are on the same logical footing as spin flip, and do not constitute change of identity as we are using the term here.

The quark model also provides a way of thinking of the high-mass particle families discussed earlier in this chapter. All of the particles we have discussed so far involve three quarks that are

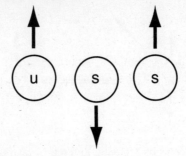

22a. The cascade, in terms of quarks.

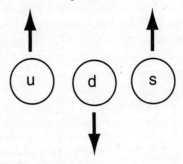

22b. The Λ°, in terms of quarks.

stationary with respect to each other. This is certainly the simplest possible configuration for them, but the question naturally arises as to whether we could have a situation where the quarks are in orbits around each other. The answer to this question is yes. Not only are such states possible, but they account for many of the dozen nucleon resonances we listed earlier. Each resonance corresponds to a different set of orbits for the quarks. Since these orbits can get bigger and bigger, there is no limit, in principle, to how many resonances there can be. In this sense, it is a good thing the quark model was formulated when it was. Otherwise, we would now be absolutely flooded with particles.

THE QUARK MODEL FOR MESONS

In the previous section we treated mesons as particles that are emitted when quark spin flip occurs. Actually, we know that mesons are hadrons and therefore must also be composed of quarks. The quark structure for mesons will be discussed in this section.

By definition, mesons have baryon number 0. The only combination of quarks that will produce this result is a quark plus an antiquark. For them, we will have $B = \frac{1}{3} - \frac{1}{3} = 0$. This leads to another rule for the quark model: *Mesons are made from a quark and an antiquark.*

Using this rule, we can find the right combinations for the different mesons just as we did for the baryons. Take the positive rho-meson as an example. This is a meson with $S = 0$, $Q = 1$, and $J = 1$. If there are to be any s quarks in the particle, they would have to appear in an $s\bar{s}$ combination. Such a combination could not have an electrical charge. Consequently, the ρ^+ must be made of u and d quarks along with their antiquarks.

Referring to the table of quark properties, we see that a u quark (charge $\frac{2}{3}$) and a $\bar{d}$ antiquark (charge $\frac{1}{3}$) will give us the proper charge for the ρ^+. It must therefore look something like Illustration 23a. A similar argument shows that the π^+ must be as shown in Illustration 23b. As was the case with the proton and the positive delta, the only difference between the two particles is in the alignment of quark spins.

In our discussion of the decay of the baryons, we treated the pions that were emitted as single particles. Clearly, we cannot treat the process $\rho^+ \rightarrow \pi^+ + \pi^\circ$ in the same way. We have to look a little more closely at the process by which a quark emits a pi-meson.

For this discussion, we can consider the quark that is going to be active during the decay process as being isolated in space, and that the other constituents of the particle will play the role of observer in whatever happens. In the simplest case, it should make no difference whether the quark we are looking at is attached to an antiquark (and, hence, is in a meson) or to a pair of

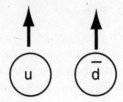

23a. The ρ⁺, in terms of quarks.

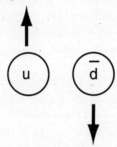

23b. The π⁺, in terms of quarks.

quarks (making it part of a baryon). In either case, we can consider the active quark to be isolated in space.

From the uncertainty principle and the discussion of virtual particles in Chapter III, we know that on a very short time scale it is possible to have processes occur that seem to violate the conservation of energy. One such process would be the creation of a quark-antiquark pair in the empty space near our isolated quark. Any quantum system can fluctuate to a state where energy is increased by an amount ΔE, provided that it returns to normal in a time Δt, where $\Delta E. \Delta t > h$. There is therefore no reason why quark-antiquark states should not appear. In fact, there is no way we could tell whether they had appeared or not. In this way, the pair is very similar to the virtual particles responsible for the strong force. It is further possible that the isolated quark will combine with the virtual antiquark from the pair, forming a meson, and that the virtual quark will take the place of the original quark in the particle. If the original quark happens to be in a meson, such as the ρ⁺, then we could have a process similar to that shown in Illustration 24. We see that the original ρ⁺ is transformed into two particles with zero spin and baryon number. One of these is positively charged and the other neutral, so it is quite evident that this process describes the decay of the ρ⁺ into two pi-mesons.

Actually, this same process of meson creation by the combining of original quarks with quark-antiquark pairs is the mechanism for the baryon decays we talked about in the previous section. What we described as "quark spin flip with pion emission" is actually a quark spin flip accompanied by the recombination of a quark-antiquark pair, which is the same process that occurs in rho decay.

Once we understand the picture of meson decay suggested by the quark model, we can see that all of the systematics of the decay process discussed for baryons can also be applied to the mesons. Mesons containing the strange quark will decay slowly if that quark has to change into a u or d, but will decay quickly if it does not have to change. As with the baryons, there should be a large number (in principle, an infinite number) of higher mass mesons corresponding to situations where the quark and antiquark circle each other in expanding orbits. For example, the A_2 meson, which has a mass of 1310 MeV and $J = 2$, would be a state in which a u and an anti-d quark orbit each other.

Finally, the fact that the decay process is identical in baryons and mesons suggests that it ought to be possible to find relationships between decay rates for mesons and baryons. After all, the process by which a quark combines with a pair should not depend very much on whether the original quark is attached to an

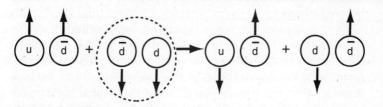

24. The decay of a rho-meson.

antiquark (as it would be in a meson) or to two other quarks (as it would be in a baryon). Thus, even at the theoretical level we see that the distinction between different particles begins to become less important, and the essential fact is that every hadron we have discovered is composed of the same kinds of constituents.

PHILOSOPHICAL CONSEQUENCES
OF THE QUARK MODEL

The question we posed at the beginning of the book was this: "Is there an underlying simplicity in nature?" We have now followed the trail of the answer through three periods when the answer seemed to be yes. First, there was the atomic picture of the chemical elements, in which the tremendous variety of materials was perceived to be composed of a relatively small number of atoms. This simplification was followed by a period of increasing complexity, during which the number of known chemical elements swiftly increased. The periodic table of the elements provided a measure of order among the elements, but why it worked was not understood.

The next major simplification was the development of the modern nuclear atom, in which all chemical elements were believed to be composed of three basic building blocks—the proton, neutron, and electron. When this picture was investigated more closely, however, a new kind of complexity developed, a development that we have been following in this book. By the early 1960s, the three elementary particles of the nuclear atom had been superseded by hundreds of elementary particles with various new properties. Once more our picture of the world had become complex.

The quark model therefore represents the most recent step in this dialectic of complexity leading to simplification leading to complexity again. The hundreds of elementary particles are now replaced by three quarks and their antiparticles, and all the particles are thought of as combinations of these three basic building blocks. With the development of the quark model, the answer to our basic question seems to be a resounding yes.

Whether we would still answer that way in the future remains to be seen.

Whether it is a permanent simplification or not, the quark model is a tremendous improvement over the previous chaos. It suggests a way of picturing the elementary particles that has been extremely productive in research and that lends itself easily to teaching. I can recall seminars on the quark model where the speakers made their point by employing colored wooden blocks with arrows painted on them to illustrate hadronic structure, which could hardly have been done using something as abstract as the eightfold way.

X

Evidence
for the Quark Model

Perry Mason turned. "Circumstantial evidence is the best evidence there is, Paul. You just have to interpret it properly."
—ERLE STANLEY GARDNER, *The Case of the Queenly Contestant*

THE QUARK MODEL AND PARTICLE STATES

ONE OF THE MAJOR ARGUMENTS in favor of the quark model is its enormous success in explaining and codifying the hadrons. If we think of the hadrons as particles, all of the simplifications and classifications we have discussed do not reduce the confusion to the point where a pattern can be discerned. Once we have the concept that each of these particles corresponds to a different arrangement of the quarks, however, the situation changes. We can see that the families of particles (such as those associated with the delta and the nucleon, for example) are indeed related to each other in a fundamental way: They are all composed of the same quarks. The patterns of decay speeds associated with strangeness are also understood in terms of the number of s quarks present before and after a decay.

In this chapter we will go over some of the early evidence that convinced physicists that there was something to the quark model. While such evidence is important, I should emphasize that today most physicists accept the existence of quarks because of the enormous success of the so-called Standard Model, which we will discuss in Chapter XIV.

Historically, there were two types of investigations of the ramifications of the quark model. One is to see whether any of the predictions of the model are wrong; the other is to see what new information about the particles can be gained.

To understand how the first type of investigation is carried out, we can consider the nucleon. In the quark model, the nucleon is a composite system made up of three particles, each with spin and isospin $1/2$, which is not an unfamiliar system in physics. There are nuclei whose constituents are nucleons (which also have spin and isospin $1/2$) that are very similar. Tritium (chemical symbol ^3H), an isotope of hydrogen with a proton and two neutrons, and ^3He, an isotope of helium with two protons and a neutron, are both very similar to the quark picture of the nucleon. This similarity was exploited by theoretical physicists to predict what sort of particles one would expect to find in the nucleon family. Calculating the various combinations and permutations of the quarks is not, after all, very different from calculating the permutations and combinations of protons and neutrons that would make up the excited states of ^3H and ^3He. When this calculation is made, an amazing fact emerges: *There is no particle predicted by the quark model that has not been found.* Perhaps even more surprising is the converse statement: *No particle has been found that does not fit into the quark picture.*

Here is probably the main reason why physicists today so readily accept the quark model. We can get some of the flavor of the process of prediction and discovery that went on in the 1960s if we look at the first (and probably best known) particle discovery that has resulted from this process—the particle called the Ω^- (omega minus).

Although the particle was first predicted from the eightfold way symmetry, it is probably easier to understand the prediction in terms of quarks. We have seen that there are particles with no strange quarks, particles with one strange quark, and particles with two strange quarks. These correspond to strangeness 0, –1, and –2-type particles, respectively. Is it not possible that there might be a baryon made up of three strange quarks? What would the characteristics of such a particle be?

From the quark table in Chapter IX, we see that such a particle would have a charge

$$Q = -\tfrac{1}{3} - \tfrac{1}{3} - \tfrac{1}{3} = -1$$

and a strangeness

$$S = -1 - 1 - 1 = -3$$

along with the usual baryon number of 1. It would also have its spins aligned. The mass of the particle was estimated to be about 1,680 MeV, and it was christened the Ω^- (omega minus) even before it was discovered.

Once the prediction was made, the race was on to find the new particle. In February 1964, a group at Brookhaven Laboratory reported the first production of the Ω^- in a liquid-hydrogen bubble chamber. It is characteristic of the scale of modern high-energy physics that there were no fewer then thirty-three persons listed as authors of the paper.

The event they saw occurred when a K^--meson entered the chamber, striking a proton target. The following string of events was seen in the chamber:

$$K^- P \rightarrow \Omega^- + K^+ + K^\circ$$
$$\quad \rightarrow \Xi^\circ + \pi^-$$
$$\quad \rightarrow \Lambda^\circ + \pi^\circ$$
$$\quad \rightarrow p + \pi^-$$

Each of the three decays was slow, allowing it to be seen in the bubble chamber. We would expect three slow decays for the Ω^-, of course, since there are three strange quarks to be converted to u and d quarks.

This exciting discovery meant that the new simplicity in particle physics was borne out by a crucial experimental test. I was a graduate student at Stanford at the time; I can remember how rapidly the news spread through the department, with groups of professors and students standing around trying to figure out all that the discovery implied.

As I have said, the original prediction of the existence of the Ω^- was made by using the eightfold way, rather than directly from the quark model. For the sake of completeness, here is how that original prediction was made.

The eightfold way predicts that the low-mass, low-spin particles will group themselves in octets. For higher spins, it predicts that other groupings may occur as well. Consider, for example, the following spin $3/2$ particles, which are members of the delta, sigma, and cascade families: $\Delta(1236)$, $\Sigma(1385)$, and $\Xi(1532)$. The last particle in this grouping has not been introduced before, but it is one of the members of the cascade family that we mentioned briefly in Chapter IX. If we make an eightfold way plot of these particles, a plot in which each particle represents a point on a chart whose axes are $(B + S)$ and I_z, we will get the graph depicted in Illustration 25. The mathematics predicts that there should be ten particles in this grouping, and it does not

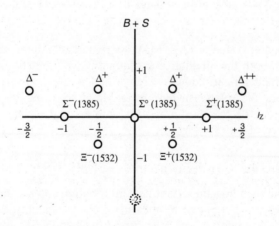

25. The eightfold way grouping of the spin $3/2$ baryons.

take much imagination to see where the missing particle should go. From the graph we see that it should be a particle with $B + S = -2$ and $I_z = 0$, which means that if it has $B = 1$, it must have $S = -3$ and $Q = -1$, which is, of course, what we require for the Ω^-.

If we consider the prediction for a moment, we can see an analogy with the development of the periodic table of the elements. There were "holes" in the table that led to the discovery of new elements, and in the same way, a "hole" in the elementary particle table led to the discovery of the Ω^-.

We can even understand how the prediction of the mass of the new particle could be made. A little arithmetic on the masses of the particles listed above reveals the following facts:

$$M_\Sigma(1385) - M_\Delta = 1{,}385 - 1{,}236 = 149 \text{ MeV}$$

and

$$M_\Xi(1532) - M_\Sigma(1385) = 1{,}532 - 1{,}385 = 147 \text{ MeV}$$

In other words, the difference in masses between the members of the grouping is roughly 150 MeV. It would be reasonable to suppose the Ω^- would have a mass about 150 MeV higher than the $\Xi(1532)$. This, in turn, leads to a predicted Ω^- mass of

$$M_{\Omega^-} \approx 1{,}680 \text{ MeV}$$

The modern value is 1,672 MeV—a remarkable agreement.

Thus, both the qualitative and quantitative predictions of the quark-model-eightfold-way were verified by the discovery of the Ω^-. Since 1964 physicists have by and large regarded this discovery as one of the most important pieces of evidence for the existence of quarks.

THE QUARK MODEL AND SCATTERING

We have seen how important new information about elementary particles can be gained by measuring the probability that two particles will interact, or, as discussed in Chapter VII, the interaction cross section. It was measurements of this type that led to the discovery of the delta and, ultimately, of the whole family of nucleon and delta resonances. A large body of data on scattering cross sections exists, and it should not be a surprise that the quark model can be used to deal with it.

To understand how this can be done, we can consider the scattering of two baryons, which is a situation we could produce by directing a proton beam at a hydrogen target. From the point of view of the quarks, this would correspond to one object made of three quarks scattering from another object that was also made of three quarks. One way to imagine the scattering is to think of one quark from the projectile scattering from one quark in the target and both moving off, dragging their respective partners with them. In proton-proton scattering there are precisely nine ways in which this kind of interaction can happen, corresponding to each of the three quarks in the projectile scattering from each of the three quarks in the target.

The cross section for scattering, you will recall, is related to the probability that a reaction will take place. The picture we have just described relates the probability that a proton will interact with another proton to the probability that a quark will interact with another quark. If we denote the cross section for one quark to scatter from another by σ_q, then in this simple picture of the scattering process, the proton-proton scattering cross section must just be $\sigma_{pp} = 9\sigma_q$.

So far, all we have done is to replace one number we are not able to calculate, σ_{pp}, by another number we are not able to calculate, σ_q. But we can go farther if we consider the scattering of a meson from a proton as conceived in the quark model. The meson is made up of two quarks (we ignore for the moment the differences between quarks and antiquarks), and each of these can scatter from each of the three quarks in the target proton. An argument similar to the one given previously tells us that the meson-proton cross section must be $\sigma_{Mp} = 6\sigma_q$, where again σ_q characterizes the scattering of one quark from another.

At first glance, it would seem that we have just obtained another relation between two quantities that we cannot calculate. But, if we divide the equation for σ_{pp} by the equation for σ_{Mp}, we get

$$\frac{\sigma_{pp}}{\sigma_{Mp}} = \frac{9\sigma_q}{6\sigma_q} = \frac{3}{2} = 1.5$$

and this does *not* depend on anything having to do with the quarks. It is, in fact, a clean prediction that tells us that proton-proton cross sections ought to be half again as big as meson-proton cross sections.

A typical value of σ_{pp} might be 45×10^{-27} square centimeter, while a typical value of σ_{Mp} might be 27×10^{-27} square centimeter. For these numbers, the experimental ratio is

$$\frac{\sigma_{pp}}{\sigma_{Mp}} = \frac{45}{27} = 1.65$$

which is close enough to 1.5 for us to say that the prediction of the model has been verified.

This is a good example of the way one can work with the quark model. The strategy is to calculate measurable quantities (like σ_{pp} and σ_{Mp}) in terms of quantities that relate to the quarks and then to do enough algebra to eliminate the latter quantities from the equations, leaving equations that link measurable quantities. These relations can then be tested against experiment, as above.

During the middle and late 1960s many relations between quantities involved in particle scattering and production were derived by using the quark model. At the level of 20 percent accuracy, virtually all of them turned out to be correct. This is another strong indication that particles are composed of quarks.

A RUTHERFORD EXPERIMENT FOR QUARKS

The strongest single piece of evidence for the existence of the atomic nucleus was the experiment described in Chapter I. From the way alpha particles scattered from atoms, Ernest Rutherford was able to conclude that most of the mass of the atom was concentrated in a small nucleus. In 1969 a collaboration between physicists at the Massachusetts Institute of Technology and the Stanford Linear Accelerator Center (SLAC) seemed to produce a similar result for the proton. Their results suggest that the proton itself is made up of small constituents.

We have already given a brief description of the Stanford accelerator. It accelerates electrons down a 2-mile tube, bringing them to energies of over 20 GeV by the end of the trip. In an area called the beam switchyard, a series of magnetic selectors and slits are used to produce an intense beam of electrons whose momentum is very accurately known. These electrons are then allowed to strike a target in the experimental areas beyond the switchyard.

In the M.I.T./SLAC experiment, the target that the beam encountered was liquid hydrogen, so the basic interaction being studied was between electrons and protons. After the impact, the direction and momentum of those electrons that had suffered collisions were measured. The similarity to the Rutherford experiment is striking. In each experiment a projectile is directed against the target whose properties we wish to study. The idea is that by looking at what happens to the projectile after it hits the target, we can learn something about the way the target is constructed. This certainly succeeded in the Rutherford experiment, where the large number of particles coming off backward was a direct indication of the existence of the nucleus.

Of course, there are also differences between the two experiments. Rutherford's alpha particles had energies of only a few MeV, so they could not create any new particles in the collisions. The electrons, having in excess of 20 GeV, can (and do) create all sorts of particles when they collide with the proton. The debris of the collision is not measured, but by measuring the momentum and energy of the electron before and after the collision, we can determine how much energy was delivered to the proton and, hence, what the total energy of the final state of the target must be. For example, if the electron gives up 5 GeV in the collision and the end product is a proton and three pi-mesons, we can say with certainty that the sum of the energies of the four particles must be 5 GeV (plus, of course, the mass energy of the proton, which is the same before and after the collision).

A second important difference is the size of the object we

believe we can "see" with the projectile in the two experiments. We can get some rough idea of this quantity from the uncertainty principle for momentum and position: $\Delta x \cdot \Delta p \leq h$. If we use the momentum of the particles as an estimate for the uncertainty in momentum, and identify Δx as the size, D, of an object we could "see" with the particle, we see that D is just

$$D \cong \frac{h}{p}$$

In the following table we give D for some sample projectiles:

PROJECTILE	TYPICAL MOMENTUM	D
	(eV/c)	(cm)
Photon	10	5×10^{-6}
Alpha	4×10^6	10^{-11}
SLAC electron	10^{10}	5×10^{-15}

We infer that a normal photon can "see" something approximately the size of an atom, a Rutherford-type alpha can "see" something approximately the size of a large nucleus, and an electron from SLAC can "see" something much smaller than the size of a proton. Thus, if any experiment has any chance of "seeing" the quarks inside the proton, this one should.

Before the experiment was performed, theorists did not expect to see many scattered electrons with final energies around 8–10 GeV (indicating energy transfers to the proton of many GeV). When the experiment was completed, however, an appreciable number of these electrons was seen—the number exceeding even the most optimistic theoretical estimate by a factor of 40. And, as in the case of Rutherford, these excess particles were seen at large angles.

This result can be understood in analogy with the Rutherford experiment if we think of the electron as scattering off tiny, pointlike particles inside the proton. Thus, the Stanford experiment provided the first direct evidence that elementary particles are not elementary, but are composed of constituents.

Since that time, a large number of other experiments of this general type have been performed. They are called deep inelastic scattering experiments because they probe to very small distances in the proton (and, hence, to "deep" beneath its surface) and create many particles in collisions. The general result of all these experiments seems to be the same: When you work with particle structure at small distances, the particles appear to be made up of constituents, even if you cannot detect these constituents directly in the debris of the interaction.

THE FLY IN THE OINTMENT

With all of this evidence in favor of the existence of quarks, it may come as something of a surprise to discover that no quark has ever been seen in the laboratory. This state of affairs certainly hasn't arisen for lack of effort. There is, in fact, a long history of unsuccessful quark searches, as we shall see in the next chapter.

XI

Where Are They?

We seek him here, we seek him there,
Those Frenchies seek him everywhere.
Is he in heaven?—Is he in hell?
That demmed, elusive Pimpernel?
   —Baroness ORCZY, *The Scarlet Pimpernel*

INTRODUCTION

FOLLOWING THE SUGGESTION in 1964 that quarks might exist, a vast amount of effort to "bring one back alive" was expended by experimental physicists. Nevertheless, years of searching have failed to produce a single particle in the laboratory that is generally accepted as a quark. A review of some of these searches and a discussion of the response that theoretical physicists have made to this result are the subjects of this chapter. We'll see that physicists have managed to take what might look like a problem—the absence of success in quark searches—and turn it into a strength.

The property of quarks that would make them easy to find is, of course, their fractional electrical charge. The standard detection instruments for particles—the Geiger counter, cloud chamber, and bubble chamber—all depend, ultimately, on the electrical interaction of the particle with stationary atoms. In the cloud chamber, for example, the charged particle passing an atom creates an ion that then serves as a nucleus for the condensation of a droplet. In other devices, the ion is detected in different ways, but all the detectors start with the detected particle removing an electron from an atom.

The ability of a particle to create ions depends on the particle's charge. This is as we might expect—the bigger the charge of the particle, the greater the force exerted on the atomic electron. If the charge of the particle is denoted by Q, then it turns out that the number of ions that charge will produce along each centimeter of its track will be proportional to Q^2. This means, for example, that an alpha particle with two protonic charges will create four times as many ions as a single proton or electron moving at the same velocity. We will see a much more dense collection of droplets for the alpha than we would for the proton, and the droplet density can be used as a way of identifying the alpha and distinguishing it from singly charged particles.

For quarks, with their fractional charge, the same thing can be done, except that now we have to look for particles producing *fewer* droplets than would a proton or electron. A quark of charge $2/3$, for example, should produce a track whose droplet density is $(2/3)^2 = 4/9 \approx 1/2$ of the density caused by a normal particle, while a quark of charge $1/3$ should produce a track of density $1/9$. Hence, if we were looking for quarks in a cloud or bubble chamber, we would look for tracks on which we counted too few droplets or bubbles. These are called *lightly ionizing* tracks, and they form the basis for many of the quark searches that have been done in cosmic rays and on accelerators.

An alternate way of utilizing the quark's fractional charge in experiments is based on what happens when a quark is absorbed in matter. There are only two places a quark can go in an atom. It can either enter the nucleus or it can go into orbit and replace one of the electrons. In either case, the resulting "quarked atom" will have a net electrical charge. This means that we can extract the atoms containing quarks by passing the material between plates that are electrically charged. The quarked atoms will tend to collect on the plates while the normal atoms, being neutral, will be largely unaffected. The collected material is presumably rich in quarks and can then be analyzed in more detail.

A large number of searches have been made using this technique. I refer to them generically as *geological* searches, since most of them involve analyzing kinds of materials which, for one reason or another, are thought to be a likely final repository for quarks—moon rocks, sea water, and even oyster shells. The term geological is a little misleading, but it is convenient.

The fractional charge of the quarks inspires this type of experiment because once a quark has been created, it cannot decay into ordinary particles. Such a decay would necessarily violate the principle of charge conservation. Hence, it would seem that once a quarked atom is formed, it will persist and be at our disposal when we start looking for it.

Unlike the direct searches, a negative result in a geological search is a little difficult to interpret. Not finding a quark in sea water, for example, might mean that there are no free quarks, or it might mean that one of the assumptions in the chain of reasoning that led to choosing sea water for the search was wrong. This should be kept in mind as we move on to other quark searches.

DIRECT QUARK SEARCHES

By far the most straightforward way to look for a quark is to repeat the procedures that led to the discovery of so many of the other particles. We could, for example, see if quarks can be found in cosmic ray experiments, either in showers or in the primary cosmic radiation itself. Alternatively, we can try to create quarks in an accelerator and see if we can detect them.

As far as cosmic ray experiments go, there are two possibilities. Either quarks are themselves present in the primary cosmic radiation that falls on the earth, or they are created in the very high-energy collisions that primary cosmic rays make when they enter the atmosphere. To test the first hypothesis, it is only necessary to put out counters and look for lightly ionizing tracks. Because quarks lose less energy through ionization than normal particles, they ought to penetrate farther into the atmosphere and, hence, be more visible in such experiments. In any case, over twenty such searches were carried out without success, and taken together these experiments can be used to establish an upper limit on the number of quarks falling on the surface of the earth. If we call this limit N, then the number of quarks that actually fall on the earth must be less than N. From the experiments, $N \approx 10^{-10}$ quarks/centimeter/second.

To get some idea of what this limit means, we note that there are roughly 3×10^7 seconds in a year. Thus, if we picked 1 square centimeter of the earth's surface and waited for a quark to hit it, we would have $10^{-10} \times 3 \times 10^7 = 3 \times 10^{-3}$ quarks/year. This means that on the average we would have to wait about 330 years before we could expect to see a quark fall on our target. And, since this represents an upper limit, we could easily have fewer (or even zero) quarks falling, corresponding to a waiting time of hundreds, or even an infinite number, of years. Clearly, if there are quarks around they are not copiously represented in the primary cosmic ray flux.

If this is true, we can still borrow a leaf from history and look for quarks in the debris of cosmic ray collisions. That is how the positron and the mesons were first seen. This type of experiment was done by setting out small detectors—detectors that will be triggered when shower particles begin arriving at a site. This signal is then used to trigger the main detector (for example, a cloud chamber would be expanded only when the small counters indicated that a shower was present). In this way, the quark search would only take place when a major cosmic ray event had occurred. In a variation on this device, some experiments scanned particles arriving after the shower, in case the quarks should be slow and heavy.

Experiments of this type were carried out in many places. For a time in 1969 there was a flurry of excitement when a group in Australia reported a couple of events that might have been caused by quarks, but this quickly died down when an improved experiment by that group and duplicate experiments elsewhere failed to turn up any similar events. In general, the shower experiments give about the same limit as the single-particle experiments—a quark flux at the earth's surface of less than 10^{-10} quarks/second/square centimeter.

The failure of cosmic ray experiments caused physicists to turn to accelerators. In fact, there is a ritual search made for quarks whenever a new high-energy machine is turned on. These searches have the advantage of being able to control the incident beam of projectiles and the disadvantage of being limited in energy by the machine design. Quarks must be produced in pairs in order for charge to be conserved, which means that there is a limit to the mass of a quark that can be produced in any accelerator, a limit given by the mass-energy relation discussed in Chapter IV. Typically, this limit will be from a few GeV to 15–20 GeV, depending on the machine. We now look at a typical accelerator search so that we have some idea of how they are done.

In Chapter VI we saw that the radius of a circle through which a magnetic field B will bend a particle with momentum P is given by

$$R = \frac{P}{Bq}$$

where q is the charge of the particle. In Chapter VII it was explained how this fact could be used to make a magnetic spectrometer. By a slightly different chain of reasoning, it can also be used to turn such a spectrometer into a highly efficient quark detector.

Consider the situation in which a beam of protons strikes a target, and the collision products go through a magnetic analyzer. For the sake of definiteness, let the incident proton have a momentum P_p. Then the highest momentum that any particle coming out of the interaction can have is P_p (otherwise, momentum could not be conserved). If such a particle has a normal unit of charge, it will be bent through some radius by the magnet. Call this radius R_n, where the n denotes the radius through which the fastest particle with unit charge is deflected.

Suppose now we have reason to believe that a quark of charge $1/3$ and momentum P is produced in the interaction. The radius through which it will be bent (call it R_Q) will be

$$R_Q = \frac{P}{B \cdot \frac{1}{3}e} = 3\frac{P}{Be}$$

where e is the charge of the electron. It can, in other words, be larger than R_n, the largest radius through which a normal particle can be deflected. Consequently, if we build our apparatus with the detectors behind a slit set for a radius R_Q, momentum conservation tells us that no particle of normal charge will ever enter the detector. Consequently, if anything is seen, it must have less than a unit charge, and some simple ionization measurements will tell us immediately if it is a quark.

Thus, by using our knowledge of the energy of the incident beam, it is possible to design very clean experiments for quark searches. When these experiments are carried out (and dozens have been), no quarks are seen. This fact is then used to put an upper limit on the probability that a quark is produced in a proton-proton collision. The best limits now suggest that a quark will be produced in less than one interaction in a trillion (10^{12}), with a strong possibility that it will not be produced at all.

GEOLOGICAL SEARCHES

The logic for geological quark searches was somewhat different from that of the direct type of measurements discussed in the last section. We can start by assuming that quarks are present in cosmic rays, but at a level just below the limits set by experiment. Assume, in other words, that quarks are striking the earth at the rate of about 10^{-10}/square centimeter/second. The direct searches are not sensitive enough to detect this small number of quarks, and it is the function of geological searches to remedy this defect.

If quarks have been falling on the surface of the earth at this rate since the earth was formed 5 billion years ago, then roughly 10^8 quarks would have fallen on each square centimeter of the earth's surface. These quarks have to go somewhere and, since they cannot decay, they must still be on the earth. The objective then becomes one of trying to guess where they are right now.

If we think about the question, we realize immediately that we have to make some assumptions about long-term geological shifts in the earth's surface. A rock that sits on the surface today, exposed to cosmic rays, may have been deep underground a million years ago, and may be deep underground a million years from now. Hence, any quarks that strike the surface will become mixed into the ground to some depth, and what we assume about that depth makes a great deal of difference as to the number of quarks we expect to find per cubic centimeter in geological material. For example, if we assume the depth of the mixing is a few kilometers, we would expect about 1000 quarks per cubic centimeter in the earth's surface. If we assumed the mixing was more or less uniform, there would be about that number of quarks in any randomly selected piece of material. The number corresponds to about one quarked atom for every 10^{21} normal atoms, a number that is just barely detectable by the techniques we will describe in this section.

But consider the long chain of reasoning involved in this conclusion. First, we have to assume that there actually *are* quarks hitting the earth. Then we have to assume that these quarks are incorporated uniformly into surface atoms. Then we have to assume something about the geological processes that carry these atoms around. A negative result could arise from any one of these assumptions being wrong. This is what I meant earlier when I said that geological quark searches are harder to interpret than direct ones.

Most people who carried out geological searches did not assume a uniform distribution of quarks in the earth's materials, but reasoned from the properties of quarks to estimate their concentration. For example, in 1968 David Rank at the University of Michigan, while working with the theory for "quarkium" (an atom in which a single electron circles a quark), concluded that this particle ought to be soluble in water and behave somewhat like lithium. From this inference he concluded that quarks ought to congregate in the sea, where biological processes might concentrate them further. Consequently, he analyzed sea water, seaweed, oyster shells, and plankton to see if he could find any quarks. Since this method is rather typical of geological searches, it is worth examining in a little more detail.

First, the various materials were ground up and heated until they vaporized. The vapor was passed through an electric field so that any charged atoms were concentrated on the charged plates. This concentrated material was then tested by heating it again in an electric arc and analyzing the light given off. Electrons around a quarked nucleus will have different orbits from those in a normal atom because the nucleus has a different charge. Therefore, from the discussion in Chapter I, we know that such atoms should emit different photons, and these photons, if detected, would be one kind of evidence that we could use to deduce the presence of quarks in the sample. From this sort of analysis, Rank concluded that if there were any quarks in his sample, there were fewer than one for each 10^{18} atoms of sea water, and fewer than one for each 10^{17} atoms in the seaweed, oysters, and plankton.

Another common method of analyzing concentrated materials is called the oil drop experiment. This is an old technique that was first used in 1910 by Robert A. Millikan to measure the charge of the electron. In Rank's work at Michigan, various organic oils (such as peanut and cod liver) in drop form were tested directly for quarks. It would also be possible to make a sample by dissolving some concentrated material in oil. In any case, the oil is sprayed into the air between plates whose voltage can be regulated. There are then two forces acting on the drop: gravity (tending to pull it down) and electricity (tending to lift it up). The electrical force depends on the charge of the drop. By adjusting the voltage on the plates and watching the drops move up and down, it is possible to determine this charge with very high accuracy; certainly well enough to tell whether it is fractional or not.

Rank was able to conclude that if quarks existed in his oil samples, there was less than one for each 10^{19} ordinary atoms. In the following table we give a sampling of the materials that have been tested and the number of normal atoms in which a single quark would have to be diluted to explain the negative result of the experiment.

MATERIAL	NUMBER OF NORMAL ATOMS
	(rounded off to nearest factor of 10)
Lava	10^{22}
Grand Canyon rock	10^{22}
Sea water*	10^{25}
Ocean sediment	10^{21}
Meteorites	10^{17}
Iron	10^{21}
Moon rocks	10^{22}

*This is a more recent experiment than the one discussed in the text.

Unfortunately, in this short section we cannot do justice to all the clever techniques that were devised to find geological quarks, but all of them have produced limits on the quark concentration similar to the ones in the table.

WHAT DOES IT MEAN?

The brutal fact of the matter is that there is not a single generally accepted piece of evidence that a quark has been isolated in a laboratory. What are we to make of this?

There is an old saying to the effect that when life gives you lemons, you should make lemonade. In a sense, this is what happened when physicists realized that their quark searches were coming up empty. The absence of quarks in the laboratory, a difficult situation on the face of it, became a virtue as a new class of theories was developed in which quarks were postulated to exist inside elementary particles, but could never be shaken loose for laboratory observation.

Now I have to emphasize that this wasn't a situation in which theories were patched together to cover up an embarrassing experimental problem. Instead, as the new unified field theories (which we'll talk about in Chapter XIV) started to be developed, a rather amazing feature began to emerge. In the more successful theories, quarks were, in effect, locked in to the elementary particles. In the language of the theoretical physicists, the quarks were "confined." Furthermore, the confinement of quarks also explained many of the simple successes of the model that we discussed in Chapter X.

There's an easy way to visualize how quark confinement works. Think of elementary particles such as a meson as being like a rubber band, with the quarks themselves as the ends of the band. If you reach into the particle to pull a quark out, two things are evident. First, the harder you pull, the stronger the force holding the quark into the particle becomes. Second, if you pull very hard and break the rubber band, you still won't get a free quark, but two shorter rubber bands. Each of these will have two ends, and will, therefore, be interpreted as an additional meson.

In fact, this is a process we have described in detail earlier—the creation of new particles in the collisions of elementary particles. Thus, in this sort of theory, attempts to shake quarks loose will just result in the production of sprays of mesons and other particles. And although I have used the analogy of mesons to make the discussion simpler, it's obvious the same results can be derived for baryons. The only change would be to make the rubber band Y-shaped (so it has three ends) instead of straight.

There is an additional bonus to confinement theories. In Chapter X, we talked about how particles seem to behave as if they were made up of quarks that acted more or less independently of each other. Yet if the quarks are really locked together so strongly in the particles, how could they possibly be independent? How, for example, could one quark scatter without affecting the others?

The confinement theories gave a rather elegant solution to this problem. The force that holds quarks together in these theories do, indeed, get very strong when the quarks are widely separated (think of the rubber band again). But when quarks are close together, this force drops dramatically. In these theories, you can picture a proton as being something like three marbles inside a tin can. As far as the behavior of the marbles near the center of the can is concerned, they can be regarded as free, noninteracting particles. Each marble just rolls around as if the others were absent. But if a quark tries to get too far from the others, it runs into the wall and is turned back.

Thus, the failure of the quarks searches turns out to be something of a plus. The confinement theories explain not only the failure, but the success of all the indirect evidence for quarks we've already discussed.

One final comment about confinement: Our current notion is that during the first fraction of a second in the life of the universe, the temperatures and energies were so high that matter existed in the form of free quarks. When the universe was about 10 microseconds old, however, these quarks "froze" into the elementary particles, and ever since then they've been confined.

Still, the idea that something can exist and yet not be visible in an experiment, even in principle, raises some tricky philosophical questions, to put it mildly. At present, the confined quarks are discussed because the data seem to favor them. The whole business was put into perspective by Lawrence Jones of the University of Michigan, who carried out a quark search himself and authored a recent review of the data: "It is," he said, "a question on which wise men can differ."

Enough said.

XII

Charm and the Proliferation of Quarks

Be fruitful, and multiply . . .
  —Genesis 9:1

THE DISCOVERY OF THE ψ/J

WE HAVE SEEN that the strongest motive for using the quark model is the simplicity that it introduces into our picture of elementary particles. By the mid-1970s, however, some developments had taken place that cast serious doubt on the ultimate simplicity of the whole quark picture. One set of developments was experimental and rather unexpected. It involved the discovery of new particles. The other was theoretical and, in a sense, was part of the model from the very beginning. We can begin with the experimental situation.

   In mid-1974 two experiments were being conducted several thousand miles apart. The two were completely different and the physicists involved in one had no idea of what was going on in the other. Yet each of these efforts resulted in the discovery of a new particle, and these discoveries were made so close together that they were reported in the same issue of *Physical Review Letters.*

One experiment was being done at Brookhaven National Laboratory on Long Island under the direction of Samuel C. C. Ting of the Massachusetts Institute of Technology. A beam of protons was directed against a target made of beryllium (a lightweight metal whose nucleus contains nine protons and neutrons). A series of magnetic selectors, scintillators, and Cerenkov counters were set up in a two-armed arrangement, as shown in Illustration 26. Its purpose was to look for reactions of the type

$$P + B_e \rightarrow e^+ + e^- + \text{anything}$$

In other words, the M.I.T. experiment was designed to look for electron pairs produced in the proton-nucleus collision.

This is a very difficult experiment to carry out. In Chapter VII we saw how extreme care had to be taken in order to distinguish between rare antiproton production and the copious production of pi-mesons. In Ting's experiment the difficulty is much greater, since electron pairs are produced very seldom, particularly at the large angles where the experiment was designed to make measurements. There could be billions of hadron pairs produced for each electron pair.

To get some idea of what this means, think of a rain shower falling on an average-sized city. There might be several billion raindrops in a typical shower. Finding the electron pair among all of the hadrons is analogous to finding one raindrop falling on a city during a shower—not the easiest of tasks.

On the other hand, Ting and his group had spent over 10 years perfecting the apparatus and, in the process, acquiring a solid reputation as extremely careful and precise workers. I can testify to the validity of this reputation personally, because I spent two summers with the group when it was working in Hamburg, Germany. My most vivid recollection of the "check–double-check–triple-check—and then check again" operation was that each computer program used to analyze data was written independently by two physicists in the group, and the results of the independent programs were then compared at the daily

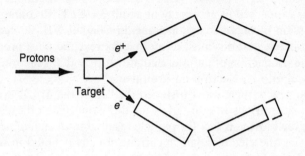

26. Diagram of apparatus used in the experiment when the ψ/J was discovered.

staff meetings. Only when everything checked out would the data analysis go ahead.

Using this careful experimental technique, the group began collecting data in the late summer and early fall of 1974. When they plotted the number of electron-positron pairs as a function

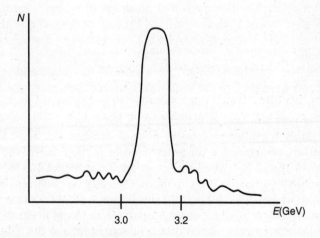

27. Graph of electron-positron pairs.

of the energy of the pair, they got a result like that pictured in Illustration 27. A large peak (corresponding to about 250 events) appeared at an energy of about 3.1 GeV. Recalling our discussion of phase space diagrams in Chapter VII, we realize immediately that this must mean that the reaction must proceed via the production of an intermediate particle that subsequently decayed into an electron-positron pair.

The new particle was christened the "J" by the M.I.T. group. There are various reasons given for this name. The official reason given by Ting is that the particle resulted from an interaction involving the electromagnetic current, usually denoted in theoretical physics by the letter "J." Others have pointed out, however, that there is a striking resemblance between the letter "J" and the Chinese character for Ting. In either case, it was clear that the group had discovered a new particle.

While this was happening at Brookhaven, another group at Stanford, led by Burton Richter of SLAC, was coming to a similar conclusion by a different route. Since the early 1960s, Richter had been working on the design of an electron-positron storage ring that could take particles from the linear accelerator and, after storing them, cause them to collide head-on. By 1972, just 21 months after construction funds became available, this facility was in operation. Electrons and positrons of energies up to 3 GeV were stored in the ring where, because of the fact that they have opposite electrical charges, they circulated in opposite directions.

The two circulating beams were allowed to collide and then counters were used to see what was produced. Schematically, we are dealing with a situation in which an electron and a positron collide and produce an assortment of hadrons. In general, the reaction will produce a spray of pions, K-mesons, electrons, and positrons, but evidence for a new kind of particle was found in some of them. But when the experimental evidence started to accumulate, a "bump" was found in the electron-positron cross section, which indicated that the interaction was taking place via an intermediate particle, as seen on the right in the illustration. The Stanford group named their new particle the Ψ (Psi).

The two groups learned of each other's work in the fall of

1974, when Sam Ting paid a visit to Stanford. By that time, both groups were ready to announce their results. Two completely independent experiments, using completely different techniques, had discovered the same particle at about the same time. The particle is now called the Ψ/J, and its properties are listed in the table. What these properties mean will be discussed in the next section.

PROPERTIES OF THE Ψ/J	
Mass	3,098 MeV
Spin	1
Isospin	0
Parity	—
Width of resonance	67 keV
Strangeness	0

THE IDEA OF CHARM: AN EXTRA QUARK

In Chapter XI we saw that one of the major pieces of evidence for the quark model was that by it every known particle was accounted for. This, in turn, implied that there were no "vacancies" in the hadronic scheme of things into which the Ψ/J could be fit. The discovery of a new particle clearly showed that some major overhauling of the quark model was needed—or that it had to be scrapped.

A hint at a resolution of this problem can be seen by considering the width of the Ψ/J. In Chapter VII we saw how the uncertainty principle could be used to relate resonance width to a lifetime for a particle. A width between 100 and 200 MeV, for example, corresponds to a lifetime typical of the strong interactions, such as, 10^{-23} second. The Ψ/J has a width 1000 times smaller than this, so its lifetime must be over a thousand times greater, or around 10^{-20} second. As far as the strong interactions are concerned, the Ψ/J is virtually a stable particle, like the lambda and the sigma.

This analogy suggests a solution to the puzzle posed by the

discovery of the Ψ/J. Just as the slow decay of the lambda and the sigma is associated with the presence of the strangeness quantum number and the s quark, perhaps the slow decay of the Ψ/J is associated with a new quantum number and a new kind of quark. In that case, the general rule for decays we discussed in Chapter IX would tell us that converting the new quark into a u or d could not be done in 10^{-23} second, but would have to involve a time similar to that required to make the same conversion for the s quark.

As it happened, theorists had suggested that a fourth quark might exist. The argument was based on some concepts of symmetry between hadrons and leptons that seemed to require that quarks come in pairs. The u and d quarks clearly do this, but the "partner" of the s quark had been missing when the prediction was made. In addition, the existence of such a quark would explain certain properties of weak interactions that had been puzzling up to that time. Thus, when the Ψ/J was discovered, a theoretical groundwork had already been laid for an addition to the roster of quarks.

The new quantum number had been labeled *charm*, and was denoted by the letter C. The quark that carried it was called the c (for charmed) quark. The relation between charm and the c quark is exactly analogous to the relation between strangeness and the s quark. A particle that contains one c quark would have a charm of $+1$, a particle containing two c quarks, a charm of $+2$, and so on. Each time a c quark has to be converted to a u or d quark, the corresponding particle reaction takes a time long compared to the time associated with the strong interactions.

All of the particles we considered before the discovery of the Ψ/J were formed from u, d, and s quarks and, hence, had $C = 0$. This is similar to the fact that the nucleon and delta families, being made from u and d quarks, have $S = 0$. Like the other quarks, the c quark has spin $\frac{1}{2}$ and baryon number $\frac{1}{3}$. It has strangeness 0, $C = 1$, isotopic spin 0, and charge $\frac{2}{3}$. Knowing these facts, we can use the techniques of the ordinary quark model to determine what new kinds of objects we should see when we open the charm dimension in the elementary particle world. But before we start this job, let us consider the Ψ/J itself.

Since it is made from an electron-positron interaction, it must have $B = 0$, and hence be a meson. It must, therefore, be composed of a quark-antiquark pair. And since all the states involving u, d, and s quarks are used up, the process of elimination leads us to the conclusion that the θ/J is made up of a c quark and an anti-c quark.

If this is the case, then two predictions can be made. First, there ought to be other particles with $C = 0$, corresponding to the c and $\bar{c}$ in different orientations. Second, there ought to be particles that have $C = 1$, corresponding to a c quark and an anti-u or d quark. Both of these predictions have been borne out by experiment.

The first point was quickly perceived by theoretical physicists. If the Ψ/J is really a c-$\bar{c}$ system with the quark spins aligned, there ought to be another particle corresponding to the quark spins antialigned, and still another corresponding to the situations where the c and $\bar{c}$ are moving around each other in orbits. In short, the theory predicts that there ought to be many particles, each decaying into lower mass members of the group by emitting gamma rays. It should be possible to detect these decay gamma rays in the laboratory. They were, in fact, seen at accelerators in Hamburg and in Italy (as well as at SLAC) shortly after the discovery of the Ψ/J itself. In Illustration 28 we show the present c-$\bar{c}$ family of particles as they are known at present. Obviously, the fact that all of these predicted companions of the Ψ/J exist lends a lot of credibility to the theory of a fourth quark.

A much more direct bit of evidence would be the existence of a particle that had $C \neq 0$—a particle, in other words, that contained a single c quark. There ought to be mesons, for example, in which a c quark is matched with a $\bar{u}$ quark to produce a particle with $Q = 0$ and $C - 1$. The same theoretical considerations that led to the prediction of charm in the first place indicate that this meson ought to decay into a K-meson and a few pions. Thus, it is a straightforward matter to look at the electron-positron reactions that give rise to K-mesons and pions to see if any peaks are seen in the phase space diagram. In 1976, experiments at SLAC showed a bump in the number of pi- and K-mesons produced in electron-proton collisions. By now, you

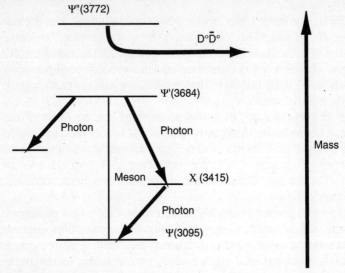

28. The currently known c-$\bar{c}$ family of particles.

should be used to interpreting such bumps as evidence for a new particle. In this case, the particle was named the $D°$-meson. It has a mass of about 1.8 GeV and is, in fact, the particle whose quark structure we pictured above. Since its discovery, other particles in the D family have been found, as would be expected. These include a D* (similar to the $D°$, except that the quark spins are aligned) and a set of charged D-mesons made from various combinations of c and u quarks.

From these developments, it is clear that the addition of the charmed quark simply adds a new dimension to the quark model, but does not change it in any essential way. The D*, for example, bears the same relation to the $D°$ as the rho-meson does to the pi-meson. All of the other games we learned to play in constructing particles from quarks can be played with the addition of charm. For example, there ought to be particles (made from c and s quarks) that are both charmed and strange. We would also expect to see baryons made with one, two, or three c quarks in combination with the others. In fact, there should be a whole new set of particles waiting to be found in accelerator experiments—enough to keep experimental physicists happy for a long time. As an example of how these newly predicted particles might look in a standard eightfold way plot, we show in Illustration 29, below, the predicted baryons associated with the spin $3/2$ baryons. This is, you will recall, the grouping of particles that led to the prediction and discovery of the omega—one of the first great triumphs for this approach to elementary particles.

It should be understood that being forced to add charm to our list of quarks suggests a disturbing possibility. What if it is not the last quark to be forced on us, but only one of many? If this were so, then the original hope of being able to use the quark model to create a simple picture of nature would be in serious trouble. Charm by itself, of course, does not imply any such conclusion. A world composed of four basic particles is as simple as a world made up of three. But remembering that the proliferation of chemical elements and elementary particles is what led

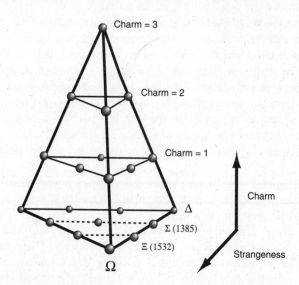

29. Standard eightfold way plot showing predicted baryons associated with the spin $3/2$ baryons.

to the idea of quarks in the first place, we recognize that any hint of a proliferation among the quarks must be taken very seriously indeed.

MORE QUARK CANDIDATES

In 1977 a group under the direction of Leon Lederman at the Fermi National Laboratory near Chicago announced a result that was to have profound implications for the quark model. The experiment they performed was similar in design to the Brookhaven experiment that led to the discovery of charm. A beam of high-energy protons from the accelerator was allowed to strike a nuclear target (either copper or platinum), and then detectors were set up to look for pairs of oppositely charged mu-mesons. The main differences between this experiment and the one that Ting had carried out were (1) that the Fermilab accelerator has a much higher energy proton beam, and (2) that the detection of mu-meson pairs rather than electron-positron pairs makes certain parts of the work a little easier to do.

In a sequence of events that seems almost a replay of the 1974 discovery of charm, this experiment produced data for the number of events in which a muon pair is seen as a function of the energy of that pair. (See Illus. 30.) Again, a series of bumps in the otherwise smoothly falling data indicated the presence of a new particle (or set of particles). As before, there was no place for these "extra" particles in the quark model, and, as before, there were theoretical arguments based on studies of weak interactions that suggested that there ought to be a new kind of quark.

Data from Fermilab and other laboratories have pretty well shown that there are, in fact, two particles causing wiggles on the curve in Illustration 30, and there may be others as well. These particles are denoted by the Greek letter Υ (upsilon) and Υ′; they have masses of 9.5 and 10.0 GeV, respectively. These particles are now universally accepted as being states similar to the Ψ/J, in which a new quark and its antiquark are bound into a meson. The new quark is called the *b*. The Υ′ is presumably the same set of quarks with spins aligned differently, and we can

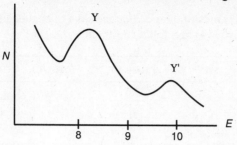

30. Graph showing probable existence of the Y and Y′ particles.

expect a set of new particles to be found corresponding to the ones that were discovered after the Ψ/J. Eventually, we can expect to see particles created that exhibit the *b* quantum number explicitly, just as the *D*-mesons were found to exhibit charm. Hence, the discovery of the Y seems to start the whole process of building up new particles from a new quark that we went through with charm, and all of the comments we made in the previous section should apply here as well.

The theoretical arguments that suggested the existence of another quark besides *u*, *d*, *s*, and *c* actually predict a pair of quarks. These are labeled *b* and *t*. The *b* has charge $-\frac{1}{3}$ and the *t* has charge $+\frac{2}{3}$. Otherwise, each is similar to the *c* quark, except that it carries a new quantum number. Although a valiant effort was made to name them "beauty" and "truth," Keats eventually lost out to the more prosaic *bottom* and *top*. Thus, we would say that the Y particle is made of a bottom quark and antibottom quark, and that eventually we will see particles that explicitly exhibit the bottom quantum number. As I write (winter 1993), there is no evidence to indicate that particles made from a top quark have been seen. However, historical precedent certainly seems to indicate that we shall eventually see such particles. The most likely sequence would seem to be the appearance of a narrow bump in a cross section, followed by a rapid proliferation of states and the eventual discovery of particles carrying new quantum numbers.

We expect that in the near future we shall be confronted with six different kinds of quarks. For the sake of convenience, we list their properties in the following table:

QUARK	SPIN	CHARGE	OTHER QUANTUM NUMBER
u	$\frac{1}{2}$	$\frac{2}{3}$	none
d	$\frac{1}{2}$	$-\frac{1}{3}$	none
s	$\frac{1}{2}$	$-\frac{1}{3}$	$s = -1$
c	$\frac{1}{2}$	$\frac{2}{3}$	$c = +1$
b	$\frac{1}{2}$	$-\frac{1}{3}$	$b = +1$
t	$\frac{1}{2}$	$\frac{2}{3}$	$t = +1$

All, of course, have baryon number $1/3$.

THEORETICAL PROLIFERATION:
THE IDEA OF COLOR

In the preceding sections we have seen how experimental results have forced us to increase the number of quarks we want to call basic. While this process was going on, a parallel development was taking place in the theory of quarks which, in essence, had the same effect. To understand the reasoning, we can go back for a moment and think about an ordinary atom.

The first three elements of the periodic table are hydrogen, helium, and lithium. Hydrogen has a nucleus composed of a single proton, and a single electron moves in the lowest orbit. Helium has a nucleus made up of two protons and two neutrons, and, hence, must have two electrons. Both of these can be found in the lowest orbit. Lithium has a nucleus composed of three protons and four neutrons, and must therefore have three electrons. Two of these electrons, like the two electrons in helium, occupy the lowest orbit, but the third electron resides in the next higher orbit. It is almost as if the first two electrons "fill up" the lowest orbit, so that the third has to find room for itself elsewhere. The idea that electrons can fill up orbital spaces follows from something called the Pauli exclusion principle. Although the derivation of this law from quantum mechanics is rather abstract, the rule itself can be stated quite simply: *No two identical spin $1/2$ particles can be in the same state.*

One way to perceive this principle is to imagine that each orbit in an atom is a kind of parking lot for electrons. Each electron in the orbit occupies one space, and when all the spaces are taken, the parking lot is full. Additional electrons must then go to the next highest orbit in which the parking lot is still empty.

Perhaps the easiest way to see how the Pauli principle works in practice is to consider electrons in the lowest orbit. The "state" of the electron in that case corresponds to the direction of its spin. Since this can have two values (up or down), we say that there are two electron states in the lowest orbit or, from the parking lot analogy, that there are two "parking spaces" there. This means that the orbit will be completely filled when it contains two electrons with their spins oriented in opposite directions. The Pauli principle guarantees that in the lithium atom, the third electron must be in some other orbit.

In passing, we should note that the same laws of quantum mechanics predict that there should be eight states in the next orbit. Since the chemical properties of an atom depend on the outermost electrons, this is the explanation of the periodic table of the elements we talked about in Chapter IX. There are only two elements (helium and hydrogen) in the first row of the periodic table. Lithium, the third element, starts a new row and is placed in the same column as hydrogen. They are chemically similar because they both have a single electron in the outermost orbit. Similar reasoning predicts that the next such element (sodium) should have eleven electrons—two in the lowest orbit, eight in the next, and one in the third. These kinds of electron distributions account for the entire structure of the periodic table.

Since quarks are spin $1/2$ objects, there is a strong prejudice on the part of theoreticians to assume that they, too, must obey the Pauli principle. Yet there is at least one elementary particle whose very existence seems to imply that they cannot. In Chapter IX we saw that the doubly charged delta (Δ^{++}) was constructed from three d quarks with spins aligned. No other arrangement will match the spin $3/2$, $Q = 2$ that we observe in this particle.

If the quarks obey the Pauli principle, this particle should not exist, since all three are identical spin $1/2$ particles and all three are in the same state. There are only two ways out of this dilemma—either quarks do not obey the Pauli principle, or the quarks in the Δ^{++} are not identical.

If we explore this last possibility, we have to suppose that the reason the Δ^{++} can exist is that its three constituent quarks are not identical in the sense implied by the Pauli principle. Since they have the same charge, spin, parity, and so on, the difference between them must lie in some characteristic that we have not yet discussed. Suppose, for example, that some subatomic gremlin had come around and painted the quarks in three different colors. Then, provided that the three quarks in the Δ^{++} were different colors, there would be no problem with the Pauli principle. Although the three are all in the lowest state and spin up, they would not be identical particles because they would have different colors.

The idea of *color* was introduced (under another name) by O. W. Greenberg at the University of Maryland almost as soon as the quark model was put forward in 1964, but it was a decade before the concept was widely accepted. The delay was inevitable because no particles have ever been seen that exhibit the color quantum number explicitly. Had they been seen, they would have been recognized in the same way that charm and the b quantum number were. This amounts to a rule which states that although individual quarks have color, the net color (a quantity derived from the sum of the colors of the quarks) of known particles must be zero.

At this point, we ought to pause for a word of caution. When I introduced the idea of strangeness, I commented on the problems that physicists create for themselves when they use common words to designate new properties of particles. The term "color" is a perfect example. No one thinks that there is any gremlin running around with brushes and buckets, daubing paint on the quarks. No one really thinks that it is necessary to picture quarks as little spinning spheres of different colors. Yet, such a picture is encouraged by the adoption of the term.

For the record then, the term color as it is applied to quarks is not the same thing as the term color as it is applied to everyday objects. In physics it refers to a property of the quarks in the same way that spin, parity, strangeness, and charge signify properties. And just as we can determine the charge of a particle by adding up the charge of each constituent quark, we can find the color of a particle by combining the colors of each quark in it. The rules for combining color turn out to be a little more complicated than the simple addition that we apply to charge, but the principle is the same.

The rule for making hadrons from colored quarks amounts to requiring that when the colors of the quarks are added the result is zero net color. If there were a similar rule for electrical charge, it would require that only those particles for which the charge of the quarks added to zero could actually exist, and if this were true, we would only see electrically neutral particles in nature. No such rule exists for charge, of course, but it seems to exist for color. As a consequence, only particles with zero color are actually seen in the laboratory.

There is some experimental evidence that color actually is a property of quarks. Two experimentally measurable quantities depend on the number of quarks. One of these is the rate at which the $\pi°$ decays into two photons. The other is the probability that electrons and positrons will create hadrons when they collide. Both these numbers seem to indicate that there are three times as many quarks as would be expected without color. These results would be in agreement with the model if each different color for a quark were counted separately.

In addition, the rule about only zero color (or "color neutral") objects being seen in nature also has a relation to the theory of quark confinement discussed in Chapter XI. If the individual quarks have color, then the zero color rule would say that individual quarks cannot be seen in the laboratory. Therefore, unless the rule can be broken, we should never expect to see quarks except in those combinations that have zero color; that is, in the combinations that give us the particles we already know about.

A SUMMING UP

So there is now clear evidence for five kinds of quarks, each coming in three colors, each with its corresponding antiquark. In addition, as we will see in Chapters XIV and XV, physicists expect that a sixth quark will join this list—it's already in the theories, and a major search is under way to see evidence for it in the laboratory. That means that there are 6 quarks × 3 flavors × 2 (to count antiparticles) = 36 fundamental building blocks of the hadrons. Yet we started down the road to the quark model because there were too many elementary particles. In the words of the well-known social commentator Yogi Berra, "It's déjà vu all over again."

Or is it? There is no limit to the number of elementary particles that could, in principle, be discovered, nor is there any limit in principle to the number of chemical elements. There are, however, two important pieces of evidence that argue that there aren't going to be any more quarks.

One of these comes from cosmology—the branch of science devoted to the description of the evolution of the universe. About three minutes into the life of the universe, protons and neutrons came together to form some light elements, most notably helium. (Heavier elements like carbon and iron were made later, in stars.) Knowing the rate of helium-forming reactions and the density and temperature of matter at three minutes, it is possible to calculate how much primordial helium there ought to be in the universe today. The fact that the theory predicts a number (around 22%) very close to what is actually observed is taken to be one of the strongest pieces of evidence for the entire Big Bang theory.

But here's the catch: For various technical reasons, if there were more families of particles containing new kinds of quarks, the amount of helium produced at three minutes would have been much lower than what we see without telescopes. It is barely possible that there could be eight quarks, but by far the most comfortable situation is one on which there are only six—the five we have already talked about and one other that we'll discuss shortly.

The other piece of evidence that argues against a further proliferation of quarks comes from experiments done at accelerator laboratories. In the next chapter we will talk about a particle called the Z_0. Its importance will be discussed at that time, but for the moment I want to talk about only one of its properties—its decay width. We have seen that the width of a peak in a plot of number of particles vs. energy can be related to the lifetime of the particle being produced. The shorter the lifetime, the wider the peak.

Now most very massive particles can decay into many different final states (they're usually called "decay channels"). Each extra channel shortens its lifetime, just as each extra risk can shorten the life expectancy of individuals. For the Z_0, the width of the peak has been measured closely, and all of the decay modes have been observed. When the books are balanced, it turns out that the decay modes we see account fully for the decay of the Z_0. If there were extra channels (associated, for example, with families of particles containing new kinds of quarks), the lifetime of the particle would be shorter than it is observed to be. There simply isn't room for any more quarks!

So there is at least some evidence that we're not in for another orgy of proliferation, that we are getting near the end of the road for elementary particle physics. To follow the road down this last little bit, however, we have to leave the hadrons for a moment and talk a little about recent developments in studies of the leptons.

XIII

Leptons and the
Weak Interactions

As I was going up the stair
I met a man who wasn't there.
He wasn't there again today.
I wish, I wish he'd stay away.
   —Hughes Mearns, "The Psychoed"

THE DISCOVERY OF THE NEUTRINO

IN CHAPTER II we discussed the beta decay of the neutron as an example of a weak interaction. We saw that in order to preserve the laws of energy and momentum conservation in this decay it was necessary to assume the presence of an unseen particle in the interaction—a massless, uncharged particle that was called the neutrino. We also saw that this hypothetical particle was so difficult to detect that it could quite literally pass through a block of lead several light-years thick without disturbing a single atom. But so successful was the theory of beta decay which Enrico Fermi put together that physicists were willing to accept the neutrino as a genuine particle despite the fact that it had never been seen in a laboratory. Indeed, it would not be an overstatement to say that there were probably a good number of physicists in the early 1950s who had more faith in the existence of the neutrino than they had in the new strange particles that were starting to turn up in cosmic ray experiments.

In 1956 this faith was justified when two physicists, Clyde L. Cowan, Jr., and Frederick Reines of Los Alamos, managed to produce laboratory evidence showing that the neutrino also existed in the real world, and not solely in the minds of theoretical physicists. Considering the miniscule effects that the neutrino has on its surroundings, this was no mean feat. It took five years of searching and refining the experiment before Cowan and Reines finally announced a definitive result.

Before going into the details of the experiment, we must digress briefly in order to define some terms. We have seen that an elementary particle can be characterized by its spin, and that the direction of the spin obeys the so-called right hand rule (see Chapter VIII). We have also seen that for every particle in nature there is an antiparticle. Consequently, we can conclude that there ought to be an antineutrino, which differs from the neutrino in the direction of its spin (since the neutrino has no charge, spin is the only property that can be different in the particle and antiparticle). By convention, the particle whose spin is in the same direction as its velocity is called the antineutrino, while the particle whose spin is directed opposite to its velocity is the neutrino.

If we use this convention, then the particle that is the invisible partner in the beta decay of the neutron is actually the antineutrino. It is customarily denoted by $\overline{v_e}$, where the subscript e refers to the fact that it is produced in concert with an electron. In keeping with this convention, we now write the neutron beta decay as $n \rightarrow p + e^- + \overline{v_e}$.

Since the difficulty of detecting the presence of a neutrino or antineutrino arises because of the very small probability that the particle will interact with the nuclei it passes, the only way to see such interactions is to find a source of the particles copious enough to overcome the small probability of interaction. You can see how this line of reasoning works by noting that the expected number of interactions that will be seen each second will be

events seen = number of neutrinos or antineutrinos ×
probability of an interaction

The fact that the probability of interaction is small can be overcome if the number of neutrinos or antineutrinos is large
enough.

Nuclear reactors, as a by-product of fission, produce large
numbers of antineutrinos—perhaps as many as 10^{18}/second. By
placing a large target and counter apparatus near the reactor at
Savannah River, South Carolina, Cowan and Reines calculated
that they should be able to get one reaction of the type

$$\overline{\nu}_e + p \rightarrow n + e^+$$

every 20 minutes. Not a prodigious number of events, of course,
but enough to do the job.

The apparatus they used is sketched in Illustration 31. It consisted of layers of water (which contains hydrogen atoms whose
nuclei served as the target for the antineutrino projectiles) inter-
leaved with layers of a scintillating liquid. The entire block was
then surrounded by detectors that would see the photons given
off in the scintillators.

When one of the rare antineutrino interactions occurs in the
water, there are two products—the positron and the neutron.
The positron will annihilate with an atomic electron in the water
within a millionth of a second or so. When this happens, two
energetic photons are emitted, each of which enters a scintillating layer and produces a shower of photons that will be seen by
the detectors. The neutron does not interact so quickly, since it
has no electrical charge. Instead of losing its energy through an
electrical force, it undergoes a series of collisions that slow it
down to the point where it can be captured by a nucleus. The
water in the target is mixed with a small amount of cadmium
chloride to augment the absorption process, cadmium being an
element that absorbs slow neutrons very efficiently. When the
neutron is absorbed in the cadmium, the new nucleus gives off
one or more photons as the protons and neutrons rearrange
themselves to accommodate their new partners. These photons
are also converted into a signal in the scintillator.

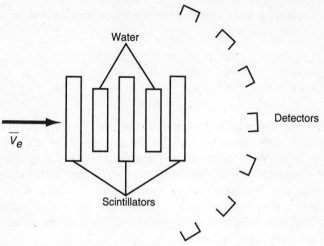

31. Apparatus used to find the antineutrino.

Thus, the sequence of events that would signal the presence of an antineutrino interaction would be (1) two photons from the electron-positron annihilation, followed in a few millionths of a second by (2) one or more photon characteristic of the cadmium nucleus. When Cowan and Reines finally satisfied themselves that they were seeing these events at the expected rate of one every 20 minutes, the physics community had the proof it needed for the existence of the antineutrino—a proof the physics community had waited for for 20 years.

THE MU-NEUTRINO AND A NEW CONSERVATION LAW

Once the antineutrino associated with beta decay had been seen in the laboratory, it was only natural that attention should turn to the problem of providing neutrino beams in accelerators. As we discussed in Chapter VI, this can be done by taking a beam of pi-mesons of known charge and energy from an accelerator and allowing pions to decay into a muon and a neutrino. Our convention on neutrinos and antineutrinos tells us that the two decay reactions are

$$\pi^+ \rightarrow \mu^+ + \nu_\mu$$

and

$$\pi^- \rightarrow \mu^- + \bar{\nu}_\mu$$

where the subscript μ is used to remind us that these particular neutrinos are associated with the production of the mu-meson. If the mixed beam of muons and neutrinos is then taken through a large shield (e.g., a pile of steel plates or a long earth mound), we can, by choosing a thick enough shield, ensure that the muons and any other particles that might be in the beam by accident will be absorbed in the shield and only the neutrinos will pass through unaffected.

In 1962 Leon Lederman, Melvin Schwartz, and a group at Columbia University reported on experiments designed in that way. Their purpose was to look for reactions of the type

$$\bar{\nu}_\mu + p \rightarrow n + \mu^+$$

and

$$\nu_\mu + n \rightarrow p + \mu^-$$

which would be initiated by the neutrinos and antineutrinos. The signature of such an event would be the sudden appearance in the apparatus of a charged particle (the muon). Both the incident neutrino and the final neutron, being uncharged, would be invisible.

The most striking feature of the experiments was that *only* the reactions given above were observed. No reactions of the type

$$\nu_\mu + n \rightarrow p + e^-$$

and

$$\nu_\mu + p \rightarrow n + e^+$$

were seen at all. In other words, it appears that neutrinos resulting from the decay of muons cannot create electrons or positrons, even though the Cowan-Reines experiment showed conclusively that neutrinos emitted from beta decay could do so. There are, in other words, two different kinds of neutrinos, one associated with electrons and the other with muons.

This, of course, is the reason we referred to neutrinos as v_μ and v_e. Not only does each neutrino appear in association with a particular lepton, but it can initiate reactions only if that same lepton is involved. It seems, therefore, that the leptons come in pairs, the electron with its neutrino and the muon with its neutrino.

The two neutrinos also cast some interesting light on a conservation law. By our choice of definition of the neutrino and antineutrino, we have arranged things so that there is a law for weak interactions similar to the conservation of baryon number. This is the law of conservation of leptons: *The net number of leptons cannot change in any interaction.*

For example, in neutron beta decay we have no leptons in the initial state, and an electron with its antineutrino in the final state. If we assign a *lepton number* of +1 to the electron, then we have to assign −1 to the antineutrino (just as we would assign a baryon number −1 to an antiproton). The net lepton number after the decay is therefore zero, the same as it was initially. You can, if you check the other weak interactions we have discussed so far, see that this law holds.

The existence of the two neutrinos suggest further that a more stringent form of this law may apply. If we assign an electron number to the electron and its neutrino and a muon number to the muon and its neutrino, then it appears that electron number and muon number are each conserved individually. In other words, if we consider the reaction

$$\mu^- \rightarrow e^- + \overline{v_e} + v_\mu^-$$

we start with a muon number of 1 and electron number 0. The neutrinos in the final state are arranged to give the same value for each of these. We then note that if electron and muon numbers are each conserved, the law of conservation of lepton number is automatically satisfied. These concepts are summarized in the table on page 194, in which the four leptons are given with their various quantum numbers.

PARTICLE	LEPTON NUMBER	ELECTRON NUMBER	MUON NUMBER
e^-	1	1	0
e^+	−1	−1	0
μ^-	1	0	1
μ^+	−1	0	−1
ν_e	1	1	0
$\overline{\nu_e}$	−1	−1	0
ν_μ	1	0	1
$\overline{\nu_\mu}$	−1	0	−1

Two important inferences follow from the two neutrino theory. First, if it turns out that the electron and muon are not the only massive leptons in nature (and we shall see later that they are not), then we would expect a new kind of neutrino to accompany each new massive lepton that is found. Second, if, as some theorists believe, there is some fundamental connection between leptons and quarks, then the fact that leptons always come in pairs (particle plus neutrino) would imply that quarks should always come in pairs as well. This type of reasoning is what led to the first predictions that there ought to be a charmed quark as a complement to the strange quark. Pushing the analogy a bit further led to the prediction of the b and t quarks after another lepton, the τ, was found. This discovery will be discussed later in this chapter.

PARITY IN WEAK INTERACTIONS

In Chapter VIII we introduced the idea that we could describe particles by a property called parity. In essence, the parity of a particle tells us how its wave function will look if the space coordinates are inverted. A simple way to visualize this is to say that the parity operation corresponds to looking at the particle in a mirror. If the reflection is the same as the particle, we say that the particle has positive parity.

Imagine this mirror-image operation being applied to processes as well as to particles. For example, a collision in which a proton approaches from the right and an electron from the left, when looked at in a mirror, would appear to be a collision in which a proton approached from the left and an electron approached from the right. Intuition tells us that what happens in the collision should not depend on whether we view the process directly or see it in a mirror. In physics, the idea that this particular operation should not affect anything in nature is called the principle of parity invariance. The principle holds for strong and electromagnetic interactions, and, until the early 1950s, was believed to hold for weak interactions as well.

In 1956 two young physicists at Columbia University, Tsung-Dao Lee and Chen Ning Yang, were studying some problems in weak interactions. As part of their study, they took a close look at the reasons why parity invariance was believed to hold in the weak interactions. After some careful thought, they came to the conclusion that there was no actual evidence that indicated that it must, but that there was simply a strong prejudice on the part of theoretical physicists that the weak interactions had to be like the rest of physics. On the strength of this realization, they modified the theory of beta decay to see what would happen if the idea of parity invariance did not hold. They found that there were a few situations in nature where this lack of mirror symmetry could actually be seen and measured.

In 1956 Madame Chien-Shiung Wu, also at Columbia, performed an experiment that verified beyond question that the weak interaction was not invariant under parity. Although technically difficult, the idea of the experiment was quite simple. A piece of material containing cobalt (actually, the isotope ^{60}Co) was cooled to within a few tenths of a degree of absolute zero $(-273\,^{\circ}\text{C})$. At this very low temperature a magnetic field can be applied that will lock all of the cobalt nuclei into an orientation where their spins are all pointing in the same direction. The low temperature is required so that the normal oscillations that atoms perform, and that we perceive as heat, are reduced to the point where they cannot upset the nuclear alignment. Although this procedure sounds simple, it took over 6 months of work to turn the idea of an aligned sample into reality. Even then, the sample could be kept aligned for only 15 minutes, so it could be said that this experiment involved 6 months of preparation for a 15-minute run.

^{60}Co is a nucleus that undergoes beta decay spontaneously, and has a half-life of about 53 years. As is well known, the element is widely used as a source for radiation in cancer therapy in modern hospitals. Madame Wu, however, used the decay property for a different purpose. Since she knew the direction in which the nuclear spin was pointing, she could observe the number of electrons that came off in the "up" direction and the number that came off in the "down" direction. We can understand the significance of this measurement if we look at Illustration 32. At the left is a cobalt nucleus in which the electron is emitted along the direction of the spin—the direction we are calling up; on the right, we show the mirror image of the same process. The spin of the nucleus is reversed, but the electron is still moving toward the top of the page, albeit in a different direction. Thus, if parity invariance is valid in weak interactions, we would expect to see as many electrons emitted along the direction of the nuclear spin (up) as along the opposite direction (down). In 1956 most physicists would have predicted this result.

When the experiment was actually performed the results clearly showed that the electrons were emitted preferentially in the up direction. Thus, as far as the weak interactions are con-

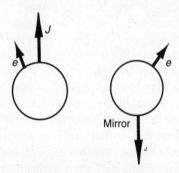

32. Representation of the experiment that disproved the validity of parity invariance in weak interactions.

cerned, nature is *not* ambidextrous. There is a differentiation between right and left in a way that is totally unexpected. For predicting that this might be so, Lee and Yang shared the Nobel Prize in 1957—the youngest men ever to do so. At the time they actually did the work, Lee was 29 and Yang was 33.

THE HIERARCHY OF CONSERVATION LAWS

The discovery that parity is not conserved in weak processes brings out an important point about the interactions we have studied so far. These have, of course, different strengths (we will come back to this point shortly), but they also seem to be arranged in a hierarchy; that is, the stronger the force, the more conserved properties there are.

For example, the strong interaction does not depend on the electrical charge of the particles being examined. In more technical language, we say that the strong force conserves the isotopic spin in any process. But if we go down one step in interaction strength to the electromagnetic force, this statement is no longer true. The electromagnetic force on a proton is much different from that on a neutron, since the former is electrically charged and the latter is not. Thus, we say that the electromagnetic force does not conserve isotopic spin, or, equivalently, that the electromagnetic force breaks isotopic symmetry.

In the same way, both the strong and the electromagnetic interactions conserve parity, but the weak interaction does not. These differences between the interactions are highly significant, since they tell us something about the way the fundamental interactions operate. In the present state of our knowledge, however, it is not possible to do more than state what the differences between the interactions are, for we have no way of showing why they should be so.

To summarize the properties that are conserved in the three interactions we have discussed so far, we list in the table below a number of conserved quantities and indicate on the right which interaction conserves them.

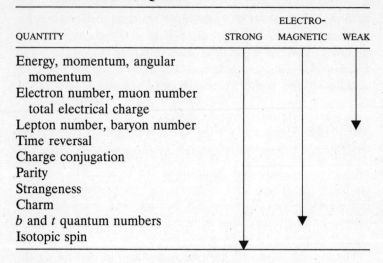

QUANTITY	STRONG	ELECTRO-MAGNETIC	WEAK
Energy, momentum, angular momentum			
Electron number, muon number total electrical charge			
Lepton number, baryon number			
Time reversal			
Charge conjugation			
Parity			
Strangeness			
Charm			
b and t quantum numbers			
Isotopic spin			

WHY IS THE WEAK INTERACTION WEAK?
THE W BOSON

In Chapter III we saw that the strong interaction could be thought of as being generated by the exchange of virtual particles. This way of thinking about the interaction is useful, and it led Yukawa to the prediction of the existence of the mesons before they were actually seen in the laboratory. Similarly, we can think of the electromagnetic force as being generated by the exchange of photons. For each of these two forces, there is a particle that can be identified with the force, and which we could say "mediates the interaction."

Starting with Yukawa, physicists have often pondered the question of whether such a statement ought not to be true for the weak interaction as well. None of the particles we have discussed so far have the requisite properties to play the role of mediator for the weak interactions, but perhaps there is some other that can. This particle has been given a name. It is called the *vector boson,* or the *W* particle. From the properties of the weak interactions themselves we can deduce a good deal about the properties such a particle ought to have.

To picture the role that such a particle would play in a process like beta decay, look at Illustration 33. A neutron emits a virtual negatively charged W particle, changing into a proton in the process. The W then turns into an electron and an antineutrino. This picture, in which the weak interaction is mediated by a virtual particle, puts beta decay on the same conceptual level as strong and electromagnetic processes.

From the diagram, we see that the baryon and lepton numbers of this particle must be zero. Somewhat more technical considerations show that the spin of the particle must be 1. Particles with even spin are sometimes called bosons (after S. N. Bose, who, together with Albert Einstein, first investigated some of their theoretical properties). Particles with spin 1 are called vector particles, so the name vector boson, or "intermediate vector boson," is sometimes used to describe the W.

There is another important statement we can make about the vector boson. When we talked about forces being mediated by the exchange of virtual particles, we saw that the higher the mass of the particle, the shorter the distance it can travel without violating the uncertainty principle and conservation of energy. This means that short-range forces, those that cannot be exerted over long distances, are mediated by heavy particles—the heavier the particle, the shorter the range of the associated force.

The weak force has an extremely short range, since it acts within a single nucleus or particle. You can contrast this with the electromagnetic force, which has infinite range. In the case of the electromagnetic force, the large range has to do with the fact that the exchanged particle, the photon, has zero mass. The W particle, on the other hand, must be many times heavier than the proton to produce the short range of the weak interaction.

In the "classical" (i.e., pre-1967) theory of weak interactions, the W particle had to have either a positive or a negative charge. This requirement was imposed by the fact that in experiments it always appeared that particles undergoing weak decay also underwent a change of electrical charge. In beta decay, for example, the neutron changes into a proton, and the W particle that is exchanged has a negative charge.

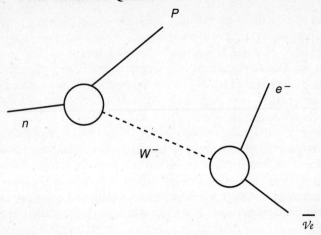

33. Theoretical role of the vector boson in beta decay.

One extremely important feature of this way of looking at the weak interactions can be seen by comparing the two reactions in Illustration 34. On the left we see two electrons interacting by exchanging a photon. This, of course, is that standard way that modern physicists look at the electrical repulsion between these two negatively charged particles. On the right we show the beta decay of the neutron in a slightly rearranged form. Seen this way, the two interactions, as different as they might seem to be, appear to have an underlying similarity. Both forces are generated through the exchange of virtual particles, and it becomes reasonable to say that the differences we observe between the electromagnetic and weak forces in nature are actually due to differences between those particles. This insight, in fact, is what lies at the heart of the unified field theories that we will discuss in the next chapter and which are, arguably, the most important insight in physics since the advent of quantum mechanics.

In 1967, Steven Weinberg, then at the Massachusetts Institute of Technology, developed the first of these theories. He predicted that there would be not only the positive and negative W bosons that physicists had discussed previously, but a third, uncharged particle called the Z. Furthermore, the theories made

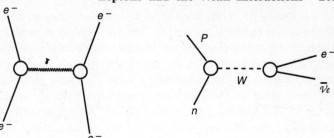

34. Representation of two interactions.

definite predictions about what the masses of these particles had to be. The predictions, which had the masses in the 80–90 GeV range, put these particles outside the reach of the accelerators that were available in the 1970s.

But as we have seen, physicists have gotten to be very good at pushing the limits of their accelerators. At the European Center for Nuclear Research (CERN) in Geneva, Switzerland, Carlo Rubbia and Simon van der Meer convinced the directors to invest in one of the first antiproton storage rings so that they could use the full energy available in head-on collisions to make these extremely massive particles. In 1983, the two physicists, together with a team of 130 others, announced that they had seen the intermediate vector bosons for the first time. Today, the accepted mass of the W (both plus and minus) is 81 GeV, while that of the Z is 93 GeV.

With this discovery, for which Rubbia and van der Meer shared the Nobel Prize in 1985, the idea that the weak interaction had to be considered as something distinct from others in nature could be dropped. The way was open to look for deep unity among nature's different forces.

THE HEAVY LEPTON: ROUNDING OUT THE PICTURE

Since the discovery of the muon in the 1930s and the two neutrino experiment in 1962, it had been assumed that there were four leptons—the electron, the muon, and the two neutrinos. This fact was the basis, as we saw, for the theoretical prediction of charm. But the question of whether there were any other leptons more massive than the muon was still left unanswered. In 1975 a group at Stanford made a search for such particles. It was performed on the same electron-positron storage ring on which the ψ/J had been discovered earlier. What was seen were some 64 events of the type $e^+ + e^- \rightarrow \mu + e +$ undetected particles. According to conventional physics, there is no way such an event could occur. If we imagine that there is a third lepton (we will call it the τ (tau) lepton), then it ought to be able to decay into either the muon or the electron via the interactions

$$\tau^- \rightarrow e^- + \overline{\nu_e} + \nu_\tau$$

and

$$\tau^- \rightarrow \mu^- + \overline{\nu_\mu} + \nu_\tau$$

The ν_τ is the neutrino that (presumably) is associated with the new particle. We could then imagine a reaction in which the electron and positron annihilate to form a pair of the new leptons, one of which decays into a muon and the other into an electron.

The report of such events is now accepted as evidence for a third lepton. The new particle is called the τ^- and has a mass of 1.8 GeV. In all other respects, it is similar to the muon and electron.

At the present time, therefore, we know of three massive leptons, and no one doubts that a third neutrino associated with the τ^- will be eventually seen. Like the quarks, the number of leptons seems to be proliferating. Just as the idea that there should be a connection between quarks and leptons led to the prediction of charm, the fact that there are now six leptons has led theorists to suggest that there should be another pair of quarks to match them. The upshot is that the growing complexity of the hadrons has its counterpart in a growing complexity of the weak interactions.

XIV

The Standard Model: Bringing It All Together

E pluribus unum.
—Motto of the United States

WE HAVE COME A LONG WAY on our quest for the ultimate nature of matter. We have broken the atom down into electrons and nuclei, broken the nucleus down into hadrons, and broken the hadrons down into quarks. We have seen that the forces that operate between elementary particles can be thought of as arising from the exchange of yet other particles. The time has come to start putting all of this information together—to weave a unified picture of the universe and its workings. The best approximation to a coherent theory we have today goes by the name of the *Standard Model*. You already have all the pieces you need to understand it, so let's start seeing how they fit together.

QUARKS AND LEPTONS

Throughout this book we have repeatedly referred to quarks as the "building blocks" or "bricks" of the universe. This is a particularly apt metaphor, for we can now see that everything in the universe is made from just two types of particles. Quarks can be put together to make elementary particles which, in turn, can be put together to make the nuclei of the atoms. The electrons, which are one type of lepton, can be put into orbit around the nuclei to make atoms, and atoms can combine chemically to make all the materials around us. The entire known universe, then, is made from quarks and leptons. And as we saw in Chapter XII, there is strong evidence that this is it—that there are no more layers to the onion, no more kinds of quarks waiting to be found. So we can begin our discussion by recognizing that this part of the quest—the search for ultimate constituents—is over.

But there is a rather interesting regularity among the quarks and leptons: they both seem to come in pairs. The ordinary particles of matter—the protons and neutrons that are the main constituents of ordinary atomic nuclei—are made from the up and down quarks, for example. Like the proton and neutron themselves, the up and down quarks form a linked pair. In the same way, there are three leptons that have mass—the electron, muon, and tau—and each of these has its own associated neutrino. The lepton and the neutrino form a linked pair, too.

And just as the building blocks come in pairs, the pairs themselves can be grouped together. There are three pairs of quarks and three pairs of leptons, and physicists have come to group them into "generations" or "families" as follows:

FIRST GENERATION
   up quark   electron
   down quark   electron neutrino
SECOND GENERATION
   strange quark   muon
   charmed quark   muon neutrino

THIRD GENERATION

bottom quark	tau
top quark	tau neutrino

There are certain regularities in this arrangement of quarks and leptons. For example, each generation involves heavier particles. The masses of the up and down quarks, for example, are thought to be around 5–7 MeV, while the strange quark weighs in at about 150, the charmed at 1500, and the bottom at 5000. The mass of the top, as yet unknown, may well be much greater than any of these. Similarly, the electron has a mass of about 0.5 MeV, the muon about 105 MeV, and the tau about 1800. Again, as we go up through the generations of leptons, the massive particles get heavier.

As was the case with Mendeleev's periodic table and the eightfold way, physicists recognize the regularities of the three generations, but do not have a clear idea about *why* nature should be arranged in this way. And as happened with both the periodic table and the eightfold way, there is a gap in the arrangement that points to a missing particle—in this case, the top quark. We will return to the search for the top quark in the next chapter.

With the delineation of the generations of quarks and leptons, we have finally laid out the basic building blocks of the universe. Everything in the universe is made from some combination of the particles in the above list. But just as a building is not just a collection of bricks, so too is a universe more than just a collection of particles. To make a building, you need mortar to hold everything together. To make a universe you need some way for the particles to interact with each other. We have called the "mortar" of the universe the fundamental forces, and it is to a discussion of this other half of the picture that we now turn.

THE FUNDAMENTAL FORCES

Before we turn to our final catalog of the forces that operate in nature, we have to clear up one point. In Chapter III we described the strong force in terms of the exchange of particles like the pi meson between protons and neutrons. But now we know that this picture was, in some sense, too simple. The protons, the neutrons, and the pi mesons are now understood to be composite structures made from quarks. How do we describe the strong interaction at this more fundamental level?

The theory that physicists have developed to describe interactions between quarks is called quantum chromodynamics, or QCD. The "chromo" refers to the fact that quarks have color, so QCD is a theory that describes the dynamics (motion) of quantum particles that have color (i.e. quarks). Without going into a lot of details, QCD describes the strong interactions between quarks on the same basis as the other forces we've discussed—as arising from the exchange of particles. In QCD, the exchanged particles are called *gluons*—they "glue" the particles together. There are eight massless gluons, and they carry color. (Technically, each gluon carries a color and an anticolor.)

In QCD, then, quarks exchange gluons to generate the strong force, just as electrons exchange photons to produce the more familiar electromagnetic force. The exchange of gluons binds the elementary particles together, and the exchange of these particles is what generates the "old" version of the strong force.

There is, however, one important difference between the exchange of photons and the exchange of gluons. When two electrons get farther away from each other, they exchange fewer photons and the electromagnetic force gets weaker. When two quarks get farther apart, however, the rate of exchange of gluons goes up and the force gets stronger. This is what leads to the quark confinement we talked about in Chapter XI. Similarly when quarks get close together, the rate of exchange drops and the force gets weaker. This is why elementary particles often behave as if they were made up of independent, noninteracting quarks.

QCD is a very powerful theory, with a great deal of experimental backing. Let me describe just one of the early (and therefore crucial) pieces of evidence for it. In 1978, scientists were monitoring electron-positron collisions in a collider called PETRA in Hamburg, Germany. Most of the reactions they saw were like the one shown in Illustration 35(a). The electron and the positron annihilated, producing a single photon of extremely high energy (up to 30 GeV at this particular machine). The energy of this photon, in turn, was converted into a quark-antiquark pair, each member of which produced a jet of ordinary particles that could be detected in the lab.

Occasionally, however, an event like the one in Illustration 35(b) was seen. The quark (or the antiquark) emits a gluon before it creates its jet. In this case, experimenters would expect to see not two, but three jets of particles in the laboratory—one from the quark, one from the antiquark, and one from the gluon. When these triple jet events were detected, it was a great triumph for QCD.

Having described the strong force at the fundamental level, we can now turn to an examination and comparison of all the forces in nature. As a matter of fact, we can now see that, as was the case with the building blocks, there are only a limited number of such forces in existence. There are, of course, gravity and electromagnetism. These are forces of which we have direct experience in our lives—I think of them as nineteenth-century forces. Then there is the strong force that holds nuclear matter

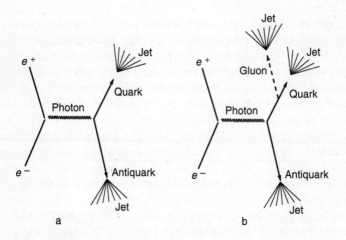

35. The production of a quark-antiquark pair (a) compared to the production of the same pair with the addition of a jet of particles (b).

together and the weak force that (sometimes) tears it apart. That's it. Everything that happens in the universe happens because one or more of these forces is operating.

The electromagnetic force is associated with the exchange of photons, the weak force with the exchange of the W and Z, and the strong force with the exchange of gluons. There is not, as we shall see in the next chapter, a generally accepted theory of gravity at the quantum level, but such a theory would describe the gravitational force as arising from the exchange of massless particles called gravitons. The differences between forces, both in strength and in range, is a result of the different particles that are exchanged. The table summarizes the four forces.

The Fundamental Forces

FORCE	STRENGTH	RANGE	PARTICLE EXCHANGED	MASS
strong	1	10^{-13} cm	gluons	0
electromagnetic	1/137	infinite	photons	0
weak	10^{-13}	10^{-16} cm	W, Z	81, 93GeV
gravity	10^{-38}	infinite	graviton	0

The forces of nature, then, differ in the sort of particles that are exchanged to generate them. Gravity and electromagnetism are familiar to us because they have a long range (in principle, an infinite range), and this has to do with the zero mass of the photon and graviton. The strong and weak forces are less familiar because they operate only on the scale of the nucleus, and are not readily available to our senses.

Although the gluon has zero mass, it also carries color and therefore, like the quarks, cannot move very far from its source. It is this fact that limits the range of the strong force. The range of the weak force, of course, is limited by the high masses of the W and Z.

The particles listed in the table—the gluon, photon, W and Z, and graviton are called *gauge* particles. The historical origin of the name will be explained in the next section, but for the moment we simply note that gauge particles are responsible for all the forces in the universe. If quarks and leptons are the bricks from which everything is built, then the gauge particles are the mortar that holds the bricks together.

On the one hand, the idea that you can reduce every process in the universe to the action of one of four forces is a major triumph of the human intellect. On the other hand, it raises a very deep question—"Why four?"

To see why this is puzzling, ask yourself this question: What is the minimum number of forces required to build a universe. With no forces, nothing can happen. Thus, a universe requires at least one force to exist. But does it need more? If you really believe that nature is as elegant and beautiful as possible, you might want to entertain the notion that the four forces, as different as they appear to be, are simply different manifestations of a single, underlying force.

When two forces are seen in this way, they are said to be *unified,* and the theory that describes them is called a *unified field theory.* There are two very important examples of the development of unified field theory in physics. The first occurred in the seventeenth century, when Isaac Newton showed that the same force that makes an apple fall to the ground also keeps the moon and the planets in their orbits. This insight unified earthly and heavenly forces—forces that had been considered to be as separate as the strong and weak forces appear today.

The second great episode of unification took place in the nineteenth century when a series of discoveries showed that the seemingly different forces of electricity and magnetism are, in fact, the same force. We have implicitly used this result in this book by talking about the force of *electromagnetism,* rather than of electricity and magnetism separately. We now understand that the force that holds notes to your refrigerator and the force that makes papers coming out of your xerox machine stick together are both manifestations of the exchange of a single gauge particle—the photon.

Having made this point, I should inject a note of caution: Not every attempt at unification has been successful. Albert Einstein, arguably the greatest scientific mind of the twentieth century, spent the last half of his career in an unsuccessful attempt to find a theory that unified gravity and electromagnetism. Thus, when physicists in the late 1960s began to think about developing unified field theories, they had examples of both successes and failures in front of them.

The tool that eventually led to the modern unified theories was one we have not yet discussed—the tool of symmetry. Symmetry is a familiar concept from our everyday life, but in the hands of theoretical physicists it takes on a special meaning. The general use of symmetry in physics is straightforward. It goes like this: First we assume that a particular kind of symmetry holds in nature (I'll give some examples below). Then we construct a theory that incorporates that symmetry, use the theory to make predictions, and ask whether those predictions match the world we actually inhabit. If they do, then we say that the universe must be built according to the symmetry and incorporate it into future theories.

Let me give you a simple example of how this procedure might work. As I write these words, I am sitting at a word processor in Red Lodge, Montana, some 5000 feet above sea level. I am not worried about falling down to sea level because of my elevation because I believe that the universe operates according to a certain kind of symmetry. The gravitational force that acts on me should not depend on what I define to be the zero of the altitude scale, whether it be sea level or something else. The only thing that matters is the *difference* in elevation between two points. Thus, if I want to go down one footstep, there will be an unbalanced force acting on me during the transition, but that unbalanced force is virtually the same whether I am at sea level or on top of Mount Everest. (They are not exactly the same force, of course, since the pull of the earth's gravity is slightly less on a mountaintop because the mountaintop is slightly farther from the earth's center.)

This argument tells me that any theory that I make for the earth's gravity cannot depend on what I choose for the altitude I call "zero." The theory must, in other words, be symmetric with respect to this choice. Another way of saying this is that in nature only forces produce change, and that in this case the force can depend only on differences in elevation, but not in the elevation itself.

Electromagnetism has a similar kind of symmetry. A bird can stand with both feet on a high-voltage line without being electrocuted because the electrical force depends only on the difference in voltage, not of the level of the voltage itself. Current will flow only if the bird's feet are at different voltages—if one is on the wire and one on the ground, for example.

In the world of elementary particles there are symmetries as well. It should, for example, make no difference if we change our definition of electrical charge, so that every negative charge becomes a positive one and vice versa. All the forces operating should be the same, since those forces depend only on whether the electrical charges involved in any interaction are the same or opposite. We would (rightly) reject any theory that did not have this sort of symmetry built into it.

During the late 1950s, theorists began experimenting with a strange kind of symmetry. Suppose, they asked, we could go from point to point in the universe, and at each point make a decision about what particle we were going to call a proton and what particle we were going to call a neutron (for example), and suppose further that the decision and point A could be completely different from the theory at point B. Our gut reaction to this suggestion would be that such a procedure would have to affect the universe—that changing all the protons in a nucleus into neutrons could not leave that nucleus unchanged. But the theorists persisted until they found that it was possible to construct a theory which had this sort of symmetry.

As it turned out, in the early 1900s, some studies of theories that were mathematically similar had been made. These early investigations looked at situations in which distance measurements ("gauges") could be stretched (think of a grid marked out on as sheet of rubber). Thus, theories of this type were called *gauge* theories, and the name persists even though today's theories have nothing to do with rulers.

The gauge theories, as you might guess from the flow of this discussion, were the breakthrough that led to unified field theories. But because gauge symmetries are not manifest in nature—it certainly does matter whether a particle is a proton or a neutron, for example—I should say a word about hidden symmetries. It is possible—even common—for a theory to have a symmetry that isn't manifestly present in the state of the universe that results from the action of the theory. Consider a roulette wheel as an example. The forces that act on the ball—gravity and the electromagnetic forces between atoms—are completely symmetrical as far as the position of the ball is concerned. They have the same description at 20 degrees as at 90. Yet these perfectly symmetric forces acting on the billiard ball produce a situation that is clearly asymmetrical. The ball winds up in one slot on the wheel, off to the right or left, despite the fact that there is no right or left hand preference in the theory. We say that the final situation "breaks the symmetry" of the theory. In the same way, elementary particles may exhibit broken symmetries in the real world.

I have to admit that I have always been fascinated by the notion of gauge symmetry. Here's why: One way of talking about the sea level "symmetry" I discussed earlier is to say that the only possible universe is one in which the zero of altitude doesn't matter. What gauge symmetry seems to be telling us on a more serious level is that the only possible universe is one in which our state of mind—the names we assign to particles—doesn't matter. In a sense, gauge theories are a continuation of a revolution that began when Copernicus showed that human beings are not at the center of the universe—that the universe doesn't care where we live. The gauge theories tell us that the universe doesn't care what we think, either.

THE ELECTROWEAK UNIFICATION
AND THE HIGGS PARTICLE

It is a matter of historical curiosity that Albert Einstein failed to produce a unified field theory because, unwittingly, he tackled the most difficult part of the problem. As we shall see in the next chapter, the challenge of unifying gravity with the other forces has still not been met by modern physicists. During the late 1960s and early 1970s, however, three physicists working independently—Steven Weinberg, Sheldon Glashow, and Abdus Salam—put together a theory that dealt with another part of the problem. This theory, based on the gauge symmetry described above, showed that the electromagnetic and weak forces are unified, that they are simply different aspects of a single underlying "electroweak" force. Furthermore, the theory predicted that at energies that would soon be available at accelerators this unification would be seen.

There are two keys to understanding the electroweak unification. First, as we have seen repeatedly, different forces are generated by the exchange of different particles. Second, although in the theory all gauge particles are massless, there is a symmetry breaking that results in the W and Z having large masses while the photon does not.

Let's start with the second point. At first glance, you might think that there could be nothing in common between the massless, ubiquitous photon and the heavy, elusive W and Z. If the universe is built to have gauge symmetry, however, then the theories tell us that all these particles ought to be massless. If the W and Z are actually observed to have a mass (as the short range of the weak force indicates), then that mass must result from a broken symmetry. The W and Z masses, in other words, must come from a process analogous to the ball falling into a slot on a roulette wheel.

But how can mass arise from a symmetry breaking? Let me give you an analogy that may make this concept palatable. Suppose you had a sprinter who was going to do the 100-yard dash, but suppose that the track had been designed in an unusual way. The first and final 40 yards were made from standard material, but the middle 20 yards was a muddy, boggy swamp. Suppose further that you couldn't see the track, but just the runner. When the gun went off, you would see the sprinter start off at a high speed, slow down, and finally, speed up again. How would you describe this sequence of events?

One way (the most obvious way) would be to say that he had encountered some sort of retarding force in the middle of the race. You could, though, imagine describing the slowdown by saying that for a brief period the sprinter was more massive and therefore moved more slowly. This may seem a strange way to talk, but in fact it's a language quite familiar to theoretical physicists. When electrons move through solids (in a computer chip, for example) they are subject to all sorts of forces from atoms and other electrons. Physicists often lump the effect of all these forces together by assigning the electron an "effective mass" that is different from its rest mass.

The electroweak theories describe the acquisition of mass by the W and Z in much the same way. Instead of a swamp, they tell us that the universe is filled with a new kind of particle, called the Higgs particle after Peter Higgs, the Scottish physicist who first suggested its existence. The Higgs particle has one unusual property. At normal temperatures it interacts with gauge particles, slowing some down (i.e., giving them mass). In effect, it creates a "swamp" that slows the W and Z down. This is the way that the W and Z, which are "really" massless like the photon, break the gauge symmetry and acquire a mass. At higher temperatures, however, the Higgs particles do not have this effect and the true symmetry can be seen. The W and Z, quite literally, lose their mass and become massless like the photon.

Another analogy that may help you think about this process is with an ordinary magnet. At high temperatures, the atoms of iron move at random and there is no magnetic field—no preferred direction. This is a situation where the true symmetry of the atoms can be seen. When the temperature is lowered, the atoms spontaneously arrange themselves so that there is a magnetic field—a preferred direction in space. The interaction of the atoms, which is responsible for forming the magnet in the first place, breaks the original symmetry. To see the true symmetry of the system, you'd have to add energy to the magnet again, heat it up, and destroy the magnetism.

In the same way, to see the true symmetry of the electroweak force, you have to add enough energy to the particles to undo the symmetry breaking associated with the Higgs interaction.

And it is in the process of adding energy, of course, that accelerators come in. In the CERN experiment described in Chapter

XIII, the collision of a proton and an antiproton in a collider supplied enough energy so that the true symmetry of the electroweak system could be seen. The predictions of the theory as to the masses and rates of production of the W and Z were verified, and the theory became the cornerstone for further work in high-energy physics. For their work, Weinberg, Glashow, and Salam shared the Nobel Prize in 1979. As I tell my students, the electroweak unification has to be right, because the Nobel Prizes have already been given out!

The upshot of all of this is that when energies are sufficiently high—when they exceed 100 GeV or so—there are no longer four fundamental forces in nature, but three. These are the strong, gravitational, and electroweak forces.

GRAND UNIFIED THEORIES
AND THE COSMIC CONNECTION

When physicists found that they had a method that unified two of the fundamental forces, they were not slow to see if the method would get them farther along the road to the ultimate unification. The next step forward made possible by gauge symmetry was the unification of the electroweak force with the strong force. The theories that describe this unification are called *Grand Unified Theories*, or *GUTs*.

There are many different versions of the GUT, corresponding to different assumptions about the way the details of the theory play out, but they all agree on one crucial point. At energies above 10^{15} GeV, there are only two fundamental forces in nature: the strong-electroweak, and gravity. This energy is truly astronomical—if we imagined scaling Fermilab up to reach it, the main ring would enclose several nearby stars! There is, then, no way of seeing the strong-electroweak unification in the same way we did the unification of electromagnetism and the weak force.

But that doesn't mean we can't test the theory at all. Although we may not be able to produce GUT-style energies, there was a time in the early history of the universe when everything was at this temperature. In fact, when the universe was 10^{-35} seconds old, the temperature was so high that collisions between ordinary particles would have had this sort of energy. The universe, in the words of Nobel Laureate Leon Lederman, is "an accelerator with an infinite research budget."

This connection between the study of elementary particles and the early stages of the universe is one of the great pieces of serendipity in twentieth-century science. The way we use the universe to test our theories is this: We take a theory and use it to predict how the Big Bang evolved in its earliest stages. We then follow that evolution through to the present and ask whether the universe we live in is actually like the one predicted by the theory. If it is, we have evidence (albeit indirect) to support the theory. If it isn't, we can discard or modify the theory and try again.

There are many triumphs of the GUT in cosmology—too many to go into in detail here. Let me, therefore, tell you about just one, so that you get a sense of how the system works. In Chapter IV we pointed out that one of the great mysteries of the universe is the absence of antimatter. Although matter and antimatter enter the fundamental theories in completely symmetrical form, and although you might therefore expect the universe to be made of equal parts of the two, in fact, there is absolutely no evidence for large concentrations of antimatter anywhere. Until the advent of the GUT, this absence was simply a mystery.

In the early 1980s, however, physicists applied the GUT to the early evolution of the universe. Adjusting the parameters of the theory to match the results of experiments at (the relatively low) energies, they then extrapolated up to GUT energies and calculated how the universe would evolve. What they found can be explained simply, even if the details are far from simple. When the universe was 10^{-35} seconds old, there was enough energy available to create all sorts of particles in collisions. As a result of these creation processes, there were about 100 billion and 1 particles of matter made for every 100 billion antiparticles. In the subsequent maelstrom, the 100 billion particles of matter found the 100 billion particles of antimatter and annihilated, converting their mass into high-energy photons. The lone particle of ordinary matter couldn't find a partner with which to annihilate, and so it survived to the present. It (and its brothers and sisters) makes up everything that we call the material universe.

This picture of the early universe explains why antimatter is conspicuously absent in terms of a fundamental symmetry, a fundamental theory (the GUT), and some phenomena observed at modern-day accelerators. Similar applications of the GUT have solved most of the other nagging problems associated with the Big Bang theory—so much so, in fact, that it is getting to be difficult to tell particle physicists from astronomers these days.

But although the GUT take us a long way toward the ultimate unification of the four forces, they do not include gravity. The development of a theory in which *all* the forces of matter are unified—a so-called "Theory of Everything" (TOE)—remains an unsolved problem at this time. It is, in fact, one of the last hurdles to be cleared before we arrive at the end of our quest to understand the basic structure of the universe. We'll turn to the TOE and other remaining fundamental problems in the next chapter.

XV

Unsolved Problems

There was a veil past which I could not see
There was a door for which I had no key
  —Rubáiyát of Omar Khayyám

S O NEAR and yet so far! We have come a long way from Greek theorizing about the existence of atoms, and even from Rutherford's primitive experiments on the nucleus. We now know that the basic building blocks of the universe are the quarks and leptons, and we know that the seemingly different forces that act between them can, in all likelihood, be unified into a single universal force. The general outline of a final theory is in sight, but we haven't yet reached the point where we can retire to our front porch rocking chairs. There are some real problems that face elementary particle physicists as they approach the millennium. In this chapter I'll give a brief description of the problems that I see as being crucial to the future of the field. As with any list of important scientific questions, there is a certain amount of subjectivity in my choice of topics. I don't apologize for my choices, but you should be aware that others might have put together a different list.

THE SEARCH FOR THE TOP QUARK

If you talk to theoretical physicists, you will hear them talking about the top quark as if evidence for it had already been seen in the laboratory. In fact, as I write this (fall 1993), the search for the top quark has been going on for some time without a definitive resolution as to whether it exists or not. Obviously, all the theories I talked about in the last chapter would be in deep trouble if the last member of the quark-lepton family failed to show up, so the search is of more than just academic interest.

In Chapter VI we described the Fermilab collider as an example of a large modern accelerator installation. In that apparatus, you will recall, protons and antiprotons move in opposite directions around the ring and are accelerated to energies up to 900 GeV. The beams are then brought together and allowed to collide head-on, providing the most energetic collisions available to physicists anywhere in the world. It is to Fermilab that the physics community is looking for success in the search for the top quark.

The instrument (if you can call it that) used in the search is a far cry from Rutherford's simple tabletop apparatus. The collider detector at Fermilab (CDF) is an example of the new generation of particle detectors—as good an example as you'll find anywhere of high-tech virtuosity. Weighing in at over 5000 tons with a cost of over $60 million, the CDF fills a two-story building and occupies the attention of over 200 physicists and technicians.

The detector has to be large because it must wrap around the proton-antiproton collision point so that virtually all the particles produced in each collision pass through it. So important is it that nothing from the collision be missed, in fact, that particles emerging from the collision area encounter not one, but multiple layers of detectors. In the CDF, for example, particles first go through a chamber in which thin gold wires are spaced a fraction of an inch apart. An electrically charged particle passing one of these wires will generate a small electric current. Depending on where the particle is, the electrical signals will arrive at the ends of the wires at different times. This sort of detector is called a Charpak chamber, after the French physicist Georges Charpak, who received the Nobel Prize in 1992 for his invention.

With modern techniques, a single plane of wires can locate a particle's position to within a millimeter or so, and tell when it went by to within a hundred millionth of a second. In the CDF and other collision detectors, the particle passes through multiple layers of such counters, and as a result, its trajectory can be traced back so that the location of the collision can be specified to within about .05 millimeter (a distance much less than the width of a human hair).

Once the trajectory of the particle has been established by the first set of counters, it enters a second set. These are typically composed of alternating layers of lead and scintillation counters. In this detector the particle loses energy rapidly (electrons, for example, are completely absorbed). The surviving particles enter a third tier of detectors made of alternating counters and iron plates. In this tier, the energy of the particle is measured and everything except muons is absorbed. The fourth and final layer of detectors simply registers the presence of the particles that have survived the trip through all the other detectors.

Not only does this layering scheme allow physicists to detect particles coming directly from the collision region, but it makes it possible to detect the presence of short-lived particles that decay as they emerge from those collisions. In the latter case, the detector will see the decay products, and will be able to trace their trajectories back to the position of the decay—a position that will, in general, be different from that of the collisions itself.

The most amazing thing to me about detectors like the CDF is not its size, its cost, or even the number of physicists needed to run it, but the amount of information that has to be processed when the accelerator is running. In the Tevatron, for example, there are about a million collisions per second, each producing particles whose trajectories and energies could, in principle, be recorded for later study. But at this point we run into a serious problem, because modern computers can record only a few events per second. Fortunately, in a typical high-energy experiment only a small fraction of the events are interesting. This means that the computers, instead of simply recording each event, examine each to see whether it's worth keeping.

In general, there is a multistage logic system that examines each event in the millionth of a second before the next one comes along. If the event seems to be one where quarks or gluons collided head-on and produced jets or other interesting phenomena, the data is passed on to be recorded. Otherwise it is simply thrown away (except for one in a thousand or so that it recorded for "just in case" insurance). It is these rare events that are then passed along to the experimenters to be examined and analyzed.

In the search for the top quark, for example, the theory makes some definite statements about what will happen if a top quark (or antiquark) is produced. The top quark will decay quickly into a bottom quark and leptons, and the bottom quark will, in turn, produce a spray of ordinary particles. This means that there are two likely ways for the top quark to be produced: (1) the collision produces a W boson, which subsequently decays into a top and bottom quark, and (2) a top and antitop are produced directly. In the first case, the signature of the event will be a lepton along with two jets of particles (one for each bottom quark). In the second case, the signature will be two jets and two leptons. As the search progresses, it is events like these that are winnowed out by the computer logic systems.

As I write this (fall 1993), large amounts of data have been collected at the CDF. It may be that the events that will be hailed as the discovery of the top quark are already stored on magnetic tape somewhere, awaiting analysis. This is certainly what everyone hopes will be the case. But the longer the search goes on, the more nervous everyone seems to get.

The problem is this: The failure of earlier experiments to see the top quark already shows that its mass must be more than 100 GeV. This is already significantly higher than the mass of the bottom quark. Since the two are supposed to be part of the same family, there are limits to how high the mass of the top can be without causing serious difficulty for the Standard Model. The general consensus is that if the mass of the top is more than

220–250 GeV (a limit CDF scientists expect to reach around 1995), there is something wrong with the theory.

So, although most theorists find the search for the top quark somewhat boring—it is, after all, just filling in a blank everyone knows has to be filled—the longer the search goes on, the less boring it becomes.

THE SUPERCONDUCTING SUPERCOLLIDER

Our understanding of the basic structure of matter has always been intimately tied to our ability to construct accelerators capable of probing ever more deeply into the fundamental building blocks of nature. This has been true since the discovery of the nucleus, and is equally true today. The problem, of course, is that as you probe deeper and deeper, you need more and more energy to get the job done, and the machine you have to build grows as well. Rutherford discovered the nucleus with an apparatus that would, in essence, fit on a table top. In contrast, the discovery of the quarks requires an accelerator like the one at Fermilab, with the main ring a mile across and equipment costing hundreds of millions of dollars. The next step, unfortunately, is going to be even more expensive.

Assuming that the top quark is indeed discovered at Fermilab, the present generation of accelerators will have accomplished its task. The Standard Model has been developed, and all the basic structural members of the universe are known. The next step, as we saw in the previous chapter, is to understand the unification of the fundamental forces. This process has already begun with the discovery of the W and Z particles and the electroweak unification. Yet although current accelerators can, indeed, reach energies at which the unification can be seen, they cannot produce the Higgs particle. As we saw in Chapter XIV, this particle not only produces the unification of the electromagnetic and weak forces, but is responsible for the mass of everything in the universe. Clearly, finding and understanding the Higgs particle and its interactions is the logical next step on the path we have been following.

In 1982, a group of physicists under the auspices of the American Physical Society began a study of what sort of accelerator would be required to understand the electroweak unification and find the Higgs. The proposal that eventually resulted from their work was for a machine called the Superconducting Supercollider (SSC). The "superconducting" part of the title refers to the fact (see Chapter VI) that all the magnets in the main ring of the accelerator will be made from superconducting materials, thereby lowering the energy requirements of the machine. The "supercollider" refers to the fact that the machine will have two beams of protons circulating in opposite directions, with the beams brought together to produce head-on collisions at two points on the ring.

So far, so good—the SSC seems to be a rather obvious extension of the kinds of machines we've discussed before. The difference comes in the scale of the thing. To fulfill its mission, the SSC would have needed to accelerate each proton beam to 20 TeV (a TeV, you will recall, is 1000 GeV). This would have required a main ring no less than 54 *miles* in circumference! It will also be very expensive—current estimates put the cost of the facility at between 8 and 10 billion dollars.

The SSC was designed to have three booster accelerators to take the protons up to energies of 2 TeV, at which point they were to be injected into the main ring and accelerated up to the full 20 TeV. The main ring would be housed in a twelve-foot tunnel about 100 feet underground. Each ring would have over 4000 magnets to bend the beam, and there would be two collision areas with large detectors at diametrically opposite ends of the ring. The beams themselves would contain over 100 trillion protons each, but would be only about as wide as a human hair. The rate of collisions was expected to be about 100 million per second—significantly higher than the rate now handled at Fermilab.

In January 1987, President Reagan authorized the beginning of the SSC project. Originally, the plan called for having the SSC up and running by the end of the 1990s. In November 1988, the site for the SSC was chosen—an area of rolling prairie in Ellis County, Texas, about 30 miles south of Dallas. By the time the project was killed by the House of Representatives in October

1993, some 14 miles of tunnel had been drilled through the Austin chalk, the geological formation underlying the town of Waxahatchie, Texas.

The decision to scrap the project had little to do with science. Everyone agreed that the SSC was the right machine being built at the right time. But there is more involved in this sort of decision than simple arguments of scientific progress. Ten billion dollars is a lot of money, even in a town like Washington. It became a symbol of budget-busting profligacy, and as such became a highly visible scalp that congresspersons could take home.

One development that I found disturbing in the SSC debate was a kind of dog-in-the-manger argument by some prominent physicists. In effect, they argued that instead of spending 10 billion dollars on the SSC, Congress should fund other kinds of research (usually their own). But of course, this isn't the way the political system works. Money not spent on the SSC will not automatically go to other scientific research, but will just stay in the general budget. Scientists who want additional funding will still have to compete with entitlement programs, highways, defense, and the myriad of ways the government spends its revenues. I wish them luck.

It will, I suppose, come as no surprise to the reader that I think the abandonment of the SSC was a major mistake. In a sense, the entire story in this book points to the inevitability of building a machine like this. It represents the culmination of a 2000-year quest, a culmination that will now belong to another generation, another country.

If it were just a matter of adding a feather to the cap of American science, I suppose it might be possible to argue that the exigencies of the federal budget force us to forgo the SSC. But I think this decision will also prove to be shortsighted even in straightforward economic terms. For the fact of the matter is that the promise of the SSC was that it would have allowed us to understand mass, one of the fundamental properties of matter. The last time human beings made such a discovery was in the nineteenth century, when the nature of electrical charge was uncovered. The economic benefits that flowed from that discovery are incalculable—try to imagine modern society without electricity, if you can. It seems to me that, by killing the SSC, Congress has cut future generations off from the equivalent of the use of electricity. That's a high price to pay for a highly visible scalp!

But the SSC is all water under the bridge now, I suppose. Where do we go from here?

THE THEORY OF EVERYTHING

We seem to be very close to a completely unified picture of the universe. The electroweak unification was one of those watershed events, it seems—an event that casts a long shadow into the future. Once we understood how two forces could be unified, it was only a matter of a few years until the strong force was added to the unification list. At first glance, it seems that it should be a relatively simple matter to incorporate the fourth force, gravity, into this scheme.

Unfortunately, things aren't so simple. The fact of the matter is that the strong, electromagnetic, and weak forces can all be described by theories that incorporate quantum mechanics and the exchange of virtual particles. The best theory of gravity we have—General Relativity—couches its description in terms of concepts like warped space-time fabrics, rather than in terms of the quantum. For my entire professional life, I have watched as some of the most brilliant minds in theoretical physics have tried to overcome the technical difficulties involved in making a quantum theory of gravity. None has succeeded, and as of this writing, developing a quantum theory of gravity remains one of the great unsolved puzzles of science. And until gravity is put on an equal footing with the other forces, it seems unlikely that we will be able to produce any sort of unified theory that includes it.

We can, however, get some sense of what a unified theory of all the forces might look like. For example, we have seen that there are two basic kinds of particles in the universe—quarks and leptons (the "bricks") and gauge particles (the "mortar"). And although we haven't made a point of it, you will note that the quarks and leptons are fermions (see Chapter VIII), while the gauge particles are all bosons. These two classes of particles seem to play very different roles, a difference reflected in their different spins. Yet at the level of gravity, we expect them all to behave pretty much the same. The gravitational force any particle exerts depends only on its mass, not on its spin or any other characteristic. Gravitationally, bosons and fermions must be equivalent. Thus, a fully unified theory would have to have a way of changing fermions into boson and vice versa—reactions that are not seen in any laboratory today. In a sense, at temperatures characteristic of the unification of gravity with other forces, there would be only one force acting and only one basic kind of particle. This universe would be as simple as it could possibly be.

With their characteristic humor, theoretical physicists have taken to calling theories that unify gravity with the other forces "Theories of Everything" (TOE). Such a theory, if it existed, would truly mark the end of our quest for the basic structure of the universe.

The general unification scheme, then, says that at progressively higher energies more and more forces will unify until the full unification is reached. Remembering the cosmic connection we discussed in Chapter XIV, this means that early in the life of the universe temperatures must have been high enough that the unification would have been seen directly. For reference, in the table I list the unifications, their energies, and the time when we think that the universe had a temperature corresponding to those energies.

Unifications

UNIFICATION	ENERGY (GeV)	TIME (sec)
electromagnetic-weak (Standard Model)	100	10^{-10}
strong-electroweak (GUT)	10^{15}	10^{-35}
gravity-everything (TOE)	10^{19}	10^{-43}

For reference, the energy and time associated with the gravitational unification are called the *Planck energy* and *Planck time,* respectively, after Max Planck, one of the founders of quantum mechanics. He was the first to realize that at those energies it would be necessary to treat gravity according to the rules of quantum mechanics.

One point should be obvious from the table: We are not going to be exploring the GUT and TOE unifications in the laboratory anytime soon. The SSC, for example, would have supplied energies 10 billion times too small to reach even the lowest of the requisite energies. And although some visionaries have suggested that extrapolating current increases in accelerator energies into the future will bring us to Planck energies sometime in the twenty-first century, for the foreseeable future we will have to use the universe itself as the ultimate test of our theories.

Just to give you a sense of what a TOE might look like, let me briefly describe one popular type of theory—the so-called "Supersymmetry Theories" (SUSY). One way of thinking about these theories is to note that the basic idea of the gauge symmetry (see Chapter XIV) that lies at the base of our understanding of unification is that there are some properties of particles that can be regarded as arbitrary definitions, and can therefore be changed arbitrarily from one point in space to another. In the case of the electroweak unification, this property was electrical charge (or more precisely, isotopic spin). In the case of the GUT, the property was color. In the case of the SUSY theories, the property has to do with particle spins. In these theories, we can go from one point to another, changing bosons into fermions and vice versa, without affecting anything that can actually be measured.

One interesting outcome of these theories is that they predict a whole new slate of elementary particles, each a partner of a known particle except for its spin. There is, for example, a spin $1/2$ particle called the photino that is identical to the photon except that it is a fermion while the photon (with spin 1) is a boson. There are selectrons (spin 1), sneutrons, and sprotons, and so on. In fact, the supersymmetric particles in this scheme are somewhat like antimatter—they provide a kind of mirror image universe (provided, of course, that they exist). One argument given for building the SSC, in fact, is that it might make it possible to produce the SUSY particles.

But whether it is supersymmetry or some sort of theory as yet undreamed of, it seems pretty clear to me that we shall someday have our Theory of Everything. We will be able to write down a single equation—perhaps one that will fit on the back of an envelope—that will, in a very real sense, explain everything about the basic structure of the universe.

And then what?

THE END OF SCIENCE?

In his excellent book *Dreams of a Final Theory* (Pantheon Books, 1992), Nobel Laureate Steven Weinberg discusses the question of the TOE. He likens the progress of science to a series of theories in which some arrows always point backward—in which there are always some terms or concepts that must be defined and clarified at a deeper level. Indeed, the progression from materials to atoms to elementary particles to quarks can be cast in this general scheme. In Weinberg's language, a TOE is a theory in which all the arrows point forward, in which there are no further fundamental questions to ask.

If such a theory was developed and tested, if we really could work out and test all its ramifications, then that would indeed be an end to the 2000-year quest for the ultimate understanding of the universe. Particle physicists could turn the equation over to the engineers and take a well-earned vacation.

But would that mark an end to science? Of course not. There are many questions in science, many ancient quests besides the search for the fundamental structure of the universe. The development of a TOE would do very little for these parallel endeavors. To give one example, telling a paleontologist that a trilobite was made from quarks and leptons gives him or her very little interesting information. The paleontologist wants to know how the trilobite evolved, why it became extinct, how it fit into its ecosystem. The TOE will shed no light whatsoever on these questions.

Furthermore, over the past few decades we have come to realize that systems made up of simple constituents can have surprising properties. The human brain, for example, is made up of the same sort of neurons as those of other animals, including squids and worms, yet it reaches a level of complexity that produces cathedrals, symphonies, and Theories of Everything. In fact, I think that the study of complex systems like the brain may very well mark the beginning of a new quest in science. It is only recently begun, but it promises to be every bit as interesting as the one started by those Greek thinkers two millennia ago.

But that, of course, is a story for another book.

Appendix A

SOME COMMON SYMBOLS

A—The number of nucleons in a nucleus
B—A magnetic field
c—The speed of light
h—Planck's constant
I—Isotopic spin
J—Spin
m_p—Mass of the proton
m_e—Mass of the electron
Q—Total charge of a particle
S—Strangeness
σ—Cross section
Z—Total charge of a nucleus

Appendix B

A CATALOG OF PARTICLES (Unstable)

NAME	SYMBOL	CHARACTERISTICS	Chapter Discussed
Mu-meson (muon)	μ	lepton	V
Pi-meson (pion)	π	meson	V
Lambda	Λ°	strange baryon	V
K-meson (kaon)	*K*	strange meson	V
Delta	Δ	nonstrange baryon	VII
Rho-meson	ρ	nonstrange meson	VII
Sigma	Σ	strange baryon	VII
Cascade	Ξ	doubly strange baryon	VII
Sigma 1385	$\Sigma(1385)$	strange resonance	VII
Omega minus	Ω^-	triply strange baryon	X
Psi/J	Ψ/J	meson containing charmed quarks	XII
D-meson	D°	charmed meson	XII
Upsilon	Y	meson containing bottom quark	XII
Tau-meson	τ	heavy lepton	XIII

A CATALOG OF PARTICLES (Stable)

NAME	SYMBOL	CHARACTERISTICS	Chapter Discussed
Electron	e	lepton	I
Photon	γ	light	I
Neutrino	v	lepton	II, XIII
Proton	p	baryon	I
Neutron	n	baryon	II
Positron	e^+	antielectron	IV

In general, an antiparticle is denoted by a bar over the particle symbol. Thus, the antiproton is written $\bar{p}$, and so forth.

Glossary

The following terms are important in any discussion of elementary particles. In addition to a short definition, the chapter in which the term is introduced or discussed in the text is generally given in parentheses immediately after the term itself.

ATOM (I)—the "indivisible" smallest piece of matter first postulated by the Greeks, but now known to be composed of a nucleus circled by electrons.

BARYON (VIII)—any hadron that contains one proton in its final set of decay products.

BETA DECAY (II)—the process by which a neutron decays into a proton, an electron, and a neutrino. If the process occurs when the neutron is inside a nucleus, we speak of nuclear beta decay.

BOOTSTRAP (XIV)—a theory of elementary particles in which logical consistency is the ultimate requirement.

BOTTOM QUARK (XII)—one of the new quarks whose existence is shown by the discovery of the upsilon particle.

BUBBLE CHAMBER (VII)—a device in which the track of a particle crossing the chamber is marked by a string of bubbles condensing on ionized atoms.

C QUARK (XII)—the quark whose existence is demonstrated by the discovery of the Ψ/J particle. It carries the charm quantum number.

CATHODE RAY (I)—an old term for electron.

ČERENKOV COUNTER (VII)—a device that identifies particles

passing through it by observing a flash of light generated in a manner similar to a sonic boom.

CHARGE CONJUGATION (VIII)—the mathematical operation that turns a particle into its antiparticle.

CLOUD CHAMBER. See WILSON CLOUD CHAMBER.

COLOR (XII)—the property of quarks that allows them to be arranged in ways that seem to violate the Pauli principle.

CONSERVATION LAWS (XIII)—any observed regularity in nature which indicates that a particular quantity (electrical charge, for example) is the same before and after a reaction.

COSMIC RAYS (IV)—energetic particles (primarily protons) that are created in stars and enter the earth's atmosphere.

CYCLOTRON (VI)—a device that accelerates protons to high energies.

EIGHTFOLD WAY (IX)—a way of grouping the elementary particles that reveals regularities in their properties.

ELECTRON (I)—the small, negatively charged particle that normally circles the nucleus of an atom.

ELECTRON VOLT (abbreviated eV) (V)—The energy acquired by one electron falling through one volt of potential energy.

ENERGY (IV)—the ability to do work. It comes in many forms (kinetic, potential, and mass) and is conserved.

FERMI (abbreviated as F) (I)—A unit of length equal to 10^{-13} cm, about the distance across a proton.

FNAL (VI)—the Fermi National Accelerator Laboratory—the world's largest accelerator, located near Chicago.

FLAVOR (XII)—the aspect of a quark that tells which of the six kinds of quark it is.

GAUGE SYMMETRY (XIV)—a symmetry in which no measurable property of the world changes if protons and neutrons can be substituted for each other at each point in space independently.

GAUGE THEORY (XIII, XIV)—any theory that incorporates gauge symmetry.

GEIGER COUNTER (I)—a device that detects ions created by a charged particle.

GEOLOGICAL SEARCHES (XI)—searches for quarks presumably trapped in various materials on the earth.

GeV (V)—abbreviation for giga (10^9) electron volts of energy.

HADRON (VII)—any particle that participates in the strong interactions.

HALF-LIFE (II)—the time required for one-half of a given sample of particles or nuclei to decay.

HEAVY LEPTON (XIII)—a particle having properties similar to the electron or mu-meson, but more massive.

ION (I)—an atom from which one or more electrons have been stripped, leaving it with a net positive charge.

ISOTOPIC SPIN (VIII)—a mathematical quantity related to the number of different charges in a particle family.

keV (V)—abbreviation for kilo (1000) electron volts of energy.

LEPTON (VII, XIII)—a particle (like the electron, muon, and neutrino) that participates in the weak, but not the strong, interactions.

LEPTON CONSERVATION (XIII)—a rule which states that the net number of leptons before and after an interaction must be the same.

LINEAR ACCELERATOR (VI)—an accelerator in which the particles move in a straight line as they gain energy.

MAGNETIC MONOPOLE (XIV)—a hypothetical particle that carries a single magnetic pole.

MAGNETIC SPECTROMETER (XII)—a device that uses the bending of particles in a magnetic field to separate out those of a given momentum.

MATHEMATICAL QUARKS (XI)—a term for quarks that might "exist" only in theory, but never be found in the laboratory.

MESON (V, VIII)—originally, any particle whose mass is between that of the electron and proton—in modern terms, any particle whose decay products do not include a baryon.

MeV (V)—abbreviation for mega (10^6) electron volts of energy.

MU-NEUTRINO (XIII)—the neutrino given off in the decay of the mu-meson.

NUCLEAR DEMOCRACY (VII)—the idea that every particle is equally "elementary."

NUCLEUS (I)—the heavy positively charged center of the atom, composed of protons and neutrons.

NUCLEON (II)—a term referring to both the proton and neutron.

NEUTRAL CURRENT (XIII)—the uncharged object exchanged when a neutrino scatters from a hadron.

NEUTRINO (II, XIII)—a zero-mass uncharged particle emitted in the process of beta decay.

NEUTRON (II)—a particle of approximately the same mass as the proton, but uncharged.

PARITY (VIII)—a mathematical operation which exchanges right and left.

PARITY VIOLATION (XIII)—refers to the observed fact that in some beta decay processes electrons are emitted preferentially in the right-hand direction.

PARTICLE-WAVE DUALITY (III)—a pseudodilemma brought about by the fact that elementary particles behave neither as waves nor as particles.

PAULI PRINCIPLE (XII)—the principle that states that two spin ½ particles such as electrons or quarks cannot occupy the same state.

PHOTON (I)—the "particle" associated with light.

POSITRON (IV)—the antiparticle of the electron.

PROTON (I)—the massive, positively charged particle that is the nucleus of the hydrogen atom.

QUARK (IX)—the hypothetical particle that is believed to be the basic constituent of the elementary particles.

QUARK CONFINEMENT (XI)—the theory that there is some reason why quarks might exist inside of elementary particles but may not be seen in any experiment.

PLANCK'S CONSTANT (I, III)—a fundamental constant of nature important in quantum mechanics.

SCINTILLATOR (I)—a type of particle detector that emits a flash of light when a particle strikes it.

SPIN (VIII)—the property of an elementary particle analogous to the rotation of the earth on its axis.

SLAC (VI)—acronym for Stanford Linear Accelerator Center, the world's largest electron accelerator.

STORAGE RINGS (VI, XII)—devices in which accelerated particles are kept moving in circles by magnetic fields.

STRANGENESS (V, VIII)—the property of elementary particles that governs the speed at which they decay.

STRONG FORCE (INTERACTION) (II, III)—the force or interaction responsible for holding the nucleus together.

SYNCHROTRON (VI)—an accelerator in which magnetic fields and acceleration are synchronized to keep the particles moving in a narrow ring.

TACHYONS (XIV)—hypothetical particles that move faster than the speed of light.

TeV—abbreviation for tera (10^{12}) electron volts of energy.

TIME REVERSAL (VIII)—the mathematical operation analogous to running a movie film backward.

TOP QUARK (XII)—the as yet undiscovered quark that is the partner of the bottom quark.

UNCERTAINTY PRINCIPLE (III)—the principle that states that it is impossible to measure both the position and momentum of a particle with infinite accuracy.

UNIFIED FIELD THEORY (XIV)—a theory in which two or more interactions are seen to be different aspects of a single process.

W BOSON (XII)—the hypothetical particle that is supposed to be exchanged in beta decay and other weak interactions.

WAVE FUNCTION (III)—in quantum mechanics, the mathematical function that gives the probability of finding the particle at a given point.

WEAK INTERACTIONS (II)—processes, like beta decay, that proceed slowly on the nuclear time scale.

WILSON CLOUD CHAMBER (IV)—a device used in early work that records the passage of charged particles by the presence of droplets formed on ions left by their passage.

Y^* (IX)—an old expression for higher mass resonances associated with strange baryons.

Index

(Figures in italics denote illustrations or tables.)

250 Index

ABOUT THE AUTHOR

Coauthor of the bestselling book *The Dictionary of Cultural Literacy* and the highly acclaimed *Science Matters,* and author of *1001 Things Everyone Should Know About Science,* James Trefil has published more than a dozen books on science. A former Guggenheim fellow and a regular commentator for National Public Radio, he is the Robinson Professor of Physics at George Mason University, and science consultant to *Smithsonian* magazine. He lives in Annandale, Virginia.

THE LITTLE BOOK OF
Coffee

Alain Stella

Flammarion

Coffee is one of the most popular drinks in the world.
How did coffee drinking begin? Who invented it?
What did coffee taste like in its earliest days?

On the surface of it, there's nothing simpler than preparing
a cup of coffee. Yet traditions evolve and tastes vary from
one country to the next. What different methods are used
to make it and what are the great traditional flavors?

Arabicas, robustas, gourmet beans and blends—each type
of coffee has its particular characteristics. How do you choose?
And what is the best way to get the most from each one?

A N S W E R S

Orientation p. 6

The alphabetical entries have been classified according to the following categories. The categories are indicated with a small colored rectangle.

■ Raw Material:
Superior Beans
Cultivation
Processing

■ Practical Information:
Preparation
Flavors
Traditional Types

■ Context:
History
Sociology
Economics

The information given in each entry, together with cross-references indicated by asterisks, enable the reader to explore the world of coffee.

The Story of Coffee p. 11

The Story of Coffee provides a detailed overview of the themes and information provided in the alphabetical entries.

Alphabetical Guide p. 27

The entries, arranged in alphabetical order, tell you all you need to know to find your way around this fascinating world. The information is enriched with detailed discussion of all the major varieties of coffee and text boxes which put the different flavors in context.

O R I E N T A T I O N

I. COFFEE: AN ADVENTURE

A. Coffee the Conqueror

Follow coffee through its prehistoric and legendary phases, from its various uses in Ethiopia, its voyage across the Red Sea to Yemen, and its spread throughout the Arab world.

- *Etymology*
- *Invention*
- *La Roque, Jean de*
- *Legend*
- *Opponents*
- *Treatises*
- *West Indies*

B. From Camel's Back to Ship's Hold

The Western world took so strongly to coffee that camel caravans couldn't cope with demand. In the seventeenth century, the mercantile trading companies of the European powers resorted to diplomatic efforts and underhand tactics to guarantee the development, and their monopolization, of the coffee trade. These days, the routes coffee takes from plantation to table are filled with less adventure and risk than in earlier centuries, but its commercial trajectories are still quite complex.

- *Caravans*
- *Expansion*
- *Coffee Exchange*
- *Maritime Trading Companies*
- *Ports*
- *Route*

C. Coffee Slaves

Not everyone involved in the coffee trade became rich. For a long time production was ensured by unscrupulous planters who often resorted to slavery. Today's market for this worldwide commodity is based on a more equitable distribution of profits.

- *Consumption Statistics*
- *Max Havelaar Foundation*
- *Plantations*
- *Production Statistics*
- *Roasters*
- *Slavery*

II. FROM PLANTATION TO SUPERMARKET

A. The Coffee Tropics

Arabica and robusta coffee plants both thrive in tropical climates. While every plantation has its own carefully run nursery system, plantations and their methods differ greatly from country to country. Coffee plants are productive for about fifteen years and yield six to eight harvests per year.

- ◼ *Arabica*
- ◼ *Coffee Plant*
- ◼ *Cultivation*
- ◼ *Organic*
- ◼ *Robusta*

B. A Question of Skill

The "cherries," or fruit, of the coffee plant each hold two beans. These have to be cleaned, dried, hulled, sorted and roasted. The execution of these steps contributes to a coffee's quality and taste. Part of the processing takes place on site, generally by hand. Large-scale distributors and local roasters take care of the rest.

- ◼ *Blends*
- ◼ *Coffee Shops*
- ◼ *Decaffeinated*
- ◼ *Processing*
- ◼ *Roasting*
- ◼ *Selection*
- ◼ *Sorting*
- ◼ *Tasting*

C. Pure Coffees

Superior beans can be enjoyed unblended. They are all arabicas and include Blue Mountain, Kona, and Sidamo Mocha. Some of these have only been available for a few years. Connoisseurs appreciate their distinctive tastes. Their natural characteristics evoke subtle idiosyncrasies for a maximum of enjoyment.

- ◼ *Best Beans*
- ◼ *Blue Mountain*
- ◼ *Brazil*
- ◼ *Caracoli*
- ◼ *Colombia*
- ◼ *Costa Rica*
- ◼ *Kenya and East Africa*
- ◼ *Kona*
- ◼ *Mocha*

III. TASTE AND COLOR

A. Coffee Making

There are a number of ways to make and enjoy coffee. But without good water, the right bean or blend, and the correct grind, results will be disappointing.

- *Bedouin*
- *Boiling*
- *Coffeepot*
- *Espresso*
- *Grinds*
- *Instant Coffee*
- *Melitta*
- *Pressure Percolation*
- *Rules of Thumb*
- *Steeping*

B. Traditional Flavor

It has been a few centuries now since coffee first made its way from eastern Africa and the Arabian peninsula to the rest of the world. Traditions, practices, inventions, serving styles and combinations have developed everywhere.

- *Alcohol*
- *Austria*
- *Britain and America*
- *Cappuccino*
- *Chicory*
- *Chocolate*
- *Ethiopia and Yemen*
- *Extracts and Liqueurs*
- *Fortune Telling*
- *France*
- *Iced Coffee*
- *Irish Coffee*
- *Italy*
- *Milk*
- *Northern Europe*
- *Turkish*

C. At Home or in a Café?

Whether at home or at a café bar, slowly savored or quickly gulped down, coffee drinking should always provide that liquid bite that you're after. Coffee brings refined pleasure to everyday life, a small miracle repeated every day, sometimes several times a day.

- *Café life*
- *Caffeine*
- *Coffee Set*
- *Coffeehouses*
- *Health*
- *Insomnia*
- *Literature*
- *Painting*
- *Percolator*
- *Waiter*

THE STORY OF COFFEE

It is impossible to fully comprehend the extent to which coffee has changed the world. The moment Europeans first tasted it in the fifteenth century, it began to conquer the globe. It is the second most consumed beverage after water, a fact bitterly contested by tea drinkers. In addition to its wonderful flavors and aromas, coffee offers stimulation to both mind and body. From America to Japan by way of the Middle East and Europe, coffee warms the spirit and fosters friendship. Whether at the family breakfast table, in a Bedouin tent, in a busy office, or at a café, drinking coffee means human contact, hospitality and the sharing of ideas.

I. Coffee: An Adventure
A. Coffee the Conqueror

Coffee has its own prehistory, the exact phases of which are impossible to establish with precision. Before its widespread consumption as a beverage, the fruit of the coffee bush was boiled and utilized for medicinal purposes in Ethiopia,* where it grew in the wild. It is also likely that Ethiopia is the site where coffee became what it is today, with the possibly accidental invention* of roasting.* For centuries, coffee stayed put where it got started as a drink. Then one day, it was somehow transported across the Red Sea, and planted in Yemen, where the first coffee plantations* began. This inexplicable development was thought to date back to the late fourteenth century, but a recent discovery has set it back at least two centuries. It was not until the fifteenth century that coffee really spread from Yemen, through the Sufis, who brought their beloved *qahwa* to Mecca and as far as Cairo, where it took root throughout the Muslim world. The Sufi sect used coffee to keep themselves going through their religious practices. This fact is at the heart of the legend,* told in *A Thousand and One Nights,* of coffee's discovery by a goatherd in "Arabia the Happy."

Coffee made its European debut in the mid-seventeenth century, where it was taken up in fashionable circles as part of the craze for all things Turkish. Praised for its medicinal virtues in a number of learned treatises, it was nevertheless at odds with certain political and economic priorities, and became the object of moral and other sorts of condemnations (see Opponents*), in both Europe and the Arab world. For a century, coffee drinking was a privilege for the elite. With its cultivation in European colonies in Asia and the West Indies,* it was gradually made accessible to everyone and became a

Coffee merchants in Yemen.

Bags of coffee being loaded onto a ship in Santos, Brazil.

popular beverage, finally, as a crowning mark of success, replacing soup for breakfast.

B. From Camel's Back to Ship's Hold

Long cultivated, for reasons of climate, thousands of miles from the centers of consumption, coffee was carried by camels in caravans,* then loaded onto ships and transported from Yemen to Cairo, Alexandria, Istanbul, and as far as India. Circling the African coast to Yemen, the major European maritime companies* first traded directly in coffee beginning in the eighteenth century. The companies also started colonial coffee plantations, as the European powers spread coffee cultivation right round the world between the tropics of Cancer and Capricorn. The precious bean brought wealth to such ports as Mocha,* the first major point of exportation, Marseilles, and Amsterdam.

C. Coffee Slaves

Coffee production was not a very ethical business and the practices followed have often cast a shadow over the pleasure of coffee drinking. Slave labor was used systematically by the colonial powers to clear land and cultivate* vast plantations* under abominable

Joseph Vernet, *The Port of Marseilles,* 1754. Oil on canvas. Musée du Louvre, Paris.

conditions. Today, small growers in the often poor coffee-producing countries are subject to market fluctuations and speculation, exaggerated middleman percentages, and the harsh laws of supply and demand. The back-breaking, uncertain work of coffee cultivation is often unfairly remunerated. In the generally wealthy consumer countries, some associations such as the Max* Havelaar Foundation, work to correct these injustices.

The stakes are enormous, considering the extent of the global market, which is continually expanding. Each year around seventy countries produce the ten million bags of coffee (a little more than six million tons) drunk by two out of three people worldwide. A sizeable portion of all coffee is bought, roasted, packaged, and sold by the major industrial roasters* belonging to multinational food conglomerates.

Raphaël Robin-Noiret, *Coffee Picking.* Oil on ivory. Musée des Arts Africains et Océaniens, Paris.

II. From Plantation to Supermarket
A. The Coffee Tropics

The places where coffee is grown are located in a tropical belt circling the globe and are as varied as the beans themselves. The immense Brazilian plantations cut into thousands of acres of virgin forest have nothing in common with the terraced, garden-sized plots of Yemen or the steep Andean slopes. The volcanic ash of Hawaii contrasts greatly with the red earth of Brazil.*

The two branches of the *Coffea* species, arabica* and robusta,* make for two distinct types of coffee cultivation. The fragile arabica bushes grow only at altitudes of over 3,200 feet, making large tracts of land and full mechanization impossible. The need for pesticides and insecticides to protect these delicate plants entails added costs, even though organic growing methods have been adopted in some cases. The hardy robusta bushes can be grown on flat land and allow for economical mechanized cultivation.

Coffee growing begins in nurseries. Once transplanted, coffee bushes yield fruit after three or four years, during a fifteen-year

Coffee plantation in Brazil.

15

period. The bushes flower and produce fruit after each rain. Harvest frequency and season varies with climate. Generally, there are six to eight yearly harvests. In the best plantations, only ripe fruits, called "coffee cherries," are picked (by hand). Elsewhere, in the case of inferior arabicas and robustas, all fruits are picked—ripe "cherries" or unripe "quakers"—sometimes by machine.

B. A Question of Skill

At the beginning of processing* the fruit is removed from the bean. In many countries, small growers send their yield to be processed in cooperatives. The work requires skill as well as machinery. The "wet" method calls for repeated washing and sorting. Practiced in areas with ample water supply, this entails fermentation, stripping, drying, and hulling the fruit, and makes for the superior "washed coffees" of which Colombia is the world's foremost producer. The more rudimentary "dry method" is practiced in arid and poorer regions. The cherries are simply dried, then hulled by simple machines. The resulting "natural" coffees are less consistent in quality.

Coffee "cherries" being sorted in Indonesia.

Beans are precisely graded in terms of quality and origin. This allows international buyers to select and order "green beans" that are shipped to the consumer country for the next phases of treatment. They are roasted by either industrial or smaller expert roasters.* Roasting, which varies according to national and individual taste, determines coffee's aroma and flavor. Some beans are decaffeinated.* The majority are sampled and then blended.* Blending allows for consistent quality regardless of variations in harvests.

C. Fine Beans

The best arabicas are not always blended. This is especially the case for the very best beans.* Nobody would dare try to improve upon Jamaican Blue Mountain,* the "caviar of coffee," whose delicate, fruity, chocolatey flavor commands high prices. Over the past twenty years, certain top roasters have been

The coffee
department
in the
Dallmayr
delicatessen,
Munich.

selling some of the world's finest beans at more affordable prices. These include Brazil Sul de Minas, Columbia Supremo, Costa Rica* Tarrazu, Hawaiian Kona,* Kenya* AA, Mocha Sidamo, and others whose names alone are the stuff of dreams.

A well-roasted top bean yields authentic taste that no blend can truly equal: the special flavors of the original fruit come from being raised in a specific place, with its particular soil, watered by warm or cool rains, under burning or gentle sun. No two fine coffees are alike. Some are acidulous or tangy, others present slightly earthy or mossy accents. Like wine, their quality varies from year to year, and harvest to harvest. This is why connoisseurs love them.

Coffee lovers delight in the botanical oddity known as *caracoli.** Caracoli are single beans contained in a coffee fruit, rather than the usual double bean. The flavor is doubly intense. Caracoli are sorted out from the regular beans and sold separately.

III. Taste and Color
A. Coffee Making

In the first pages of his *Memoir for Oblivion,* in one of the finest passages ever written on the topic of coffee, Mahmud Darwich explains: "There is nothing one could call 'the taste of coffee'; it is not a concept, a concrete object, a thing in itself. Everyone has his own 'coffee,' so idiosyncratic that I can judge a man, sense his inner elegance, by the kind of coffee he serves."

Coffee can be considered a sort of language. Its preparation requires a certain familiarity with more or less complex instruments and subtle differences in quantities. There are dozens of techniques, all based on either boiling,* steeping,* or percolation* (filtering and pressure percolation*), as is the case with espresso.* Each method, used correctly, yields delicious coffee with particular characteristics.

Good coffee begins with pure water at the right temperature, and beans freshly ground to the required consistency. The coffee-making procedures of modern kitchens, where quick and easy preparation is at a premium, differs greatly from the Bedouin coffee ceremony.

From top to bottom and left to right: percolator, an Italian coffeepot, plunger pot, Cona-type pot, and an espresso machine.

Instant coffee, the paper filter invented by Melitta* Benz, and convenient home espresso makers have replaced the old-fashioned long-brewing percolators of yesteryear.

B. Traditional Flavor

Coffee presents a rich array of gourmet traditions, from the sublimely concentrated Italian* *ristretto* to the light, large Nordic cup. In Italy, coffee ranges from the frothy cappuccino with its dash of chocolate,* to the cold *granita di caffè* poured over shaved ice. In Vienna, at the crossroads between Italy and Turkey, coffee comes in every nuance, from black to nearly white. The specialty known as Viennese coffee is proudly served in a glass and topped with whipped cream. Turkish* coffee, the only coffee served with its grounds, remains popular throughout the Arab world and in most of the countries formerly under Ottoman rule. Fortunes can be read in Turkish coffee grounds. The brew is savored slowly, like the passage of time itself.

Other countries were slower to appreciate fine coffee, due to historical disadvantages. These include Great Britain, long under the dictatorship of tea; the United States, invaded by barbaric instant varieties; and France, duped by the use of chicory* and the harsh robustas from their African colonies. Today, English- and French-speaking peoples are making up for lost time at an exponential rate. Credit should also be given to the delicious spiked coffee specialties (see Alcohol), including Irish* coffee, the New Orleans classic *café brûlot,* and coffee with calvados, a Normandy specialty.

Coffee lovers everywhere delight in discovering their favorite flavor in different forms, including pastries, ice cream, and cocktails. There is a special place in the coffee lover's heart for Ethiopia and Yemen, the birthplace of coffee drinking. In both countries, fine Mocha beans are still roasted at home. And coffee's ancestor, made by boiling lightly roasted coffee fruit once the beans have been removed, is drunk to this day.

Preceding
double page:
a café in
Salonika,
c. 1885.

C. At Home or in a Café?

Coffee is a convivial drink, to be enjoyed among family, colleagues, and friends alike. The first cafés* of Europe opened only twenty years after the drink was introduced there. Their ancestors were the coffeehouses of the East, which sprung up in the sixteenth century in larger towns throughout the Arab world and Turkey. Aside from Turkish coffee, musical entertainment, lively conversation, games and reading abounded there. It was often former subjects of the Ottoman Empire, including Armenians, Syrians, and Greeks, who opened cafés in Europe a century later. The first English coffeehouse was started in Oxford by a Lebanese Jew, and the second in London by a Greek. The overwhelming popularity of coffeehouses in England made England the European capital of coffee for nearly a century.

Cafés soon became social institutions in all the major European cities, playing an important role in political and cultural life at least until the mid-twentieth century. Their atmosphere, with their waiters,* terraces, and beautiful percolators*, and their conviviality made them popular with artists, and they were immortalized by many painters and writers of the eighteenth through twentieth centuries.

The café is still going strong today. Cybercafés and global chains such as Starbucks have given the café concept new impetus in the 1990s. Coffee drinking in the home has evolved, too. There are those who like to recreate a continental café atmosphere at home, with antique or contemporary accoutrements, and the latest professional-quality home brewing methods.

Alain STELLA

24

█ Alcohol

Spiking coffee is an old tradition. In Normandy, France, combining coffee and calvados is an old favorite: apple brandy is poured into a cup of hot black coffee. In Italy* it is the *caffè corretto,* in which either grappa—the intense marc brandy—or cognac is added. And throughout France a similar "cordial" can be enjoyed, an excellent source of warmth against the cold of winter.

Then there is the superb *café brûlot* (or *brûleau*), a classic New Orleans coffee cocktail. The recipe consists in setting aflame a combination of cognac and curaçao flavored with spices and lemon (or orange) juice and then pouring black coffee over the top. Imperial coffee is served in Vienna's finest cafés. Known there as *Kaisermelange,* it is made by beating an egg yolk in a shot of cognac with a spoonful of powdered sugar, adding hot coffee and then serving the mixture with warm milk on the side. The most famous hot coffee mixed drink is of course Irish* coffee.

Le Sauvignon, rue des Saints-Pères, Paris.

Café au lait with *vino santo, uvetti,* and raisins marinated in grappa.

■ ARABICA: The Subtlest Flavors

The beans of *Coffea arabica*, one of the two main species of coffee plant,* are known as arabica. The plant is indigenous to Ethiopia, and accounts for 75 percent of the coffee produced today. Unlike *Coffea canephora*, which yields robusta* beans, the arabica plant requires particular conditions and thrives only at high altitudes. Its superiority derives from this particularity: the higher the altitude the longer it takes for a bean to ripen, and the slower the ripening process the greater the aromatic and flavorful qualities. This sensitive plant thrives between 2,600 and 6,650 feet (800 and 2,000 meters) above sea level. Below these levels, temperatures are too high. At higher altitudes it succumbs to frost. But only equatorial regions offer all the conditions the plant requires, average temperatures between 68 and 77°F (20 and 25°C), regular precipitation without humidity, great quantities of not too intense sunlight, and deep, rich soil, preferably volcanic.

The best* beans available are arabicas from East Africa and Latin America. Arabica has less caffeine* than robusta, and finer, more subtle flavors. These flavors result from growing and processing* conditions. Some are spicy or acidic, while others tend toward sweetness.

Arabica was the only known variety before robusta was discovered in the nineteenth century. Robusta then supplanted arabica consumption in many countries. For instance, after becoming habituated to the robustas grown in their African colonies, which they sometimes consumed in blends with arabica beans, France only began consuming 100 percent arabica beans in the 1970s.

A Viennese café, c. 1900.

Einspänner (Viennese coffees) being served at the Landtmann café, Vienna.

■ AUSTRIA, THE CAFÉ WALTZ

An ambassador of Sultan Mehmet IV was responsible for instigating the fashion for coffee in Vienna in 1665. Legend has it that some twenty years later one Franz Georg Kolschitzky started the first café in that city. He apparently got hold of five hundred bags of coffee beans left behind by the Turks when their siege of Vienna failed. No other city has made the café into a more important social and cultural institution. In the heyday of Vienna café culture before World War I, there were more than six hundred such establishments. Austrians, who consume more than twice as much coffee as Italians,* can go to one of their fine cafés and have an excellent black coffee called *Schwarz* (black) or *Moka* (Mocha*). These are served on a silver platter, accompanied by a glass of water. A *Moka* may also be called an *Espresso*. Austrians are also fond of coffee with milk:* there is the *Melange*, with equal parts coffee and milk; the *Kapuziner*, to which a dab of cream is added; the *Verkehrt*, where the milk's whiteness is tinged slightly with a hint of arabica; the *Brauner*, which in contrast has only the lightest cloud of whiteness; and finally the *Schale Gold*, a beautiful gold colored mixture of the two. Other great Viennese specialties include coffees with whipped cream: an *Einspänner*, also known as Viennese coffee or *café viennois,* is served in a glass and crowned with whipped cream. *Eiskaffee* is cold black coffee poured over scoops of vanilla ice cream and topped with whipped cream. It is similar to an icecream sundae.

■ BEDOUIN: The Coffee Ceremony

Turkish* coffee never made its way into the desert or the sandy coasts of North Africa and the Arabian peninsula. From Morocco to Palestine, the Bedouins, nomadic or settled, have a unique way of preparing coffee in an elaborate ceremony conducted in tents. Green coffee beans are roasted* over a fire in a grill that resembles a frying pan, then ground in a mortar. The sound and smell draw friends. Three tin-lined copper coffeepots are needed, similar in shape but with different sizes and names. Grounds remain constantly steeping in the biggest pot, which stays on the fire. The brown liquid is poured from this pot into the medium-sized pot, with the grounds held back by a dried grass filter. When the water in the second pot reaches a boil, freshly ground coffee is added and boiling continues for ten minutes. After the grounds have settled, the coffee is poured into the smallest pot with a little cardamom and saffron. It is brought once again to a boil, allowed to settle, and served.

Certain gestures are part of the code. An overturned cup is the way to ask for another serving. Shaking the cup means no more is wanted. The cup is given back as soon as it is emptied, to be rinsed and reused immediately. Although today the coffee is often bought roasted and ground, the ritual still transforms a cup of coffee into an instrument of communal harmony and hospitality.

Syrian man preparing coffee, c. 1935.

■ BEST BEANS

Besides Blue Mountain,* the superior beans of Latin America (see "Brazil" and "Colombia"), Mochas* and the wonders of Hawaii (see "Kona") and East Africa (see "Kenya"), there are exceptional coffees grown in India. These include the amazing "monsoon" varieties Mysore and Malabar, full-bodied, slightly spicy and tangy gems which acquire their qualities from having been exposed to monsoon winds for at least six weeks. There are also excellent arabica plantations in Indonesia, whose beans generally have a full-bodied taste. In Java and Sumatra, a few plantations age beans for several years to sweeten them. Excellent beans come from Sulawesi, of which Kalossi is the best known.

In Central America, besides Costa Rica,* Guatemala produces a fine, strong, fruity coffee. It is sometimes a little sharp or has a slight chocolatey taste. The most famous are Antigua and Coban. Mexico produces the finest examples of a variety of bean that originated in Brazil, the Maragogype, which is typically twice the size of other arabica beans. Beans from the Liquidambar label yield a very smooth and aromatic brew. Unfortunately, Nicaragua Maragogype beans, such as Matagalpa and Jinotega, are difficult to come by, but yield more full-bodied coffee than Liquidambar. Pacamara is a hybrid of Maragogype with another arabica, the Paca bean. This large bean is blue-green in color, has a more subtle flavor and scent, and should not be over-roasted.*

This is not an exhaustive list of the best beans. Coffee quality is highly dependent on human factors. More than once, great beans have hit the market as a result of the arrival of a conscientious and impassioned grower in a region that had until then produced nothing particularly remarkable.

Arabicas from Kenya and Central America.

■ BLENDING: In Search of Harmony

Today, the majority of commercially available coffees are blends of beans of different origins. When roasters* blend beans, they are like wine makers looking for the perfect combination of grapes. The resulting cup of coffee is ideally a well-balanced mix of body (the liquid's density and its staying power in the mouth), aroma, and flavor. Although the exact recipe for each blend is a jealously guarded secret, it is based on a few basic principles. First, balance is achieved by combining milder with fuller-bodied beans, acidulous with more chocolatey or fruity ones, and letting one characteristic dominate the others for particular results. The oldest and formerly most common blend combines the subtle softness of Mocha* with the strength of Java beans.

Bean quality varies with harvest, so blends must be reconsidered on a yearly basis. Consistent quality is one of the roaster's chief responsibilities.

Mocha-Java, Mocha-Java-Costa Rica, and Napoleon's favorite, Brazil Santos-Mocha, are classic arabica blends. French author Balzac's favorite (see "Roasters"), Mocha-Bourbon-Martinique, can no longer be tried since the last two beans have become virtually unavailable. In the early twentieth century, Ali-Bab's *Gastronomie Pratique,* the gourmet's bible, recommends: "If you cannot obtain coffee beans from the islands of Bourbon and Martinique, which have become very rare, use those of Puerto Rico and Saint-Marc, carefully blended."

Wacker's coffee roasting store in Frankfurt, Germany

Boiling

Boiling was the only known method of coffee preparation until the eighteenth century. One part of the world where it continues is Scandinavia, but the jewel of boiled coffee is Turkish* coffee. This is how it is prepared in Istanbul: one small glass of water per cup of coffee is poured into a small metal crucible with a long handle called a *cezve* (pronounced "jezvay"). Two teaspoons of coffee ground as fine as flour and lightly roasted* are then added, plus as much sugar as desired (one teaspoon of sugar per cup is typical). This is then placed over medium heat. When it begins to bubble, the mixture is stirred. Then a portion of the liquid is poured into a cup and the rest is returned to the heat and brought to a boil again. Finally, the liquid, together with the grounds, are poured into the cup.

In Greece, coffee is brought to a boil three times. In Lebanon, the coffee is more heavily roasted. In Egypt and some other Arab countries, the coffee is flavored with cardamom. In all cases one should wait a couple of minutes for the grinds to settle before drinking.

Cezve used in the preparation of Turkish coffee.

33

■ BLUE MOUNTAIN

The Blue Mountain beans of Jamaica are sometimes called the caviar of coffees. They are some of the best* and most expensive in the world. And in some ways Blue Mountain deserves its reputation. For it has excellent flavor and aroma, it is very mild, agreeably tart and chocolatey. The excellent growing conditions of Jamaica's Blue Mountains are the direct causes of this. The east of the tropical island is caressed by warm humid breezes, with mountains no higher than 7,217 feet (2,200 meters) whose excellent soil is terraced and shaded by avocado and banana trees. Mountain streams provide perfect irrigation for the arabica plants, which were brought over from Martinique in the eighteenth century.

But these conditions alone did not earn Blue Mountain the reputation, for some, of best bean in the world—a bean whose price, given the small quantities produced (a few hundred tons a year), has reached dizzy heights. Its status was the result of shrewd marketing and its popularity with Japanese consumers at the end

A Blue Mountain plantation, Jamaica.

of the 1960s. In any event, thanks to the influence of the official Coffee Industry Board, extreme care is given to the processing* of Blue Mountain coffee, from picking to packaging. Unlike other coffees, which are transported in burlap sacks, the slightly bluish green beans are packed in wooden casks. The appellation "Blue Mountain" is granted only to the highest quality beans from a very small, high altitude region (Portland, Saint Thomas and Saint Andrews provinces). The next highest qualities are the appellations High Mountain Supreme and Prime Washed Jamaica. But beware, there are also many appellations containing the words "blue" and "mountain" which may come from anywhere at all.

■ BRAZIL

In a typical year, Brazil yields 20 million sacks (132 lbs. or 60 kilos each) of coffee, making it by far the largest producer in the world. It all started in 1727, when the Portuguese governor of Pará (contemporary Belem) sent the sprightly Lieutenant Palheta on a diplomatic mission to the French governor in Cayenne. The pretext was negotiating a border conflict, but Palheta's real mission was to steal away some cultivatable coffee beans from under the watchful eyes of the French authorities. Legend has it that the young lieutenant succeeded after having seduced the governor's wife. Upon his return to Pará, the seeds were planted and the first Brazilian coffee began to grow.

At the beginning of the twentieth century, Brazil provided the world with 90 percent of its coffee. Slavery* was essential in this success story, along with soil quality and climate. The initial deforestation efforts necessary for creating Brazil's vast

fazendas (plantations*) out of the jungle was a gargantuan task. Some pioneering *fazendeiros* became historical figures. For instance Martinico Prado discovered *terra rôxa* (purple earth) in 1877. The soil in the Ribeirão Preto region is rich in basalt, extremely fertile and excellent for the cultivation* of coffee. Upon his death in 1912, his Fazenda Guatapara occupied 15,000 hectares with almost two million coffee plants.

Brazil mainly produces arabica* coffee. Those plantations that produce inferior grades use mechanical harvesting methods. The best* Brazilian beans from Bahia, Santos and above all from Sul de Minas are mild, well-balanced arabicas with fine, pure aromas. The Rio de Janeiro region abundantly produces a very particular kind of coffee called *riote*. It is harsh, bitter, highly iodized and even salty. Most connoisseurs consider it the worst coffee in the world. But some people love it.

■ BRITAIN AND AMERICA

By the end of the seventeenth century, thanks to the great success of coffeehouses,* England was the European country where the most coffee was consumed. A century later, consumption more or less equaled that of tea. While it is certainly possible to find a good cup of coffee and excellent coffeeshops* in Great Britain, a large majority of people still drink instant* coffee. On the other side of the Atlantic, the colonies of the New World were drinking coffee at least as early as 1668. Documents indicate that the beverage was being consumed in New York at that time, sweetened with sugar and flavored with cinnamon. As a symbol of independence from the British, coffee soon became America's favorite beverage. In the 1920s, Prohibition also contributed to an increase in coffee consumption.* But instant coffee made its way into people's kitchens, and its practicality won over millions of Americans.

Walter Pidgeon, Greer Garson, and Donna Corcoran enjoy a coffee during a break from filming.

Among the instant and robusta* coffees served in the United States at that time, it was rare to find a good cup of coffee in the 1960s. But things have changed since the arrival of Starbucks in the 1970s. With Starbucks came a taste for espresso,* cappuccino,* and good coffee in general. Coffee flavored with vanilla, hazelnut, caramel, chocolate, mint and other syrups appeared on the west coast and became popular throughout the land. Some say its principal merit lies in attracting young people away from soda pops and other beverages. Of course, classic iced coffee is served everywhere.

Café Life

The first places that can legitimately be called cafés opened in Mecca, Cairo and Istanbul in the sixteenth century. Prices were moderate and the atmosphere was lively, with animated discussion, backgammon and chess, recitals of poetry, and the essential café function of being a meeting place for relaxation or business. The first European coffeehouses* in Europe date to the 1670s, about twenty years after the beverage first appeared there. These establishments were often owned by Armenians or Syrians. Caffè Florian in Venice, Procope in Paris, and Demel's in Vienna are among the most famous of this first generation, where aristocrats and members of the bourgeoisie, artists and intellectuals, would meet. In France,* as in other countries, important political events took place under their roofs. It was from the Café du Foy in Paris that Camille Desmoulins urged an angry

Caffè Florian, Piazza San Marco, Venice.

Terrace at the Café de Flore, boulevard Saint-Germain, Paris, 1949.

crowd to take up arms at the start of the French Revolution. In the twentieth century, Parisian cafés were famous for being frequented by artists and intellectuals. The Parisian cafés around Montparnasse, Le Dôme, La Rotonde and La Coupole saw the likes of Guillaume Apollinaire, André Breton, John Dos Passos, Ernest Hemmingway, Pablo Picasso, Amedeo Modigliani and Marc Chagall. After World War II, such figures of the city's bohemian nightlife as Juliette Gréco and Boris Vian, discovering jazz, *joie de vivre* and the

philosophy of existentialism, would start their evenings at the Café de Flore or Les Deux Magots. Simone de Beauvoir and Jean-Paul Sartre could be seen at all hours of the day in Les Deux Magots, keeping warm, writing or holding court.

■ Caffeine

Awareness of the stimulating effects of coffee goes back a long way. They are the result of the presence of caffeine, an alkaloid (trimethylxantine 1-3-7), which is only dangerous to health* at very high doses. There are several ways of reducing your intake of caffeine without resorting to decaffeinated* coffee. Firstly, choose arabica* beans, which contain half as much caffeine as robustas*—an average of 1 percent as compared to 2 percent for the latter. Secondly, drink espresso* in preference to filtered coffee, which is slightly more caffeinated. This is because a small amount of water passes so quickly over the grinds in the making of espresso that less caffeine gets dissolved. The least caffeine is obtained in what might seem to have the most, a short single shot or *serré* espresso, which uses the least water. If you find one of these too thick and full-bodied, just add hot water.

Cappuccino

Cappuccino is an international institution. The name "cappuccino" derives from the similarity of its color to the hoods worn by Capuchin monks. It is basically an espresso* to which pressure-steamed milk* is added. Real Italian cappuccino requires that the milk be well frothed, and the foamy top is sometimes even softly modeled into an appealing shape before being dusted lightly with cocoa. A pleasure for the eyes even before it is tasted, cappuccino is certainly the best way to add milk to coffee. Certainly cream offers its own set of pleasures, for instance in traditional Italian* *caffè con panna,* and Irish* coffee. The French version of the cappuccino is known as a *café crème,* a *petit crème* or a *noisette,* not to be confused with the Austrian* *Kapuziner,* to which real cream is added.

Cappuccino in Stresa, on Lake Maggiore, with foam shaped like a peacock's fan.

CARACOLI
A Mysterious Concentration

Just as some great wines are the result of late-season grape-picking, some best* coffee beans find their qualities doubled as a result of a botanical mystery. Usually each coffee cherry contains two beans (see "Coffee Plant"). When there is only one, it is called a "caracoli." These beans (whose name derives from *caracol,* Spanish for snail) are round rather than oblong, and are usually found near the ends of branches. Simple anomaly, insufficient pollination, genetic error? No one knows. But one thing is certain to connoisseurs—not having two beans creates a doubly tasty result. In Brazil* and Colombia* they are vaunted as aphrodisiacs. They were highly appreciated in Brittany, France, at the turn of the last century, where they were home-roasted at the hearth. In Germany they are known as "pearls."

Caravans

Until the eighteenth century, coffee was grown only in Yemen and Ethiopia.* Merchants from the Middle East would buy beans at the large Bayt al-Faqih market. Bayt al-Faqih was one of Yemen's main production areas and was located near the port* of

Sahara Caravan Map, 1869. Maritime Museum, Rotterdam.

Hodeidah. Jean de La Roque* described the chain of transactions that began there in his *Voyage to Arabia the Happy*. His description of 1716 is based on the account of a contemporary merchant voyage that left from Brittany. "Coffee is purchased for the whole of Turkey at Batalfaguy, where all the Egyptian and Turkish merchants come for that purpose. They load a large quantity on camels, which each carry two bales, weighing about two hundred and seventy pounds each, and head to a small port on the Red Sea, some two leagues away. There they load it on to small ships that carry it sixty leagues further along the gulf to another, larger port named Jeddah or Zieden, the port serving Mecca. From there the coffee is again loaded on to Turkish vessels, which take it to the furthest port, at the end of the Red Sea, Suez. From there it is loaded on to camels and carried to Egypt and other provinces of the Turkish Empire, by various caravans or across the Mediterranean Sea." Sixteenth and seventeenth-century merchant sailors and camel drivers from Egypt and Yemen were the harbingers of the beverage that would soon be consumed worldwide.

■ Chicory

Chicory is a herbaceous plant. It is related to the endive and has a not dissimilar flavor. It has been used as a dried and roasted coffee substitute. After Napoleon decreed its consumption during trade blockades, the French developed a taste for it. It is still frequently blended with coffee in northern France for its slightly bitter aftertaste. As with robusta* beans, which were consumed far more widely in France than the superior arabica beans, the predilection for chicory helps to explain the general ignorance about, and indifference to,

Jan van Huchtenburgh (1647–1733), *Landscape with Turkish Caravan.*

quality coffee on the part of the French. In Italy* it is not chicory but roasted barley that has been a traditional coffee supplement. These days this kind of blend, called *caffè d'orzo* has made a resurgence as a means of reducing caffeine intake. The slightly bitter and burnt taste characteristic of coffee during the shortages of World War II is back—but for new reasons.

■ Chocolate

Coffee and chocolate arrived in Europe at the same time and have always been wonderful together. There are many formulas and recipes for adding chocolate to coffee and coffee to chocolate, because the mixture produces one of the most delicious combinations invented by humankind. Cocoa has been used to flavor coffee in Italy* for

a long time. Cappuccino* is lightly dusted with it and in Turin a *bicerin* is a delicious mixture of equal parts coffee, chocolate and cream. The Italians also invented a cream-filled chocolate candy that is made specifically to be dissolved in coffee. In the United States, following a trend for flavored coffee that began in California, chocolate is one of the most frequently added flavorings. Yet, everywhere in the world, coffee-flavored chocolates are more common than chocolate-flavored coffees. Probably the most intense combination can be found at superior chocolate makers, who dip deliciously crunchy individually roasted arabica* beans in their finest chocolate, marrying tastes, materials and textures in an unforgettable mouthful of pure pleasure.

■ Coffee Exchange

The activity of coffee trading on the international market is second only to petroleum. It is subject to great oscillations and changes. Until 1960, Le Havre in France housed one of the most important exchanges. It no longer exists, and between 1962 and 1969 the International

Coffee Organization used import and export quotas to regulate the coffee exchanges in New York (arabicas) and London (robustas). Basically, if prices dropped, each country's exportation quotas were reduced as well. Since this agreement was not renewed in 1989, coffee prices have been fluctuating dramatically due to climatic and political conditions, as well as variations in consumer* demand. For example, arabicas were at $1,000 a ton at the end of 1993, but were selling for four times that a year later, after Brazilian plants suffered from freezing temperatures. Then, in July 1995, when prices were back at $1,400 a ton, all producers in South America concluded an agreement to limit exportations.

Harvest volume, supply and demand, and speculation are all important factors in creating price variations. Price fluctuations have serious repercussions in countries whose economies are largely dependent on coffee.

Prices at the Santos Exchange (Brazil) in June 1988.

45

Coffeehouses

The first English coffeehouse opened in Oxford in 1650. Students were the primary clientele. Two years later, the first London coffeehouse opened in Cornhill on Saint Michael's Alley. By the end of the seventeenth century there were hundreds of these establishments. The Grecian was one of the most famous intellectual coffeehouses, which were known as "penny universities" because a penny was the minimum charge. Two of these coffeehouses became important literary* hubs, filled with writers and editors. The writers Joseph Addison, John Dryden, Samuel Pepys, Alexander Pope and Jonathan Swift frequented Will's. In the second decade of the eighteenth century, the poet Daniel Button opened a coffeehouse bearing his name. It attracted the literary elite to his establishment, especially members of the Whig party. But around 1730, the heyday of the English coffeehouse came to an end for two reasons. The penny-a-seat policy diminished the aristocratic atmosphere, and the great era of English tea drinking began. The British East India Company had fewer dealings in coffee and chocolate than its Dutch and French counterparts, and, with strong encouragement from governmental quarters, turned to China for its trade. The nineteenth century saw the uncontested reign of British tea interests in the Indies.

Coffee Plant

Both of the main varieties of coffee plant belong to the Rubiaceæ family of the order Rubiales. There are about a dozen varieties of *Coffea arabica** and fewer varieties of *Coffea canephora* (var. *robusta**).

Arabica originated in Ethiopia.* It is a large shrub or tree with shiny evergreen leaves, attaining heights of thirty feet or more. Plants bloom three times a year, producing red or white flowers that smell like jasmine. Eight months or so after flowering, the plants produce red berries known as coffee cherries. Each cherry contains two light green oval seeds which, after processing* and roasting,* become beans. Two arabica varieties are sold commercially by their scientific names: Maragogype, which yields very large beans, and Bourbon, a variety introduced to Brazil* from Reunion (then called Bourbon Island).

Coffea canephora bushes are resistant to both insects and disease—hence its name, robusta. It grows to a height of about forty-five feet and its beans are

Anonymous painting of a coffeehouse, 1668. British Museum, London.

Coffea arabica.

47

Traditional French enamel coffeepots. Private Collection.

usually smaller and thinner than those of arabica plants. It is native to tropical western Africa and grows at lower altitudes. Another variety of *canephora*, called Kouillou, is grown in Madagascar.

■ Coffeepot

From the Bedouin copper boiler method to the most expensive espresso* machines, dozens of kinds of coffeepots have been invented over the centuries. After steeping* (the French press method is its latest descendent), the first great advance took place at the turn of the nineteenth century, when Jean-Baptiste Belloy invented the first pressure* percolation coffeepot. The Belloy model was a great success and was imitated, sometimes with variations, around the world. Traditional French enamel coffeepots follow this principle.

So do Italian stove-top models, also invented in France, but which became popular after World War II in Naples. The system is simple and ingenious. The pot is placed over heat with a bottom reservoir filled with water and the top piece acting as the decanter. When the water boils, the coffee is ready, prepared by the basic laws of gravity. Gravity is also the principle behind filtered drip and electric drip coffee making. Pressure* is another important principle that is variously applied. The most elegant and magical method is probably the Cona, with its two glass globes and burning wick. The Italian Moka Express is a well-known stove-top model made by Bialetti. The basic principle in all espresso* machines is high pressure.*

■ COFFEE SETS

Istanbul was the first place where coffee service sets became works of art in their own right. The Topkapi Museum's outstanding collection includes beautifully engraved coffeepots made of *tombak,* a fine alloy of copper and zinc, with matching ceramic cups from Iznik or Kutahya, or porcelain ones from China and Europe. These little cups have no handles and resemble egg cups. To protect the drinkers' fingers, they were placed in decorated copper, silver, or gold holders called *zarfs.*

In Europe, the finest porcelain and silver was used to make coffeepots and cups for privileged drinkers. The French king Louis XV shared his love of coffee with his mistress, Madame du Barry. In a sterling silver cylindrical roaster, the king hand-roasted beans grown exclusively for him in the Versailles greenhouses. He then prepared the coffee in one of the three gold coffeepots he commissioned in 1754 and 1755 from the court jeweler Lazare Duvaux.

Today, the enameled tin and glazed earthenware coffeepots of not so long ago are sought after by collectors. Serious coffee lovers sip their brew from simple white porcelain cups, which perfectly complement coffee's temperature and flavor.

■ Coffee Shops

You can, of course, buy coffee in a supermarket. But nothing compares to the delicious aroma of freshly ground superior coffee beans wafting through a coffee shop. Coffee roasters sell beans fresh roasted, whole or ground to order. Thus freshness and the appropriate grind for the type of coffee-maker used by the customer are assured. House blends* and the best* beans are available—along with the expertise of the roaster, with whom you can discuss your needs and preferences: whether you prefer your coffee full-bodied or sweet, whether you're buying coffee for the morning or the afternoon. Often you will be offered samples.

There are some coffee shops whose reputations extend beyond national borders. These include Verlet and Faguais, both in Paris, Higgins in London, Peet's in Berkeley, and Wijs & Zonen in Amsterdam. Most of these establishments have been in existence

for more than a hundred years, maintain solid traditions and know how to innovate with the times as well. Their clients can be demanding, ordering particular blends and degrees of roasting, and frequently have the connoisseurial fondness for discussing origins, merits and preferences for particular beans. Novelist Honoré de Balzac was more than a coffee enthusiast. He drank more than fifty cups of coffee a day and systematically devoted several hours to the purchase of the three kinds of coffee he preferred, which he blended himself. He would buy Bourbon coffee on Paris's right bank, Martinique coffee in the center of the city, and mocha* on the left bank.

Wijs & Zonen coffee store, Amsterdam.

51

■ COLOMBIA

C olombia produces approxi-mately fifteen million bags (154 lbs. or 70 kilos each) of coffee per year, making it the second largest producer in the world after Brazil.* But unlike Brazil, Colombia produces no robusta* coffee. It is the largest single producer* of mild, or what the experts call washed, coffee (see "Pro-cessing"). In this country of thirty-six million inhabitants, coffee has a strong impact on the economy. Fluctuations in prices and harvests have immediate effects on the lives of hundreds of thousands of people. For instance, lower prices can lead directly to increased cocaine cultivation.

Spanish missionaries planted arabica* here in the Andes at the beginning of the nineteenth century. The plant grows at altitudes between 2,600 and 6,650 feet (900 and 2,000 meters). Colombian coffees are among the best in the world, due to geographic, climatic and growing conditions (see "Cultivation"), as well as rigorous processing* policies. Since they are so mellow and smooth, they are used to cut the heavier characteristics of other beans (see "Blends").

The best* beans are Excelso and Supremo. Supremo is comprised of larger beans, and is usually better, with very consistent characteristics. It is smooth, sweet and aromatic; a perfect breakfast coffee. In addition to quality designations, Colombian coffees are also labeled by point of origin: Medelin, Armenía, Nariño, Bogotá, etc.

Colombia's dynamic marketing poli-cies, including advertising and spon-sorship of sporting events, ensure the

A planter's residence,
Colombia.

Transplanting young
coffee bushes.

international reputation of
its product. It also benefits
from some international
trade agreements meant to
decrease the incentive for
cocaine production. One
third of Colombia's coffee
production is purchased
by Germany, which then
serves as a conduit for the
rest of Europe.

▪ Consumption Statistics

Coffee is consumed in every country on the planet. In fact, it is the second most popular beverage after water. Two out of three people in the world are coffee drinkers, for an estimated 4 billion cups a day! Men tend to drink more than women: 1.7 cups as opposed to 1.5 cups on average per day. The USA is the world's largest consumer of coffee, importing 16 to 20 million bags annually (2.5 million pounds), representing one third of all coffee exported. More than half of the United States population consumes coffee typically drinking 3.4 cups of coffee a day. In order of per capita consumption, Finland

and Sweden come first, with 29 lbs. (13 kilos) per inhabitant per year, or the equivalent of five cups a day. In Denmark and Norway it is 26 lbs. (12 kilos) per inhabitant, or four to five cups a day. Holland follows with 20 lbs. (9 kilos), or three cups. Austria, Belgium, and Germany (the number two importer) and Luxembourg consume nearly 18 lbs. (8 kilos), or three cups. France (the number three importer) and Switzerland average 13 lbs. (6 kilos), or two to three cups. A few more statistics: Canadians consume nearly 9 lbs. (4 kilos) per year, Spaniards 6½ lbs. (3 kilos), the English 5½ lbs. (2.5 kilos) and the Japanese (the fourth largest importer) 4½ lbs. (2 kilos).

■ COSTA RICA

With three million bags (of 132 lbs. or 60 kilos) per year, Costa Rica is the eleventh largest producer of coffee in the world. This small Central American country is a paradise for coffee connoisseurs. For instance, the cultivation of robusta* is against the law! Coffee comprises 25 percent of exports and is the single greatest source of income, with a tenth of the population employed full time and a fifth employed during harvest season. School vacations are even arranged in relation to the cycles of the coffee crop.

The first coffee plants cultivated in this country came from Cuba in 1780, with large-scale cultivation beginning some thirty years later. Procedures and quality are subject to rigorous standards, and thus bean quality is excellent (see "Best Beans"), especially in the San José region. Tarrazu (there are about twenty plantations), Tres Rios and Tournon are lush, tasty, well-balanced coffees, in which body and aroma, light acidity and full-bodied flavor are united. But all the coffees of Costa Rica are excellent. There are about three hundred varieties, bearing the names of

plantations or cooperatives (see "Max Havelaar Foundation" and "Routes"). The volcanic soil is ideal for growing coffee. Thanks to fertilization and excellent irrigation, yields in the central valleys are among the highest in the world—about 1.5 tons per hectare. Beans are classified according to the altitude at which they are grown. The highest designation is SHB, for "strictly hard bean." These are grown at between 4,000 and 5,250 feet (1,200 and 1,600 meters). Drink SHBs after breakfast, during the day to fully appreciate their richness. And never pass up the opportunity of tasting Costa Rican caracoli.* These can be found at better quality coffee shops.*

Coffee harvesting, Monte Alegre plantation.

Harvest transport, 1934.

Coffee picking in Yemen.

▪ **Cultivation**

High altitude arabica* cultivation is often a family affair. Plants are first grown in nurseries and then transplanted. Beans begin to appear three or four years later, with significant bean production occurring after fifteen years. Coffee plants have a somewhat unique flowering-ripening cycle. The same plant can both flower and bear fruit (which are called coffee cherries) at the same time, because flowering and ripening are stimulated directly by rain. If it rains all year long, coffee plants will continually yield flowers and cherries. In most coffee producing regions there are one or two rainy seasons per year, each of which lasts for months. Rains falls in October in Kenya,* in April in New Guinea, in August in Jamaica—so that is when both the first cherries appear and harvesting begins.

The better coffee growers harvest only ripe beans, and the harvest season lasts months. It is done by hand and requires up to eight passes per plant. On less quality-conscious plantations, such as those in Brazil,* there is only one pass and all the cherries are picked, whether they are red-ripe or not. This method, known as stripping, is quicker and less costly. On Brazilian coffee plantations situated on plateaus, machines are used that straddle coffee plants and remove all the cherries. None of the best* beans are harvested in this way. Most of the work done in the cultivation of coffee is done by hand and by women, and sometimes by children. After harvesting, the coffee seeds are extracted from the cherries and dried.

■ DECAFFEINATED

There are two traditional ways to remove the caffeine from a green coffee bean. In the first method, beans are softened in steam or hot water to render them porous, before being submerged in a chemical bath that dissolves only caffeine. Then they are rinsed again in steam or hot water before being dried. The second method is to soak the beans in water. The caffeine dissolves in a matter of hours. Since the aromatic qualities of the bean also dissolve, the water is then treated with a chemical that absorbs only the caffeine, and then the beans are put back in the water to reabsorb the aromatic elements. Needless to say, neither method successfully retains the coffee's full aroma and flavor. If a low-quality robusta loses most of its taste in either process, what will remain of a more subtle arabica? At least a new, more effective method using carbon dioxide has been invented, and its application has been spreading. If you have to drink decaffeinated coffee, it is wise to avoid the best* arabica beans, since their particular flavors will be lost. Try 100 percent arabica blends, whose taste can be more or less improved during roasting.* Espresso* is probably the best way to go, because bean quality is less important than for filtered or steeped* coffee. The difference in taste and aroma between a regular espresso and a decaf espresso remain within acceptable limits for most, even those with relatively demanding palates.

Advertisement for Hag brand decaffeinated coffee, Germany, c. 1930.

KAFFEE HAG

An Italian
ristretto
(short espresso).

A Gaggia
espresso
machine.

■ Espresso

The Milanese Achille Gaggia invented the high-pressure espresso method of making coffee in 1948. Espresso differs greatly from all other kinds of coffee because the pressure emulsifies the coffee's oils and colloids. The result is especially full-bodied and aromatic, with an excellent taste that lingers long in the mouth. High-pressure percolation, as it is sometimes called, releases only the bean's smoothest and most aromatic components, and causes a satiny foam to appear on the surface. If the machine is sufficiently powerful (14 bars), well cared for and properly used, and if the beans have been freshly roasted* and finely ground, the result is sure to be delicious. Good espresso is recognizable by the color of its foam. It should be nut brown, with a reddish or brownish tinge depending on the type of coffee employed. The liquid should be opaque, homogeneous, and about an inch and a quarter to an inch and a half deep in an espresso cup. While the full range of a bean's qualities are not extracted in the making of espresso, it nevertheless provides the senses with a uniquely exquisite treat. With its small size and intense taste, it is more like a gourmet delicacy than a simple drink. This is why it is such a fine way to end a meal. It can even be drunk between courses, rather than as a simple morning eye-opener. To intensify the experience while at the same time consuming less caffeine,* order a "short" espresso (with less water).

■ ETHIOPIA AND YEMEN
The World's Oldest Coffees

It is commonly recognized that *Coffea arabica* originated in Ethiopia, and that its consumption began there as well. But it was not until it arrived on the Arabian peninsula, and particularly in Yemen, that its legendary international reputation began to spread. The most popular theory among Arab peoples is that the great Sufi master Ali ben Omar al-Shadili, known as the Saint of Mocha, brought coffee to Yemen from Ethiopia in the fourteenth century. This date was pushed back two centuries in 1997, when archeological excavations near Dubai (United Arab Emirates) uncovered a coffee bean from the twelfth century. The bean apparently came from Yemen.

The inhabitants of the Oromo region in Ethiopia eat the fruit of the coffee plant, which they call *boun*, raw. This is also their word for the plant itself and a drink made by boiling the skin and the lightly roasted pulp of the fruit without the seeds. In Yemen, the same brew is called *qishr*. Ethiopians from the Galla region also make it into an invigorating porridge: the fruit is peeled and pounded into a sort of flour that is mixed with animal fat. Ethiopians also use different sorts of coffee leaves and beans in various states of freshness for medicinal purposes.

Mochas* from Ethiopia are often excellent. Of course, Ethiopians also roast these beans and prepare coffee as a beverage. Ethiopia is one of the only countries in the world where almost as much coffee is consumed as is exported. Families roast their coffee themselves each morning, and the mode of preparation is triple heat steeping, making for a tasty and mild brew.

Terraced coffee plantation, Yemen.

Etymology

It was on the coast of the Red Sea in Yemen, from where it traveled to the rest of the world, that coffee acquired the Arabic name *qahwa*. There are dozens of derivitaves of this term, from *Kaffee* in German and *café* in French, to *koffie* in Dutch, *caffè* in Italian, *kafeo* in Greek, *quevay* in Persian, *kawa* in Polish, and *kahveh* in Turkish. In the language of ancient Arabic poetry, *qahweh* designated another beverage that also has its effects on mental acuity: wine. The first coffee drinkers in Yemen were mystical Sufis. They called the drink *qahweh* in honor of its capacity to help them keep alert during liturgical practices. It was dispensed by the master from a large red earthenware receptacle, usually accompanied by a prayer in which the near-homonym *qawi* ("powerful," one of the names for Allah), is repeated sixteen times.

Expansion

The first stage in the spread of coffee throughout the world was its crossing of the Red Sea, having traveled from its native Ethiopia* to Yemen. Archeological evidence places this occurrence in the twelfth

Coffee Cultivation, Bourbon Island. Watercolor. Attributed to Patu de Rosemond, first half of the eighteenth century. Musée des Arts Africains et Océaniens, Paris.

century. Coffee was cultivated in the Yemeni highlands and plants were closely guarded. To prevent foreign germination, only roasted or boiled beans were sold. In 1658, however, the Dutch began coffee cultivation on Ceylon (Sri Lanka), and then in southern India, with plants they had pilfered from Yemen some forty years earlier and had been nurturing in the Amsterdam Botanical Gardens in the interim. In 1696, plants from India were brought to Batavia (Jakarta), where they became the foundation of Dutch plantations on the Indonesian archipelago. In the eighteenth century the French took their turn, initiating coffee cultivation in the West* Indies and Guyana. Their coffee plants* also derived from the Amsterdam greenhouse, and were nurtured and guarded from the 1720s on. Brazilian* coffee growing began at the same time. Its plants were purloined from French Guyana! At the end of the eighteenth century, Spanish Jesuits had plantations in Colombia,* Central America and the Philippines, and the English were growing coffee in Jamaica. All of these plants were arabicas.* Meanwhile, the

Plantation buildings in Java, c. 1900

Portuguese had discovered indigenous robusta* plants in what is currently Angola. They began cultivating this taller, more vigorous type of plant, which soon came to be called *Coffea canephora*.

■ Extracts and Liqueurs

Ice cream makers and pastry chefs have been using coffee extracts for a long time. Natural extracts are simply very concentrated decoctions which can even be added to a cup of warm milk.* They have a long shelf life. Coffee liqueurs are used in cocktails. For example, Kahlua is made with quality Mexican arabica,* sugar cane alcohol, vanilla and caramel. Its alcohol content is not too strong (26 percent), and it can be drunk alone, as an essential ingredient in classic cocktails like a Black Russian (Kahlua and vodka), or even enjoyed with milk and ice. Of course, no artificial coffee flavoring is any match for the real thing.

Fortune Telling

The past is so full of mystery that it is not known when the practice of reading the future in coffee grinds began. But since the urge to decipher the brown marks left by or in the grinds seems so natural, such practices are probably as old as coffee drinking itself. In fact Turkish* coffee, whose method of preparation leads to grinds in the cup, is the only coffee making it possible to find out what will happen in the future. In the nineteenth century, a particular kind of coffee reading or divination by coffee grounds was quite popular. People would turn their cups over on their saucers to let the grinds slide along the sides of the cup, leaving interpretable shapes and patterns. A cross would mean that one should look after one's health; flames, that one should not trust one's initial impressions; a butterfly, that an enemy was near by. Horse-shaped splotches indicate excellent possibilities for love and a fish connotes a once-in-a-lifetime opportunity.

◼ FRANCE

When Soliman Aga, the ambassador of Sultan Mehmet IV, arrived in Paris in 1669, "Turqueries" of all sorts, including coffee, became all the rage in the city. In 1672, an Armenian named Pascal opened the first Parisian coffee shop among the other establishments at the Foire Saint-Germain. By then a coffee shop had already been in business for a year in Marseilles, and people there had been enjoying coffee at home for more than a decade. Thanks to large-scale colonial production which kept prices relatively low, coffee gained popularity outside of the upper classes, and by the 1750s it had pretty much replaced soup in the morning meal. In the nineteenth century, workers drank it to warm up and to renew their energy, as Émile Zola's *Germinal* makes clear.

There are approximately 70,000 cafés* in France today, frequented by some five million people a day for a cup of coffee, also called *café* in French, and usually in the form of an espresso.* More and more people make Italian-style espresso and cappuccino* at home. Sadly, France has suffered from some historical handicaps in relation to coffee: the adoption of a Middle Eastern tradition of nondescript and generally over-roasted coffee, corruption of the palate resulting from the use of chicory,* and the widespread consumption of inferior robusta* beans cultivated in its colonies from the turn of the twentieth century until the 1960s. But times change. More and more quality coffee shops* where better beans are sold open every year, and large commercial companies increasingly sell better beans and blends.

Above: Parisian coffee vendor at Les Halles, c. 1900.

Left: Hippolyte Benjamin Adam, *Coffee Vendor at the Corner of the Porte Saint-Denis*, 1830. Watercolor. Musée Carnavalet, Paris.

■ Grinds

The beauty of good coffee, with its full flavor potential, is realized only if the beans are ground immediately before brewing. Once roasted and ground, coffee beans begin to oxidize, rapidly turning stale and bitter. State of the art packing and refrigeration retard the process, but can never prevent it. In the olden days, coffee was ground with a mortar and pestle, or in a wooden mill with a drawer at the bottom and a crank at the top. Today, electric grinders do the job quickly. Some do it far better than others, though. Grinders with rotating blades should not be used. Only mills with burr grinders make uniformly ground coffee. They cost more, but they're worth it.

Different types of coffeepots require different types of grind. Extra fine ground coffee, with the consistency of flour, is used for Turkish* coffee; very fine for espresso;* fine for steam pressure Italian and electric paper filter* pots, medium for traditional, Neapolitan, and Cona vacuum coffeepots, and coarse for plunger pots. The right grind can only be obtained with the adjustable burr grinding wheel of a better mill. The next best thing is to have the beans ground at the roaster, whose grinder can be regulated for a dozen types of grind. Pre-ground is preferable to fresh poorly ground coffee.

■ Health

Some coffee haters, including the London women who signed a 1674 petition against it, have claimed it leads to impotence (see "Opponents"). Some coffee lovers have sworn it cures scurvy, smallpox, and gout. Today, the properties of caffeine* can be more scientifically gauged. While its effects

Preceding double page: A blend of arabicas from Kenya, Central America, and Ethiopia, in beans and ground, roasted in (from top to bottom) "medium," "European," and "Italian" styles.

on sleep vary according to individuals, it is known to stimulate the nervous system, acting on both mental and physical functioning.

send a determined abuser straight to coffee heaven.

■ Iced Coffee

Cold temperatures have an excellent effect on the taste of coffee, and there is nothing more refreshing on a summer afternoon. In Italy* coffee is drunk after being poured over stacked ice cubes, or as a *granita di caffè*, where it is served cold with shaved ice, and sometimes topped with *panna* (whipped cream*).

Title page of Nicolas de Blegny's *Le Bon usage du thé, du caffè et du chocolat*, 1687.

Caffeine affects the cardiovascular system, and is used to combat migraine headaches as well as certain forms of heart disease. It is a diuretic and speeds digestion, and is therefore recommended for some urinary and digestive problems.

Excessive coffee consumption can lead to more or less serious nervous system disturbance, from insomnia, irritability and anxiety to arrhythmia. Radically over-doing consumption can cause real damage. Ten grams of caffeine, the equivalent of one hundred cups of coffee, could

Even in winter Sicilians usually have a *granita di caffè con panna* with a roll for breakfast. You will also find the exquisitely refreshing *caffè freddo shakereto* in Italy. A sweetened espresso is combined with crushed ice in a shaker and served in a large glass. Basic iced coffee is an easier affair. Simply prepare coffee as you usually would, sweeten as much you like and keep it in a sealed container in the refrigerator. Remember that if you are going to add ice you should use less water when you make the coffee.

Iced coffees, served covered with damp napkins.

▇ Insomnia

Coffee's reputation as a stimulant is due to its caffeine* content, praised or damned by different people at different times. A little coffee may enhance mental and physical activity, while too much may lead to sleeplessness. While an exceptionally high quantity of coffee can be dangerous, a normal amount's effect usually varies according to the individual. In *The Philosopher in the Kitchen* (1825), Anthelme Brillat-Savarin claims that two cups of coffee kept him awake for forty hours. But there are people who drink three cups after dinner and sleep like babies. Even doctors and chemists are stumped by the evidence. For those given to insomnia, decaffeinated* coffee is a wise choice. So is moderation in all things, coffee included.

▇ Instant Coffee

In 1899, a coffee importer and a roaster got together and asked the Japanese inventor of instant tea, Sartori Kato, to apply his dehydration methods to coffee. Along with an American chemist, the team founded the Kato Coffee Company and went to work. Their instant coffee was sold to the public two years later, at the Pan American Exposition in Buffalo. Freeze-drying, invented in 1965, came as a great improvement. Coffee is first prepared, then frozen, pulverized, distilled by heat in a vacuum, and cooled. This process transforms the coffee into flakes. Today, to give new impetus to a market which is stagnating because of the spread of electric coffeepots, manufacturers have been improving the flavor and aroma of their products, offering for example 100 percent arabica* versions, versions with milk,* and cappuccino* versions. However, it is difficult to imagine them coming up with something that would compare with the taste of traditionally brewed coffee. The invention has at least introduced millions of people the world over to coffee, thereby enabling many coffee-producing countries to find new openings.

Freeze-dried instant coffee.

▇ Invention

Coffee grows wild on the high, cold, rainy Ethiopian plateaus, where the drink was invented. It has been keeping this region's nomadic shepherds and peasants awake for perhaps thousands of years. *Boun,* the fresh fruit of the coffee bush, is still used here for a

Yemini woman drinking *qishr,* brewed from the lightly roasted fruit of the coffee bush without the bean.

variety of potions, concoctions, and medicinal purposes. How did the custom of roasting the beans get started? It is easy to imagine a bush fire suddenly filling the air with an irresistible aroma which the people decided to reproduce. It was on the other side of the Red Sea, in Yemen, that coffee was first grown in plantations.* Recent archeological evidence dates this development to the twelfth century at the latest.

■ IRISH COFFEE
Whiskey, cream, and good coffee

One of the world's favorite spiked coffees is Irish coffee (see "Alcohol"). It was invented right after World War II by a bartender at Shannon Airport. The standard recipe calls for

two parts strong black coffee to one part whiskey. Among twenty or so brands, some experts recommend Paddy for its "bite." Add brown sugar and top with thick whipped cream.* While it requires good coffee and top-drawer whiskey, Irish Coffee's secret is in the cream, which is gently slipped into the glass on the back of a teaspoon to form a delicate surface layer. The Irish claim that their green grass and superior dairy cows produce the best, incomparably smooth, cream.

The consumption of Irish coffee is as fine an art as its preparation. The coffee and cream must be savored at the same time, not one after the other, by passing the coffee through the cream. Mixing them with a spoon would be sacrilege.

Following double page: School of Pietro Longhi, *The Coffee Shop.*

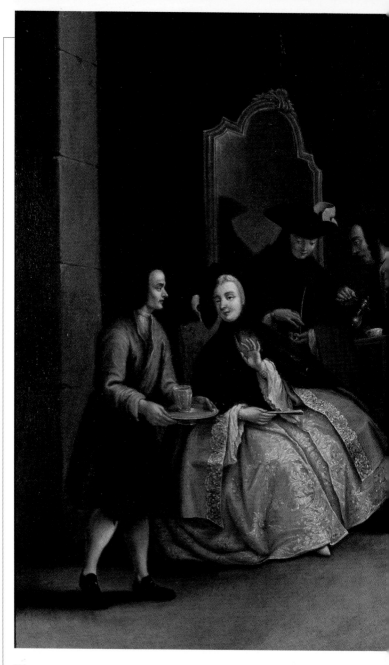

◼ ITALY: The Great Art of Espresso

Venetian merchants brought coffee to Italy around 1600. The success of the first *bottega del caffè* (coffee shop) in Venice in 1683 led dozens of others to follow suit. The elite met in luxurious cafés, such as Florian's in Venice and Greco in Rome. Today, the best way to get a feel for Italian life is to stop at one of the myriad bars where customers stand at the counter drinking espresso. The beans are more heavily roasted the farther south you go, but the coffee is universally good. Orders are for *ristretto*, made with just enough water to cover the bottom of the cup with dark coffee, regular *espresso*, which

fills it halfway, or *lungo*, made with more water for a diluted espresso that reaches the brim. Some coffee lovers order a *doppio*, or double, made with fourteen grams of coffee and the regular amount of water.

In addition to black coffee and the delicious cappuccino, there is also *caffè macchiato* ("spotted coffee"), an espresso with a couple of drops of milk, and its opposite, the *latte macchiato* ("spotted milk"), milk with a splash of coffee. *Caffè con panna* is topped with a bit of whipped cream.* Other Italian specialties include *caffè corretto*, spiked* with grappa, and *granita di caffè* (see "Iced Coffee").

■ KENYA AND EAST AFRICA

One of the world's greatest coffees grows in Kenya, at an altitude of nearly two thousand feet. This coffee is prized for its tangy flavor, its staying power on the palate, and its uncommonly elegant green fruit taste. The best Kenya beans are the biggest, classed AA. The rare AA+ beans are exorbi-tant. Kenya coffee is perfect for spring and summer, mild yet invigo-rating. Between 1914 and 1931, the Danish writer Karen Blixen owned a coffee plantation* of five hundred acres north of Nairobi. "In the wild-ness and irregularity of the country, a piece of land laid out and planted according to rule looked very

well," she wrote in her 1937 novel *Out of Africa*. "I was filled with admiration for my coffee-plantation that lay quite bright green in the grey-green land, and I realized how keenly the human mind yearns for geometrical figures."

After independence, many formerly European-run plantations were split up and redistributed to landless peasants. With technical assistance from the government, they became skilled coffee growers in their own right, producing the distinctively tangy beans loved by connoisseurs. Other east African countries produce excellent, rare coffees. Burundi coffee is full and mild, Zimbabwe has a hint of citrus and spices, and Tanzanian beans, grown on the slopes of Mount Kilimanjaro, are similar to Kenya's, but slightly milder.

Coffee plantation in Kenya.

■ KONA

Coffee bushes planted on the flanks of a volcano, Hawaii

Kona is the only coffee produced in the United States. This Hawaiian bean* is one of the most coveted in the world. The name Kona derives from the region where the plantations are located. The beans derive from Brazilian and Guatemalan arabica* plants brought over in 1818 and 1829, respectively. Some 650 farms cultivate the coffee, all located on the slopes of the Mauna Loa Volcano. Each farm is approximately two hectares large, and they are situated along what is known as the twenty mile "coffee belt" at altitudes between 820 and 2,460 feet (250–750 meters) on the 13,000-foot (4,000-meter) volcano. At times production has been interrupted by volcanic eruptions and falling ash. But nature usually cooperates and the rich volcanic soil produces gorgeous green coffee beans, which upon roasting* yield a smooth, aromatic, slightly acidulous and peppery coffee that is ideal for drinking in the evening. The quality is truly perfect and many consider it the best in the world. The only problem with Kona coffee is its price: production is only 500 tons per year, the Japanese and Philippine labor employed is expensive, demand by American tourists is strong, and exportation is minimal.

Kona is the only place in Hawaii where coffee is grown. All beans bear the Kona appellation, and coffee made with Kona beans is always excellent. But there is in fact a hierarchy depending on the size of the beans and the number of flawed beans in the bag. Kona Extra Fancy has the largest beans and fewest flawed beans, followed by Kona Fancy and Kona Prime. In any case, the coffee is so good that milk and sugar need not be added.

La Roque, Jean de

Jean de La Roque was the son of Pierre de La Roque, a merchant and traveler who introduced coffee to Marseilles' elite. His *Voyage de l'Arabie heureuse* (Voyage to Arabia the Happy) describes the first two expeditions to Mocha sponsored by French merchants (1708–10 and 1711–13).

The popularity of this chronicle inspired readers to visit faraway places, and was part of the French craze for all things exotic. It offers vivid

cannons each onto *The Curious* and *The Delight,* two raiders authorized to make seizures on the high seas. The ships were commanded by captains Walsh and Lebrun de Champloret, under the supervision of one Monsieur de la Merveille, who would retell the adventure to Jean de La Roque. They set sail on January 6, 1708, and met with many adventures before reaching Mocha a year later. The ships returned to Saint-Malo on May 8, 1710. More than satisfied with their results, the ship owners soon organized a second expedition.

Legend

Storytellers in the little cafés of Damascus and Constantinople would captivate audiences with the tale of Kaldi, the Yemeni goatherd. One of the many versions of this ancient legend is in *A Thousand and One Nights.* Kaldi's flock, grazing in the mountains, frolicked night and day without sleep, leaving the shepherd without a single moment's rest. He told his problem to

descriptions of Mocha and the coffee of Yemen, as well as the adventure of sailing around Africa to the Arabian peninsula. In 1707, a group of ship owners and merchants from Saint-Malo, in Brittany, paid 7,000 francs for the right, held by the French East India Company,* to engage in the Arabian coffee trade. They loaded fifty

some dervishes, who decided to have a closer look. They noticed that the goats fed on the tiny red fruits of a bush that grew abundantly in the region. Tasting the berries, the dervishes were filled with energy. They gathered them to take back to their monastery, where they were used from then on to keep them alert

Title page from Jean de La Roque's *Voyage de l'Arabie heureuse,* Amsterdam, 1716.

Histoire Pittoresque du Café.

during their long hours of religious practice. The legend of Kaldi is one of dozens explaining the discovery of coffee. It is based in fact, since the Sufis of Yemen were the first to make regular use of coffee, linked to their worship, spreading it throughout the Middle East.

Literature

"The coffee bean, the smell of ambrosia!" said the seventeenth-century Turkish poet Belighi. From the start, coffee's praises have been sung by writers for its delicious taste, its stimulating properties which aid creative work, and the way it unites people for intellectual discussion. Authors such as Ernest Hemingway and Henry Miller evoke the artistic melting pot of certain Parisian cafés between the world wars, while the characters in Tolstoy's masterpiece *Anna Karenina* enjoy many a cup of coffee.

Develly, *Histoire Pittoresque du Café*. Second half of the nineteenth century. Lithograph. Bibliothèque Nationale de France, Paris.

Before that there was the Italian author Carlo Goldoni's play *La Bottega di Caffè* (1750–53), a satire on scandals and gambling in a coffeehouse,* and of course before the turn of the eighteenth century, great English poets such as John Dryden and Alexander Pope were writing satires with moments that take place in coffeehouses. Even John Keats's satirical *Cap and Bells* (1819–20) portrays a soothsayer and king discussing spiking morning coffee. Coffee evokes childhood breakfast in Colette's *La Maison de Claudine* (1922). Honoré de Balzac was most prolific in his glorification of coffee, notably in his *Treatise on Modern Stimulants* (1839), and his novels *Eugénie Grandet* (1833), and *Ursule Mirouet* (1841). Gérard de Nerval (*Le Voyage en Orient*, 1848–51, Pierre Loti (*Aziyadé*, 1879), and Hippolyte Taine (*Voyage en Italie,* 1914) are among the many who brought back from their travels beautiful descriptions in which coffee appears as one of the most important institutions in the countries they visited. Perhaps the finest pages ever written on coffee, however, are by the Palestinian author Mahmud Darwich in his *Memoir for Oblivion* (1994), detailing the loving preparation of coffee in war-torn Beirut as a deeply moving reflection on suffering and death.

Louis-Auguste Bisson, *Honoré de Balzac*, 1842. Daguerreotype. Maison de Balzac, Paris.

Coffeepot made for Balzac in 1832. Maison de Balzac, Paris.

Maritime Trading Companies

At the beginning of the seventeenth century, the traditional trade routes via caravan* from Yemen were increasingly replaced by the movements of the great European trading companies. The British East India Company, founded in 1600, was the first to establish routes through the Indies. Constantly looking for new commercial opportunities, nine years later it also became the first European trading company to land at the port of Mocha, Yemen's "Coffee Coast." But it was the Dutch East India Company, founded in 1602, which succeeded in gaining the confidence of local authorities. In 1616, a Dutch merchant named Pieter van der Broecke negotiated excellent terms for some beans. Even more importantly, he also managed to get his hands on some coffee plants which he successfully brought to Amsterdam. The Dutch managed to get them to thrive in a botanical garden, and forty years later they became the basis for Dutch coffee plantations, first on Ceylon (current day Sri Lanka) and then on the Indonesian archipelago. While these plantations were maturing, the Dutch East India Company built up their trade in Yemen, bringing coffee to Holland and to Dutch-controlled territories in

Asia. The French East India Company, founded in 1664, also traded through Mocha until the middle of the eighteenth century, when it too

Mexican coffee industry. Today the foundation has branches throughout Europe. In the interests of fair trade, it serves to place coffee-growing cooperatives in

began navigating regularly around the African continent. The French also set up coffee plantations in their colonies, in Guyana, the West Indies* and Reunion (then called Bourbon Island), and then shipped large quantities of beans to Europe.

Max Havelaar Foundation

Max Havelaar is the title of a famous Dutch novel written in 1860 by a former colonial civil servant under the pseudonym Multatuli. It describes the plight of Indonesian peasants. The Max Havelaar Foundation was founded in Holland in 1988 in response to the West's outcry against unjust wages in the

direct contact with European roasters,* avoiding middleman costs. This allows 67 percent instead of 38 percent of the sale price to go to the producers, who guarantee the beans' quality. Roasters commit to a minimum price despite market fluctuations. Three hundred cooperatives in eighteen countries are part of this association.

Ludolf Backhuysen, *The Port at Amsterdam*, 1666.

"Coffee (which makes the politician wise, And see through all things with his half-shut eyes.) Sent up in vapors to the baron's brain New stratagems, the radiant lock to gain."

Alexander Pope, *The Rape of the Lock*, 1714

▨ Melitta

In 1908, Melitta Bentz was a single mother living in Dresden. Tired of the grounds her porcelain filter let into her cup, she made little holes on the bottom of a copper pot, borrowed a piece of blotting paper from her son, and used the new contraption to filter her coffee.

Paul Iribe, *Rivals*, 1902. Lithograph printed in the magazine *L'Assiette au beurre*.

The results were just perfect, and Melitta Bentz went to Berlin to patent her invention. A few months later, the M. Bentz Company, with an initial investment of seventy-three pfennigs, launched its paper filters and filter holders on the market. The firm continued to grow and diversify throughout the 1920s. Success increased with the conical filter, which appeared in 1937. Little by little, the paper filter, used either with a simple holder or in an electric coffeepot,* replaced traditional brewing methods.

Today Melitta filters are sold worldwide, and the company has expanded into roasting as well. It is run today by Melitta Bentz's three grandchildren.

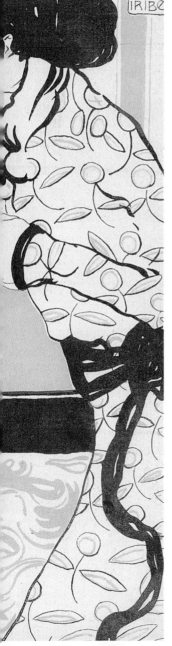

■ Milk

Westerners have been adding sugar and milk to their coffee for over three centuries. Some swear only by cappuccino.* But France's* *café au lait* and Italy's* *caffè latte* are traditional breakfast favorites, served in a bowl wide enough for dunking toast or croissants. Viennese cafés offer a wide range of coffees with milk. The record goes to Herrenhof, a coffeehouse that at one time featured twenty versions of coffee with milk. The mix of coffee's tannin with milk's casein can make for results that are hard on the digestive tract.

Coffee with milk, milder than plain back coffee, is often drunk in the morning, especially in France and Italy where a dark roast and robusta beans make for a strong brew. Milk softens the acidity of certain coffees without destroying their flavor.

"Two other serving girls hold a golden tray on which stand little coffee cups made of fine Saxon or Chinese porcelain, and the engraved golden zarfs set with precious gems."

Leila Hanoum, *Memories of the Royal Harem,* 1925

■ MOCHA

On the Yemeni shore of the Red Sea, Mocha was in antiquity the port from which the incense, myrrh, and alabaster of Arabia Felix ("Happy Arabia") made its way to Egypt and the Mediterranean. But its golden age was from 1660 to 1730, when the major European maritime companies* came in search of coffee. Decline began with competition from new colonial plantations in Asia and the West* Indies, at a time when coffee was still known throughout Europe as "Mocha."

The excellent arabicas* of Yemen and Ethiopia* are still known as Mochas today. At altitudes of up to eighteen thousand feet, the Ethiopian birthplace of coffee still produces three outstanding types of "washed"(see "Processing") Mocha: Sidamo, Yrgacheffe, and Limu, which also designate the places they are grown. All are unusually mild and aromatic, with

a slightly sharp, chocolatey taste and a subtle floral scent. These are perfect evening coffees.

Ethiopia also exports "natural" mochas. Their somewhat lower rank varies according to harvesting and processing. Under the right circumstances, they produce excellent coffee. Harrar is mellow, smooth and aromatic; ideal for any time of day. Djimah, and Lekempti are strong, slightly wild and spicy in flavor; best drunk after lunch. Sanani, a Yemeni Mocha, is also full-bodied, with a rich, spicy taste. Because qualities vary among them, Mocha appellations should be carefully checked before buying.

Coffee beans being transported in the mountains of Yemen.

"The best maxim I know in life, is to drink your coffee when you can, and when you can't, to be easy without it"

Jonathan Swift

Mocha, seventeenth-century engraving.

Gustave Taubert,
*Reading Room at
a Coffeehouse,*
1832. Gemälde-
galerie, Berlin.

■ NORTHERN EUROPE

Germans are the world's foremost coffee mavens, and among the greatest consumers. The excellence of German coffee is due to the tradition of lightly roasted fine arabicas,* and the custom of buying beans directly from coffee shops.* Germans often drink their favorite beverage—loved even more than beer—with a little milk.

Farther north, the Finns and Swedes hold the record for consumption: over five cups a day. Like the Germans, they favor good-quality, lightly roasted arabicas. The coffee is not strong, and served piping hot to warm mind and body. Traditional preparation required lengthy boiling. This has not been entirely replaced by the paper filter, especially in Norway, where a third of all drinkers still prefer boiling. Coffee is taken with cream* and often spiked (see "Alcohol"). The Swedish *kaffekask* is a mix of coffee and aquavit. Its Danish equivalent is *Kaffepunch.*

An outdoor
café in Berlin,
c. 1930.

Coffee is also the favorite drink in Lapland, where large cups of the brew from the Red Sea bring a bit of the southern sun to the polar region.

Opponents

From its first appearance, coffee has met with opposition. In Muslim countries, it has been accused of inebriating its drinkers, which is against Koranic precepts. It was provisionally banned in Mecca in 1511 and in Cairo in 1532. In Italy, around 1600, a group of clergymen vainly petitioned Pope Clement VIII to condemn the "diabolical" brew.

The overnight success of coffeehouses in England, and the freedom of expression that reigned in them, was a cause for concern. Puritans, women (who were excluded from the coffeehouses), and pub owners joined forces against the guilty beverage. In 1674, a group of angry London wives circulated *The Women's Petition Against Coffee.*

In Prussia, Frederick II, a coffee lover, worried that its popularity might threaten the Prussian beer-based economy. The Prussian state proclaimed a monopoly on roasting* which forbade roasting the beans at home, and went so far as to appoint "coffee sniffers" to patrol the streets in search of the telltale smell of infractions.

In Sweden, coffee was declared a "superfluous and noxious luxury," and outlawed several

Turkish miniature, mid-sixteenth century. Chester Beatty Library, Dublin.

times over the course of the eighteenth century. Each time bans were lifted when the government realized that taxing consumption would prove more profitable and effective. This continued through 1853.

Organic

Both for economic reasons and in response to growing demand, organic coffee is grown in several parts of the world. Plantations* in Peru, Costa Rica* and Haiti are notable examples. In fact, the higher the altitude at which plants are grown the less chemicals are needed (see "Arabica"). A stroke of entomological genius occurred in 1988 when small bees with a great fondness for beetles that are particularly damaging to coffee* plants were released experimentally in the Chiapas region of Mexico. The experiment was successful and the practice has spread to several countries in Latin America, with large-scale application in Colombia.* In addition, all over the world dead coffee leaves are left on the ground to mulch. From Kenya* to Colombia and Ethiopia* to India, the sound of crackling leaves beneath footsteps has now joined the music of small family plantations, where the voices of women and children, the songs of birds and the sounds of gushing torrents have always been heard.

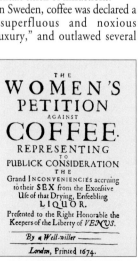

Edward Hopper, *Nighthawks*, 1942. Art Institute of Chicago.

▮ Painting

Café scenes have inspired many paintings. One of the best known and finest is a sixteenth-century Ottoman miniature painted at the time that the first *kahvehanes* opened in Istanbul. From seventeenth- and eighteenth-century Venetian artists, such as Pietro Longhi and Gianni Bertini, to Edward Hopper's *Nighthawks* of 1943, coffee drinking scenes have been represented by painters from all countries and eras.

Cups and coffeepots show up in genre scenes and still lifes from the time of coffee's first appearance in Europe. In the seventeenth century, the craze for Turkish styles led to a number of portraits of Europeans dressed in Turkish costume, holding a little porcelain coffee cup. In the nineteenth century, coffee in art symbolized the family setting and middle-class comfort. Among hundreds of others, some famous examples include Paul Cézanne's *Woman with a Coffeepot* (1890), Edouard Manet's *Breakfast at the Studio* (1868), and Henri Matisse's *Woman Reading* (1895).

■ Percolator

The term "percolation" comes from the Latin *percolare* (to filter) and any coffeepot or process in which water passes through grinds in a filter* is technically a percolator (see "Pressure percolation"). In Europe, the term is often used for the steam-pressure pots which came into use in early-twentieth-century cafés.* The first industrial percolator was invented by Angelo Moriondo, an engineer from Turin, and manufactured in Italy at the end of the nineteenth century. In 1901, Luigi Bezzera came up

Brasserie de l'Isle Saint-Louis in Paris.

with a version which could prepare and distribute coffee cup after cup. Six years later Bezzera's former partner, Desiderio Pavoni, invented the "Ideale" pot, a dazzling copper machine capable of producing up to 150 cups of espresso in an hour. The hotter the water, the quicker the coffee could be made. The trick was to strike a balance between speed and quality, since water that is too hot is known to destroy flavor. Illy, a Trieste manufacturer, came up with one of the first solutions to the problem in 1935 with the "Illetta," which created pressure with air compression instead of steam. The real revolution came in 1948, with the piston model developed by Achille Gaggia, the father of modern espresso.

Plantations

In the eighteenth and nineteenth centuries, coffee plantations spread and flourished throughout the tropics on the back of forced labor. Before mechanization, the arduous

tasks, from land clearing to bean sorting, were done by hand under grueling conditions. The introduction of machines following the abolition of slavery* at the end of the nineteenth century somewhat alleviated the hardship. At the reins of these vast growing systems were French West Indian planters and Brazilian *fazendeiros*. Wealthy and often granted noble titles by the emperor, the *fazendeiros* built *casas grandes*. A few of these sumptuous mansions can still be seen today. They were often made of granite and embellished with colonnades and wrought iron decoration. Their several dozen rooms, with fine furnishings and precious wood parquet floors, faced a vast courtyard. A battalion of servants, primarily slaves, kept the home running. The masters' children were raised by a *mae preta*, or mammy, and had a little page at their disposal to entertain them.

Brazilian coffee fortunes were sizeable. The son of a French immigrant, the engineer Henrique Dumont bought land and planted five million coffee plants* in 1879. At the time of his death, the "coffee king," as he came to be called, left his seven children enough to make their wildest dreams come true.

Candido Portinari. *Coffee Harvest in Brazil*, 1935. Museu de Arte Moderna, Rio de Janeiro.

Ports

Most coffee entering the United States today comes via the port of New Orleans, with over 2

Clipper docked in the port of Santos, Brazil, c. 1875.

of France, and the rest was exported to Italy, Switzerland, and northern* Europe. Italy's coffee port was Venice, the

million 60-kilo bags in stock in 2000. New York and Miami are in second and third place. The International Coffee Organization compiles its coffee indicator prices daily from the prices quoted at various major coffee markets (New York, Bremen, Hamburg in Germany, Le Havre and Marseilles in France). Historically, the port of Marseilles in southern France was the gateway for coffee to other European countries, all of it coming from Mocha. In 1660, 1,869 tons of Egyptian Mocha were unloaded there. One third remained in the South

first major European port to receive coffee. In Holland, it was Amsterdam where, in the seventeenth and eighteenth centuries, coffee from Yemen and the Dutch colonies in Asia was unloaded and auctioned, and the Dutch East India Company's vast ships were built. The clippers of those days have long since given way to container ships. Brazilian* coffee comes through the port of Santos, Kenyan* through Mombasa, and Jamaican through Kingston. Other major European receiving ports are Antwerp and Trieste.

▇ Pressure Percolation

Pressure and percolation are used in many methods of coffee making. Whether weak, as in the case of Cona-type vacuum processes and Italian Moka Express models, or strong, as with espresso machines, pressure serves to accelerate the passage of water through the grinds. It saves brewing time, keeps unwanted elements (such as sediments and, incidentally, caffeine*) out of the cup, and emulsifies coffee's oils, giving espresso its special density and smooth flavor.

With a vacuum coffeepot, an alcohol lamp heats the glass sphere full of water. Steam pressure causes the water to move upward through a filter-siphon into a second container filled with ground coffee. The flame is blown out, and the air in the lower container contracts as it cools. This creates a vacuum which sucks the coffee through the filter, down into

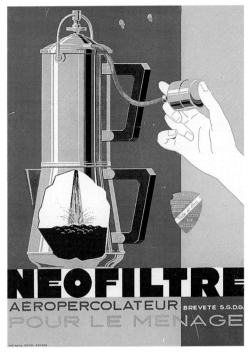

the sphere. The process allows the finest coffee to release its full aroma and flavor. In traditional percolators, heat forces water bubbles to continually pass up through a tube to sink over the top of a filter chamber filled with coffee.

The steam-pressure or Italian coffeepot consists of three pieces. The bottom portion is filled with water, the metal filter is filled with grinds and placed above it, and a top section with a spout is screwed on top. The pot is set on a heat source, creating steam pressure. This causes the water to rise up through the grinds into the top section, in the form of coffee. When used properly, the Italian coffeepot produces denser brews with longer flavors than drip methods, and retains the subtle aromas of fine coffees better than espresso.

An unsuccessful invention, the French Aeroperculator.

Conical Italian coffeepot designed by Aldo Rossi for Alessi, 1982.

Processing

There are two methods of processing coffee cherries once they are harvested, wet and dry. The wet method is used in areas with an ample water supply. Because this style requires fully ripened fruits and repeated sorting and inspection, it makes for uniform, superior-quality "washed" beans known as "milds."

The sorted and washed beans are fed into channels of running water which convey them to machines that strip the hull and pulp from each pair of beans. The beans are then placed in bins and left to ferment for several days, to remove any remaining pulp. Then they are thoroughly rinsed again. At this point,

Left:
Coffee cherries being rinsed in Java.

Below:
Coffee cherries being rinsed in Ethiopia.

the beans are still covered with a parchment-like cellulose film. They are dried for a period of one to three weeks before this layer is mechanically removed.

The dry method is simple and economical, but produces beans, called "naturals," which are less consistent in quality. The process itself is entirely natural. Cherries are harvested as soon as most of them are ripe, quickly rinsed in water and dried in the sun for two to three weeks until they wither. At this stage they are known as "hull coffee." The beans are usually unhulled by machine in the cooperative, and sorted.*

Production Statistics

As a traded commodity, coffee is second worldwide to petroleum. It is grown in some seventy countries, sixty of which export. The coffee industry employs over twenty-five million people. The standard unit of measure for coffee is the 60-kilo (132-pound) bag. A hundred million such bags are filled with coffee beans every year. Annual production is 75 percent arabicas* and 25 percent robustas.* With twenty million bags, mainly arabicas, Brazil is the world's biggest exporter. Next in line come Colombia, the

Coffee market,
Sao Paulo,
Brazil.

second-ranked exporter with thirteen million 70-kilo (154-pound) bags of arabicas; Indonesia, the third-ranked exporter with seven million bags, mainly robustas; Mexico, eighth-ranked exporter with four million bags of arabicas; Ethiopia, tenth-ranked exporter with four million bags of arabicas; Uganda, fourth-ranked exporter with three and a half million bags of robustas; Ivory Coast, sixth-ranked exporter with three million bags of robustas; India, fifth-ranked exporter with three million bags divided evenly between arabicas and robustas, Guatemala, ninth-ranked exporter with three million bags of robustas; and Vietnam, the seventh-ranked exporter with three million bags of robustas. These statistics are subject to climatic fluctuations affecting harvest. Drought and fires caused by El Niño in 1997–98, for example, greatly decreased Indonesian coffee output.

Roasters

Coffee is a product whose processing is generally completed thousands of miles from where it was born. The roasting* is traditionally carried out in the countries where it is consumed (only two coffee-producing countries are also major consumers: Brazil* and Ethiopia*). In Europe and the United States, coffee lovers of long ago roasted their own beans (a practice which continued for longer in the country). The first retail roasters were neighborhood grocers or small roasting shops (see "Coffee Shops"), but coffee, like every other consumer product, has been affected by twentieth-century global commercialization. For example, most coffee drunk in France today comes from one of two American multinationals, Philip Morris and Sara Lee. The first owns the international group Kraft Jacobs Suchard, while the second owns the Dutch group Douwe Egberts. Given the ever increasing competition from these multinationals in the high-quality sector, smaller companies have, in order to survive, devised imaginative marketing strategies which emphasize the superior quality of their products.

Roasting

The green coffee beans bagged on a plantation* have no taste or aroma whatsoever. That comes from roasting. Roasting takes twelve to twenty minutes depending on the machine, at a temperature between 355 and 480°F (179–249°C). This process changes the beans' color to brown, and releases the essential oils that hold coffee's delectable aroma. The art of coffee roasting is infinitely subtle. The roasted beans' hues range among light, medium, light French (medium brown), European (deep brown), French (very dark), and Italian

Kurukhaveci coffee-roasting establishment, founded by Mehmet Effendi in 1871, Istanbul.

Following pages, top left: A nineteenth-century cast-iron coffee roaster.

Bottom left: Coffee being roasted at Piansa, in Florence, Italy.

Right: Flowering robusta bushes in Java.

(almost black). Lighter roasting brings out mild, mellow, earthy flavors. Dark roasting makes for strong, full-bodied, and sometimes bitter coffee. The difference is as great as between raw and cooked food, or green and black tea. It is entirely possible to try roasting at home. You need a heavy frying pan, a wooden spatula to keep the beans constantly stirred, and a source of heat (a stove will do fine). You will also need some practice to get it just right. A well-roasted bean should crumble when pinched between the fingers, but must not be burnt.

■ ROBUSTA

The robusta is a hardy coffee bush that grows to a height of thirty feet in the wild. Its fruit makes for lower-quality coffee than the arabicas; far less aromatic yet strong to the point of bitterness, and highly caffeinated. Robusta beans are often used to produce low-end and instant coffees.

As the name indicates, the robusta bush is extremely hardy, and can live through almost anything, except frost, which is unheard of in its growing zone in the tropics. Robustas are impervious to disease, insects, and heat. They favor the sweltering humidity of tropical forests such as those of West Africa, where they grow in the wild and have been cultivated since the eighteenth century. Robustas flourish today in three main forested areas: Indonesia, where they were first planted by the Dutch in the early twentieth century to replace the arabica plantations wiped out by disease in 1877; Brazil, where they are called by the lovely name of *cornillon* but represent only 15 percent of the national coffee output; and the Ivory Coast, where they were introduced by the French in 1930 and grow in a forest at altitudes of between 300 and 1,200 feet (100–400 meters). Robustas are also cultivated in Uganda, Vietnam, the Democratic Republic of Congo, India, Madagascar, Thailand, and twenty or so other countries, mainly in West Africa.

Routes

Coffee beans make their way from grower to roaster* by various means. The number of middlemen involved depends on the quantities exported and roasted. High-volume industrial roasters can buy direct from the country of origin, while small retailers have to resort to brokers to make their purchases. Large plantations often process, bag, and even ship their own beans, but small growers have to sell their crops through cooperatives. In some countries, the exporter is a government agency which processes* the beans. In other countries, exporter companies are privately owned and are either independent or are subsidiaries of major international trading companies. In some cases, private, public, and commercial concerns work together. Various middlemen, agents, and brokers may play a role in a particular country's system. In Kenya,* for example, independent growers and cooperatives ship their beans to the Coffee Board of Kenya, the government agency in charge of auctioning them to brokers. The brokers then market the "green gold" internationally.

Although the beans are usually packed in shipping containers by this point, transactions are still carried out in relation to the number of 60-kilo (132-pound) bags. The beans are carefully labeled according to their quality, permitting buyers to place orders without having to sample them beforehand.

Rules of Thumb

There are a few rules that apply to all methods of preparing coffee. The first is that a good cup of coffee requires quality beans (see "Best Beans") that have been freshly roasted* and freshly ground. If you purchase prepackaged coffee, those bags that have valves on them are preferable. They guarantee a better aroma. Whether as beans or ground, keep your coffee refrigerated in a sealed container to maintain freshness. Use a grind* that corresponds to the kind of coffee maker (see "Coffeepot") you use; either have your coffee shop* grind it for you, or do it yourself. Good

Harvested robusta being taken for processing in Madagascar.

Leopold Boilly. *Young Housewife,* c. 1800. Johann Jacobs Museum, Zurich.

A broker checking the quality of beans before purchase, Sumatra

water is also important—it should not be chlorinated, so use either mineral or filtered water. During preparation, water temperature should be between 194 and 203°F (90 and 95ºC). In other words, it should be just beginning to boil. If it is not hot enough the aromatic qualities will not be released. If it is boiling, it will destroy these qualities and the coffee will be bitter. You should also clean your coffee maker after each use. Coffee dregs and acid or greasy deposits will affect the taste.

Water should be boiled only in the making of Turkish* coffee, otherwise coffee should never be allowed to get too hot. Use a thermos to keep your coffee warm, rather than leaving it on any kind of heated burner. As the French put it, "Coffee that's boiled is spoiled."

■ Selection

The advertisement showing a *gringo* expert taster* choosing coffee beans at a remote plantation* is no more than an image. In fact, the major coffee companies rely on their brokers (see "Routes") to select

and purchase the beans they will roast. Price is a function of quality, subject to an official system of labels for the beans. A trusted broker may contact a roaster* to buy a quantity of Costa Rica SHB (Strictly Hard Bean) Grade AB, European prepared—in other words beans grown at a high altitude, "screen 16" size, and machine-sorted to eliminate "stinkers." They know what they're talking about. If an appellation is not precise enough, or in the case of unusual variables—bean or harvest quality, an unfamiliar seller—the transaction may be carried out after sampling. In these cases, the sample is sent to be tasted in the buyer's laboratories.

▢ Slavery

By the end of the eighteenth century, the coffee first stolen from the Arabian peninsula was being cultivated throughout the tropics by the European colonial powers. At the end of the seventeenth century, the Dutch were the first to establish plantations* based on forced labor in Java, as well

Johann Moritz Rugendas. *Black Slaves Bringing in the Coffee Crop*, nineteenth-century engraving.

SORTING

as the first, in the next century, to organize the trade of slaves to supply their planters in Guyana with virtually cost-free labor. Hundreds of thousands of men, women and children were then abducted from Africa to work the plantations of first the West* Indies and then Brazil.*

Slaves were the essential element in Brazilian coffee cultivation from its beginning. Jungles had to be cleared, and like harvesting on the plantations, this required the labor of a great numbers of workers. The only way slaves could stop working was by dropping dead. To the blistering heat, tropical humidity, attacks by mosquitoes, snakes and indigenous peoples, was added the iron discipline of the lash. While some plantations were less cruelly violent than others, if slaves were not killed immediately, the application of the whip, plus iron leg restraints and/or chains linking hands and feet were common punishments for those who tried to escape. In general, conditions improved gradually until the abolition of slavery in 1888.

Beans being sorted in Brazil.

■ Sorting

The fruit of the coffee plant is harvested and processed* to obtain the coffee bean, known as "green gold." This green comes in infinite tones and textures, from the pure and shiny hue of Hawaiian,* and the blue-tinged Jamaican Blue Mountain,* to the gray-blue Puerto Rican, not to mention all the shades of jade, or the many yellow and brownish varieties.

After the beans tumble out of the hulling or peeling machines, they are sorted by size, a painstaking task. To begin with, currents of pulsating air strip off the remaining outer coatings, separating the plump, healthy beans from the shriveled and misshapen ones. In some processing plants, they are then placed on slanted vibrating trays that remove the rare and prized *caracoli.* The beans are next sifted and grouped according to size. As a rule, the bigger the bean, the better the coffee. A final sorting and inspection, done either by hand or machine, eliminates any discolored beans. This is a crucial step, because a single rotten bean, or "stinker," can contaminate

a whole 60-kilo (132-pound) bag. When the final sorting is done manually, it is usually performed by women.

■ Steeping

The French were the first to come up with the idea of steeping instead of boiling coffee. This method is better for retaining the beans' true aroma and taste. The ancestor of today's French Press coffeepot,* complete with an interior pocket (the "sock"), appeared in 1710. Simmering water was poured over the grounds in this filter and left to steep. The filter was then carefully removed. A half-century later, steeping had replaced boiled coffee in France. Today, the plunger or French Press method is used. This process was popularized in the early twentieth century by the French firm Melior. Plunger pots brew superior, full-bodied coffee. The glass recipient perfectly retains flavor, making this method a favorite for tasting* among industry professionals. Hands-on involvement lends special appeal to this way of making coffee as well. The plunger is manually pressed to filter the coffee after steeping. At least ten coarse-ground grams per cup are required

René Brno, poster for the Melior French press, c. 1950.

for this process. A new German version of the French Press uses a fine metallic mesh filter. The filter is filled with ground coffee, placed in the pot, and removed after steeping. The advantage of this variant is that it can be used with finely ground coffee, which further enhances the flavor of superior beans.

■ Tasting

Coffee tasting is one of the essential activities of coffee experts and coffee dealers. Coffee is tasted before purchase from producers (usually on the basis of samples), bean quality is verified upon delivery and tasting is again necessary when beans are blended.* Despite the image created by some advertisements, tasting always takes place in the laboratory of the importing company. The procedure may vary somewhat, but generally white porcelain cups filled with different varieties or blends are placed on a lazy Susan, or turning table. The experts sit around the table and dip small spoons into the porcelain cups as they come around, sniff the contents and then inhale deeply—and noisily. The tasters' comments are recorded, "Number 2 is harsh," "Number 4 is bitter," "Number 1 is grassy," etc. The table turns, opinions differ and discussion begins. After each person has tasted each brew three or four times, the cups are lifted to reveal the names of the samples. Expert tasters have a highly developed vocabulary to describe the qualities of what they drink. There are more negative terms, such as burnt, bitter, rank, moldy and musty, than positive terms. Probably because superior coffees provide sensations that are too subtle to be expressed very easily, positive designations, such as sharp, fruity, rich, full, round, delicate, mild and smooth tend to be a little more abstract.

the "Saint of Mocha" (see "Ethopia and Yemen"). In the same year, Philippe Sylvestre Dufour's copiously illustrated *Traitez nouveaux et curieux du Café, du Thé et du Chocolat* appeared in Paris. It went through several editions, in a variety of languages.

Coffee's popularity all over Europe was unchecked by protests voiced in a number of treatises and pamphlets (see "Opponents"). The widely read doctoral thesis of a Marseilles medical student named Colomb accused coffee of attacking the brain, causing paralysis, impotence, and emaciation. In his *Trèatise on modern stimulants* (1839), Honoré de Balzac's chapter on coffee discusses the best ways to achieve a constant level of alertness from coffee, even for chronic drinkers.

Illustration from Philippe Sylvestre Dufour's *Traitez nombreux et curieux du café, du thé et du chocolat*, 1671.

Treatises

Initially Europeans could only quench their thirst for coffee by reading the travelogues of two botanists, one published in 1582 by the German Leonhart Rauwolf, and the other ten years later by the Italian Prospero Alpini. It seems that the first book entirely about coffee was by Fausto Nairone, a Maronite monk from Syria. It was first written in Latin and published in 1671. This book recounts a Christianized version of the legend* of Kaldi the goatherd, and the story of

Ibria ou Pot Pour Faire Cuire — le Café

■ TURKISH

Aleppo, Damascus, Baghdad, and of course Istanbul, capital of the once vast and powerful Ottoman Empire, were one after the other conquered by coffee in the first half of the sixteenth century. "Turkish coffee," as it came to be commonly called (though in Greece it is known as "Greek coffee") is made the same way everywhere. It spread as far the Balkans and Algeria, where it is still popular today. The time-proven method involves boiling in a *cezve,* a long-handled metal pot. In the past, the coffee would sometimes be poured without the grinds into a metal or china serving pot.

Today, although the traditional storytellers and dancing girls are no longer to be found in the cafés of Istanbul, some fine establishments are still in business in the Great Bazaar in the Beyazit neighborhood, and in Salacaq, on the eastern bank of the Bosporus. Run-of-the-mill coffeehouses are everywhere, and coffee vendors weave among pedestrians and traffic. Their copper trays sway on handles that are specially hinged so that the filled cups cannot overturn, no matter what angle they are tilted. Cafés are the place for conversation, games, hookah smoking, and *keyif,* the fine art of watching time pass.

Whether *sade* (without sugar), *orta* (barely sweetened), *az sekerli* (slightly sweetened), or *sok sekerli* (with plenty of sugar), Turkish coffee is the stuff of daydreams. The beans, which are very lightly roasted and finely ground to the texture of flour, make a mellow, soothing brew. It is drunk after the glass of water often served with it, not before.

Rosalba Carriera (1675–1757). *Turkish Man.* Pastel.

Parisian waiters in 1925.

Le Masurier. *Mulatto with a White Girl Visiting Blacks in their Hut in Martinique* (detail), 1775. Oil on canvas, Ministère des DOM-TOM, Paris.

a pot of glowing coals, from which he must carefully pick a lump or two with metal pincers to place atop the tobacco plug.

West Indies

France began its challenge to the dominance of Mocha* with plantations in the West Indies. In 1720 Gabriel de Clieu, an infantry captain from Dieppe stationed in Martinique, was granted authorization to have two coffee shoots brought to the island then under his charge. The shoots derived from a plant that the mayor of Amsterdam had presented to the French king several years before. It turned out that the region's climate was perfect for coffee, which was then introduced to Jamaica by Sir Nicholas Lawes in 1728.

Ten years later, the colonists exported their first coffee shipments. By 1737, France was receiving 7,000 tons. Plantations* were then established on the islands of Guadeloupe, Dominica, and Santo Domingo, which became the leading coffee producer, with 40,000 tons in 1790. The colonial administrators brought thousands of African slaves a year to the West Indies, starting in 1730. By 1775, there were over 192,000 in Jamaica alone. There were some 500,000 on Hispanola Island in 1791, when a freed slave named François Dominique Toussaint-Louverture led the revolt against the French. Several plantations went up in flames as the slaves gradually took control of the island. When Haiti became the first independent black state in 1804, France definitively lost its position as the world's leading coffee producer.

Waiter

"Always shouting, always running, never sitting down . . ." was how French author Germain Nouveau described the often colorful figure of the waiter. Whether he be a poor kid tending a Third World stall, or one of those stylish figures that grace prestigious establishments in Paris, Vienna or Venice, the skills required for the job remain the same, even if the status doesn't: alacrity, dexterity, affability, knowledge of his clients' habits, and discretion. The waiter is supposed to be quick on his feet, adroit, affable, and an expert on his clientele's preferences. Although many waiters are students earning their way through college, in some countries the job is highly prized, particularly in major cities near tourist attractions, where tips can easily double the wage packet. In some prime sites in Paris, for example, the position of waiter is kept in the family and handed down from father to son. Waiting tables can also be a dangerous job: in the Middle East, where a cup of coffee is often accompanied by a puff on a hookah pipe, the waiter can be required to carry

12th century Wild coffee is found growing in Yemen.

14th century First documented evidence of coffee (called *qahwa* by the Sufis) consumption in Yemen.

15th century Coffee is introduced in Mecca and spreads throughout the Islamic world.

1511 Coffee is banned for the first time in Mecca.

1554 The first two *kahvehane* (coffeehouses) open in Istanbul.

1599 The first mention of coffee in English: a merchant named Anthony Sherley writes of "infidells drinking a certaine liquor, which they do call Coffe."

1609 First contract between a European maritime company, the British East India Company, and officials at the Yemeni port of Mocha.

1616 The Dutch merchant Pieter van der Broecke manages to sneak a few coffee bushes out of Mocha. They are successfully transplanted in the Amsterdam Botanical Gardens, becoming the ancestors of all Dutch plantations in Asia, and the French plantations in the West Indies.

1650 The first English café is opened in Oxford by a Lebanese Jew named Jacob.

1658 The Dutch plant coffee in Ceylon.

1660 1,869 tons of Egyptian "Mocha" are shipped to Marseilles.

1665 Kara Mehmet, Turkish ambassador to Vienna sent by the Ottoman Sultan Mehmet IV, popularizes coffee there.

1669 Soliman Aga, Turkish ambassador to France sent by the Ottoman sultan Mehmet IV, popularizes coffee in Paris.

1670 The Hoppe, Europe's oldest extant café, opens in Amsterdam.

1672 An Armenian known as Pascal opens the first Parisian coffeehouse.

1673 The first German coffeehouse is opened in Bremen by the Dutchman Jan Dantz.

1683 The first Italian café is opened in Italy, under the arcades of the Piazza San Marco in Venice.

1685 The first Viennese café is opened by the Armenian Johannes Diodato. Lloyd's Coffee House opens in London and goes on to become Lloyd's of London, the world's best-known insurance company.

1686 The Procope opens in Paris.

1689 The first "American café," the London Coffee House, opens in Boston.

1696 The Dutch plant coffee in Java.

1710 Invention in France of the "sock" method and steeped coffee.

1715 The French plant coffee on the island of Bourbon (today Réunion).

1716 Publication of *Voyage to Arabia the Happy* by the Frenchman Jean de La Roque.

1719 Coffee bushes stolen from Dutch Surinam are planted in French Guyana.

1720 Caffè Florian opens in Venice. Frenchman Gabriel de Clieu plants coffee in Martinique.

1727 Coffee bushes stolen in French Guyana are planted in Portuguese Brazil.

1728 Coffee introduced to Jamaica by Sir Nicholas Lawes.

1734 Johann Sebastian Bach writes the *Coffee Cantata*.

1780 France's West Indian plantations, primarily in the Dominican Republic, make it the world's foremost coffee producer.

1791 Slave revolt on Hispanola Island begins, led by François Dominique Toussaint-Louverture.

1800 Invention in France of the coffee filter (and percolation) with the Du Belloy coffeepot.

1825 Invention in Germany of the vacuum coffeepot, later marketed under the English brand name Cona.

1888 Slavery abolished in Brazil.

1895 The engineer Angelo Moriondo of Turin invents the first steam-pressure coffeepot.

1900 Brazil produces 90 percent of the world's coffee.

1901 Instant coffee is invented in the United States by the Japanese engineer Sartori Kato.

1908 Melitta Bentz invents the paper coffee filter.

1948 Achille Gaggia, a barman in Milan, invents the modern espresso machine.

1962 Agreement negotiated by the International Coffee Organization, signed in London by coffee producing and consuming nations to moderate coffee price fluctuations.

1965 Invention of freeze-dried instant coffee. Parisian roaster Perre Verlet markets top quality coffee in the restaurant industry.

1970 Coffee catches on in Japan.

1971 Three students in Seattle start Starbucks, America's leading chain of espresso bars.

1989 The London International Coffee Organization agreement is not renewed.

1997 A twelfth-century roasted coffee bean, apparently from a Yemeni plantation, is discovered in an archeological dig in Dubai.

2000 Starbucks decides to market Fair Trade coffee in its American stores.

SELECTED BIBLIOGRAPHY

Allen, Stewart Lee. *The Devil's Cup: Coffee, the Driving Force in History*. New York: Soho Press Inc, 1999.

Campbell, Dawn. *The Coffee Book*. Gretna: Pelican Publishing Co, 1995.

Davids, Kenneth. *Coffee: A Guide to Buying, Brewing and Enjoying*. Weybridge: Cole Publishing Co, 1994.

Dicum, Gregory, and Nina Luttinger. *The Coffee Book: Anatomy of an Industry from Crop to the Last Drop*. New York: New Press, 1999.

Kummer, Corby. *The Joy of Coffee: The Essential Guide to Buying, Brewing and Enjoying*. New York: Houghton Mifflin Co, 1997.

Pendergrast, Mark. *Uncommon Grounds: The History of Coffee and How it Transformed our World*. New York: Basic Books, 1999.

Roden, Claudia. *Coffee: A Connoisseur's Companion*. New York: Random House, 1994.

Stella, Alain. *The Book of Coffee*. Paris: Flammarion, 1997.

BLUE MOUNTAIN One of the finest coffees in the world, from Jamaica. Especially mild, aromatic and tangy, with a trace of chocolate. Rare and costly, this coffee should end a perfect meal, either lunch or dinner.

BRAZIL Whether Bahia, Santos, or Sul de Minas, Brazil's best arabicas are mild and balanced. A good choice for morning.

COSTA RICA Tarrazu, Tournon, and Tres Rios strike the perfect balance between strength and aroma. Full-bodied and slightly acidic. For daytime drinking.

DJIMAH One of Ethiopia's three exceptionally full-bodied Mochas, Djimah approximates what coffee must have tasted like when it was first discovered. Good for daytime drinking.

EL SALVADOR PACAMARA A delicate hybrid arabica from El Salvador, subtle and light in flavor. Good for breakfast or evening.

GUATEMALA Antigua, Coban, and Huehuetenango are superior, well-rounded coffees. They can be spicy or chocolatey, slightly tart, and often strong. Best for daytime.

HARRAR This Ethiopian Mocha is smooth and mild, with a floral scent. An ideal morning coffee.

JAVA This increasingly rare Indonesian arabica has a strong, spicy, earthy flavor. Good for daytime.

KALOSSI A superior coffee from Celebes (Sulawesi) that is full-bodied and fruity, and has a complex aroma. Best for daytime drinking.

KENYA AA The world's most fruity and delicately acidulous beans. To be drunk straight, preferably in the evening, for a taste of truly great coffee.

KONA This rare, expensive Hawaiian coffee is one of the finest in the world. It is mild and aromatic, lightly tangy or peppery. Perfect for evening.

LEKEMPTI A full-bodied Ethiopian Mocha, wild-tasting and slightly spicy. Best for after lunch.

LIMU One of the outstanding washed Ethiopian Mochas, which are the world's lowest in caffeine. Mild and richly aromatic. Ideal for evening.

MALABAR A great Indian "natural" arabica. "Monsooned"—exposed to monsoon winds for several weeks—it takes on a yellow color and rich flavor ranging from wild grass to spicy. A full-bodied daytime coffee.

MARAGOGYPE Grown in Nicaragua, and Mexico as "Liquidambar," this giant or "elephant" arabica bean makes a smooth, aromatic morning coffee.

MYSORE An Indian arabica low in caffeine, for afternoon or evening.

SIGRI One of the world's top coffees, grown from Jamaica Blue Mountain stock in Papua New Guinea. Well-rounded, mild, full-bodied and aromatic, with a touch of chocolate flavor. Perfect for evening.

SANANI A natural Mocha from Yemen with a wild, full-bodied flavor. It can pack a punch and is variable in quality.

SIDAMO An excellent washed, small-bean Ethiopian Mocha. Mild, smooth, slightly tart, floral-scented. Sidamo is very low in caffeine and perfect for evening. Its *caracoli* beans are highly valued.

SKYBURY (QUEENSLAND, AUSTRALIA) An outstanding washed coffee. Highly aromatic, with a hint of chocolate flavor. Good for evening.

C O F F E E S

TANZANIA A coffee with the fruity, zesty taste of other east African beans, but often slightly milder. Good for morning and evening alike.

YAUCO SELECTO One of the world's strongest, rarest, and most expensive coffees, from Puerto Rico. Full-bodied, potent, and highly aromatic. For daytime.

YRGACHEFFE An outstanding and rare washed Ethiopian Mocha. Very low in caffeine and extremely smooth, with tangy and chocolate overtones and a floral scent. Sublime for evening.

ZIMBABWE Deliciously acid and slightly spicy. Best for evening.

I N D E X

Photographic credits: ATLANTA, J. Martinez & Compagny 34–5; BERLIN, Gemäldegalerie 88; CHICAGO, Art Institute 92–3; DUBLIN, Chester Beatty Library 91; LE HAVRE, Maison P. Jobin & Cⁱᵉ/Philippe Jobin 78; Musée de l'Ancien Havre/J.-L. Coquerel 45 top; LONDON, Bettmann Archives 60, 88–9; British Museum 46–7, 90; ICO 13, 16, 26, 28 bottom, 52–3 top, 98, 106, 108–9; MILAN, Luisa Riccarini 37, 74–5; CERA 64 bottom; PARIS, archives Flammarion 42–3 bottom, 70 top, 80, 81, 83, 97, 112 bottom /F. Morellec 69; Archives Photo 38; Bibliothèque Nationale de France 86 bottom; Bios/Alain Compost 103 /Dominique Halleux 104; Dagli Orti 21, 42 top, 94–5, 106–7; Jérôme Darblay 27 bottom, 33 bottom; Giraudon 66, 115; Hoa-Qui /E. Valentin 44 top /C. Pavard 76–7; J. Laiter 44 bottom; Magnum /René Burri 14–5 bottom, 54–5, 79, 100 top, 111 top /Gilles Peress 24 /Richard Kalvar 27 top /Henri Cartier-Bresson 113 /Bruno Barbey 94; Pascal and Maria Maréchaux 10, 58, 61, 72, 86–7 top; Patrice Pascal 19, 48–9, 50, 68, 112 top; Photothèque des Musées de la Ville de Paris 82; Rapho/R. Michaux 99; Réunion des Musées Nationaux 6, 12, 14, 62–3 top; Roger-Viollet 20, 25, 30, 40, 65, 67, 114; Christian Sarramon 39, 51, 101; Top 102 top / Christine Fleurent 4–5 /Daniel Czap 73; VANVES, Explorer/J. Moss-P. Resear 36 /J.-P. Saint-Marc 56–7 top; Visa/C. Valentin 71; VENASQUE, Terres du Sud/Ph. Giraud 52–3 bottom; ZURICH, Johann Jacobs Museum 105.

Translated and adapted from the French by Chet Wiener and Stacy Doris
Copy-editing: Bernard Wooding
Typesetting: Julie Houis, A Propos
Color separation: Pollina S.A., France

Originally published as *L'ABCdaire du Café* © 1998 Flammarion
English-language edition © 2001 Flammarion

ISBN: 2-0801-0547-7
N° d'édition: FA0547-01-V
Dépôt légal: 05/2001
Printed and bound by Pollina S.A., France - n° L83256

Page 6: Eugène Girardet (1853–1907), *Café arabe* (detail).
Musée d'Orsay, Paris.

A New Birth of Freedom

A

NEW BIRTH

of FREEDOM

Human Rights, Named and Unnamed

CHARLES L. BLACK, JR.

YALE UNIVERSITY PRESS
NEW HAVEN AND LONDON

Reprinted by arrangement with G.P. Putnam's Sons, a member of
Penguin Putnam Inc.

Book design by Debbie Glasserman.
Set in Weiss type.
Printed in the United States of America.

Library of Congress Cataloging-in-Publication Number 98-88790

ISBN 0-300-07734-3 (pbk.: alk.paper)

A catalogue record for this book is available from the British Library.

The paper in this book meets the guidelines for permanence and
durability of the Committee on Production Guidelines for Book
Longevity of the Council on Library Resources.

10   9   8   7   6   5   4   3   2   1

To the sacred memory of Abraham Lincoln

Contents

Foreword

This book lays bare the legal foundation for those rights that, though not specifically named in the U.S. Constitution, cannot be dismissed if the deepest constitutional commitments of the American experiment are to be vindicated. That experiment took the European constitutional pattern of enumerated rights[1] and unenumerated powers[2] and turned it inside out. For the Americans, sovereignty itself was to be limited and the rights retained by the People were infinitely numbered. It was a source of grave concern to James Madison and Alexander Hamilton, the principal drafters of the *Federalist Papers*, that the movement to incorporate a Bill of Rights in the new U.S. Constitution would somehow becloud this radically new structure, with its list of specific powers and its implied domain of unassailable rights; if some rights were listed, they fretted, wouldn't there be a tendency to slight those rights that were not explicitly named? This problem was considerably deepened when, after the Civil War, the Constitution was amended to superimpose on the states the federal idea of limited powers and retained rights. Where were courts and other officials to look to determine the content and extent of these rights if they were never explicitly listed or described?

It is dramatically satisfying that the sword severing this Gordian knot should be wielded by Charles L. Black, Jr., now in his ninth decade. Charles Black was in his early forties when he wrote *The People and the Court* in 1960. This book was a frontal assault on the reigning dogma in elite constitutional circles whose biggest idea was that judicial review was fundamentally unsound. In this work Black endeavored to make judicial review respectable by giving it a *legal* rather than merely a *political* foundation. Thereafter he undertook much the same legitimizing task for the Supreme Court's action in *Brown v. Board of Education* in the most important article published by the *Yale Law Journal* in that decade, "The Lawfulness of the Segregation Decisions." Rereading this article now, with its mighty defense of arguments that, it must be said, are more to be inferred from than found in the *Brown* opinion, one can hardly imagine that there was a time when a different view of the Court's action widely prevailed. Most sound law professors and judges in the late fifties, however, would have thought Herbert Wechsler's 1959 "Neutral Principles" article to have said the last word on the subject: that there was simply no justification for preferring the rights of association of African Americans (who wished to integrate) to the same rights of those whites who wished to segregate themselves. Confronting this reluctant but principled stand, Black wrote, unforgettably, "Does segregation offend against equality?" After acknowledging that "[e]quality, like all general concepts, has marginal areas where philosophic difficulties are encountered," Black plainly and eloquently explains why he has no philosophic difficulty in concluding that segregation violates equality:

> [I]f a whole race of people finds itself confined within a system which is set up and continued for the very purpose of keeping it in an inferior station, and if the question is then solemnly propounded whether such a race is being treated "equally," I think we ought to exercise one of the sovereign prerogatives of philosophers—that of laughter. The only question remaining (after we get our laughter under control) is whether the segregation system answers to this description.
>
> Here, I must confess to a tendency to start laughing all over again. I was raised in the South, in a Texas city where the pattern of segregation was firmly fixed. I am sure it never occurred to anyone, white or colored, to question its meaning. The fiction of "equality" is just about on a level with the fiction of "finding" in the action of trover. [3]

In their concern with the primacy of doctrine, the law professors and, it must be added, some of the most prominent judges had simply neglected to appreciate that the system of legal segregation by color amounted to the setting up of a caste system by the State—an unquestionable violation of the State's relationship to its citizens, to say nothing of the plain words of the "equal protection" clause. The doctrine had become so severed from reality that plaintiffs and defendants, appellants and appellees had replaced flesh and blood human beings who were not symmetrically placed, like subatomic particles, but stood in historic situations unique to their time and situation.

Black was once described as "the only certified genius at the Yale Law School," and it is easy to see why. In addition to these pathbreaking constitutional works, he had coauthored with Grant Gilmore what remains the most important treatise on the law of Admiralty and published three books of highly praised poetry. I will not discuss these works as they are not directly germane to *New Birth of Freedom*, except to say that Black's inimitable prose style, which is in such fine evidence in these works, has also been important—I would even say necessary—in securing attention for his more radical constitutional ideas.

For the constitutional theorist Black's next work in 1969 was even more momentous than his earlier work: in *Structure and Relationship in Constitutional Law* Black definitively discredited the idea that the only legitimate constitutional arguments are derived from precedent or precedent itself derived from the history and text of the constitution. Just as influential in our constitutional development—as even a cursory reading of *McCulloch v. Maryland* would have indicated were it not for the blinders academic scholarship of this era had placed on lawyers and judges—were the structures set up by the constitution and the relationships among them that it mandated. This short work—only ninety-eight pages—has been so influential that now the words "history, structure and text" come as automatically to the tongue of the most chaste strict constructionist as "history and text" once did. Reflecting on this historic achievement, I wrote in

1987, "[I]f one looks across the scene of constitutional lawyers and judges of the present day, only Black can be ranked with the highest class [that would include Marshall and Story in the last century, and Brandeis, Holmes, and Hugo Black in this one.]"[4]

All comparisons are to some degree invidious; remarks like those just quoted are not calculated to endear one to one's contemporaries. Yet with each passing decade the essential truth of this paragraph seems more unassailable.

At present, Black's fame rests largely on his revitalization of the particular form of argument—structural argument—with which he is associated. Structural argument is one of half a dozen ways, or modalities, of framing legal arguments about the U.S. Constitution.[5] Although Black's name is most associated with structural argument, the present volume is, in some ways, a recognition of the limits of that form.

Structural argument reached its apogee in the early days of the Republic when its structures and relationships were first being hammered out, and then fell into disuse as other forms—chiefly doctrinal—achieved dominance. But in the turmoil following the *Brown* decision, doctrinal argument was not of much use; indeed *Plessy v. Ferguson* was a fifty-eight-year-old precedent when *Brown* effectively overruled it. Black reclaimed this form and, in his inimitable prose style, made it his own.

But there are limitations to any single form. Because structural argument derives its power from inferences arising from the political relationships ordained by the Constitution, it is of limited applicability to those human rights issues that are not, at bottom, about politics. It is a strain to suggest that the source of privacy rights lies in the freedom to assemble, or that *Roe v. Wade* is essentially a matter of federalism. While structural argument can provide useful insights into these questions, it misses something.

Rather it is in the realm of ethical argument that human rights questions must center. Like structural argument this form has been much neglected during most of this century in favor of sometimes highly attenuated doctrinal arguments. Thus it was felt that the justification of *Roe* had to be placed on *Griswold*—which provided a highly unconvincing basis—and cases like *Skinner* (involving a state program of eugenics) were absurdly tied to the vacuous, and therefore specious, jurisprudence of the equal protection amendment.

The difficulty with developing ethical argument is that it is usually confused with moral argument and thus with morality generally. The American moral ethos is simply not the same as its constitutional ethos; there is no warrant to read the moral preferences of judges—or anyone else—into the constitutional decisions of governments, no matter how pleasing this might sometimes be. Indeed the constitution is remarkably sparing in its moral preferences, choosing to leave most moral questions to the private sector. While there are many eloquent persons arguing that the Constitution enshrines their particular moral preferences, there is a profound difference between the legitimacy that moral debate confers and that offered by legal argument.

Indeed, much is sacrificed by trying to make moral and legal argument coextensive. Although there is a wide spectrum of opinion on the proper role of marriage, for example, there is a general consensus that the State cannot dictate whom we marry; while there is an active debate on whether or not couples should employ artificial means of birth control, there is a similar consensus that the State cannot simply forbid contraception; and so on with respect to procreation and the right to educate one's children and determine one's own medical care. The difficulty is that when the Supreme Court has recognized these intuitively obvious rights, there has been precious little in the way of constitutional text or original intent or political structure to which the Court could turn. The rights that we can infer from the most basic American constitutional ideas of limited sovereignty and the enumeration of powers are, after all, themselves unenumerated. Perhaps for this reason, the doctrine which has been purpose built to support these holdings—"substantive" due process—has never been able quite to shed its factitious aura and as a result has been an easy target for every critic whose ultimate argument is "Oh yeah? Where does it say that?"[6]

Charles Black, in the book that follows, has taken up the task of providing a constitutional basis for decisions that construe the unenumerated rights of the Constitution. The same man who provided legal arguments for *Brown* when these were sorely needed has now accomplished the same task with respect to *Meyer, Pierce, Griswold,* and *Skinner.* In this book, the reader will once again find the watermarks of Black's work: the powerful, lucid prose, the careful, pivotal observation, and above all, the craftsman's art of legal argument. In his own words, Black seeks "the construction of a better system of reason" for the constitutional law of the unenumerated rights.

Here, as before, Black is, at bottom, concerned with legitimacy and with providing for lawyers and judges and other public officials the means of reaching satisfying conclusions that comport with the country's deepest sense of its constitutional self.

It is a widespread misperception among the laity that law is an edifice of incomprehensible complexity and that its most magnificent chambers are reserved for those who not only tolerate but seek the counterintuitive, arcane, and contradictory. Every now and then the public reacts to this misunderstanding by demanding that law be "simplified." The great codification movements of the last part of the nineteenth century were one such reaction.

In truth, the surest expressions of law are the clearest. It is not necessary to "simplify" or to distort the law in order to bring it into line with our deepest convictions of common sense. What is required is careful and conscientious adherence to what Charles Black has called "the manner of legal reasoning." Those who have sought to assist the movement toward the recognition of human rights by abandoning close adherence to the modalities of constitutional argument have therefore done this cause no positive service.

Rather, the ones to whom we are most indebted are those who have dived down the deepest to our sense of fitness and stayed down long enough to craft distinctly legal arguments that serve that sense. Charles Black's work is not powerful because it is beautiful, having been rendered by a poet; rather it is beautiful because it is powerful—because, that is, it is the product of the arts of law, and thus is linked to our sense of rightness.

On the wall of my office at the University of Texas is a large, framed excerpt from Charles Black's "The Two Cities of Law," handwritten by him for my graduation from Law School in 1975. It is evidence of Black's long preoccupation with the subject he has taken up in the present book:

> One afternoon last fall I was on my way to my class in Constitutional Law. I was going to lead a discussion of certain technicalities having to do with the application of the Fourteenth Amendment, as implemented by acts of Congress, to voting and other rights. My head was full of section numbers in The Federal Revised Statutes. I fear I was mumbling to myself, a practice I cannot recommend to those who hold reputation dear.
>
> I happen to look up—all the way up, over the tops of the red stone buildings into the sky as the Indians of Connecticut must have seen it before the white settlers came, with the great autumnal castles of clouds as far as the imagination could reach. And somehow, very suddenly, all this illimitable expansiveness and lofty freedom connected within me the words I was tracing from the Fourteenth Amendment though the statute books— "privileges or immunities of citizens," "due process of law," "equal protection of the laws." And I was caught for a moment by the feeling of a Commonwealth in which these words had not the narrow, culture-bound relative meaning we are able to give them in the "real" world, but were grown to the vastness that is germinal within them.

This sense of the transcendent is finely fitted with Black's superb and careful craftsmanship. Perhaps neither trait alone would have marked him as the most gifted of constitutional exponents. Together they have assured his stature as the most profound. The marvelous promise of his early work—like the epic promises he writes about in this book—have been richly fulfilled in Black's contribution to the jurisprudence of the American constitution.

—Philip Chase Bobbitt

Preface

This book puts forward the thesis that a sound and satisfying foundation for a general and fully national American law of human rights exists in three imperishable commitments—the Declaration of Independence, the Ninth Amendment, and the "citizenship" and "privileges and immunities" clauses of Section 1 of the Fourteenth Amendment (as those clauses ought to have been and still ought to be interpreted). These three commitments speak in solemn organic harmony. They ought at long last to be attended to as they stand—for as they stand, in their harmony, they are all we have and all we need of prime authority for our building, by the methods of law, a never-to-be-finished edifice of human rights.

I write out of fifty years' professional thought and work and feeling on and around these things. There is an enormous and many-sided literature—from that half-century and from long before. I have been concerned in this short book to state and to support, in my own voice, my own life's conclusions. I have drawn freely from my earlier writings, improving them when I can. But I think this book has its own new unity.

I know that my chief debt, far outweighing all others, is to Abraham Lincoln; once one takes courage from his recognition of the primacy of the Declaration of Independence, the rest falls easily into place.

I have been heartened through the years by the work of Kenneth Karst on the fruitfulness of the promise in "citizenship": he and I have cultivated bordering acres of this field.

I owe thanks to Jack Greenberg, Philip Bobbitt, and Walter Dellinger, for decades of exchange of ideas with them—and, more than that, for their friendship.

And to Jane Isay, now as in the past, my Editor-in-Chief.

And to Barbara Aronstein Black (if I may be indulged in one more repetition of a thought never absent from my heart) for everything.

—C.L.B., Jr.
Columbia Law School
February 1997

A New Birth of Freedom

Chapter 1

A
GENERAL
VIEW

This work is being undertaken to set upon a firmer and wider ground of legitimacy the human-rights law of the United States. In performing such a task, one always encounters new insights suggested in part by one's earlier works[1] and by their mutual harmonies.

Something else is perhaps not too obvious to mention. For a man of full four-score winters, to whom promotion through publication is no longer a possibility or even a good, to sit down of his own free will and write yet another book—for him to take that much time away from playing with his grandchildren—he must think he has something that needs to be said:

The foundations of American human-rights law are in bad shape. They creak, they groan for rebuilding.

Ours is a nation that founded its very right to exist on the ground of its commitment to the securing of nobly envisioned human rights in very wide comprehension—a country that now bases its claim to the world's regard on a questing devotion to the securing of human rights. When we are true to ourselves, to our own beginnings and to the best of our history, we do not assert our entitlement to something called "world leadership" on the ground of our now being "the only super-power"—that is, after all, just the bully's reason for claiming dominance—but on our commitment to human rights.

This first chapter will present the outline of the fresh foundation—built of venerable materials—that I propose to defend in this book. But I believe it wise to start with a preliminary overview of the principal lacks, as I see them, of the present human-rights law of the United States.

The "lacks" in the system:

First, the "enumerated" (or textually specified) rights found in the Constitution and in our Bill of Rights—the first eight amendments—are very plainly insufficient to found a system broad and comprehensive enough for a really free people to walk around in. The validation of this statement calls for the reading of a few hundred words. I will take you through that reading later. (Maybe someday we'll compose a "round robin" to the people at Philip Morris Company, who have so reverentially eulogized the Bill of Rights as the "onlie begetter" of our ensuing liberties.)

① enumeration not sufficient enough

② post civil war amendments no substantive human ri

Secondly, the guarantees of the post–Civil War constitutional amendments, as the Supreme Court has read them, have not come near to filling this deficiency. The Fourteenth Amendment guarantee of "due process of law" can and to a large extent does combine with the same clause in the Fifth Amendment to ground a developing series of national rights to "fair *procedure*." But this does not, on its face or in the normal meaning of its words, or in its natural implications, guarantee any *substantive* human rights. It says, for example, that you cannot be tried before a bribed judge, or without being informed of the charge against you. But it does not say that you cannot be sent to prison by an unbribed tribunal for marrying before the age of forty, if a state statute sets that as the minimum age—provided only that the trial, by which you are found guilty in fact of marrying at thirty-nine, is *procedurally* fair.

Thirdly (as a desperate answer to a desperate need), a thing called "substantive due process" has been thought up (I had almost written "dreamed up") as a ground for the protection of substantive rights, such as the right to marry, to have or not to have children, to send those children to religious or military school, to teach and learn German, to listen to rock music, to travel from one State to another, not to have your property taken by a State without compensation. This paradoxical, even oxymoronic phrase—*"substantive due process"*—has been inflated into a patched and leaky tire on which precariously rides the load of some substantive human rights not named in the Constitution.

Its application follows no sound methods of interpretation (how could it, given the nature of the phrase itself?) and is therefore neither reliably invocable in cases that come up, nor forecastable in result by anything much but a guess. This kind of non-standard is not good enough for a systematic equity of human rights. It everlastingly will not do; it is *infra dignitatem*, it leaks in the front and leaks in the back. The anxiety it produces (especially in judges—see below pp. 105–106) is an ill-tuned, wavering keynote for a system of human-rights law.

Fourthly, the most comprehensive and promising words in the post–Civil War amendments—the "citizenship" and "privileges and immunities" clauses of the Fourteenth Amendment—have been put into suspended animation by a not-very-clever conjurer's trick, maladroit at best, though more serious charges than that of mere incompetence may apply (see below pp. 55–85).

Fifthly, the "equal protection" clause of the Fourteenth Amendment is of force, on its face, only against the States. Its application against the national government has to wriggle like a rabbit pulled by its ears out of a hat.

Sixthly, the guarantees embodied in the First Amendment are not stated as applying against the States and their subdivisions. He who runs may read: "*Congress* shall make no law . . ."

Now that's quite a list. It comprises only highlights. We ought not to be content to face the twenty-first century with a patchy, tacky human-rights law, so poorly legitimated, so feeble of reason.

I could expand greatly on each of these statements, and will do so further on. You can't afford to rely on my mere assertion. But at least you now can see how this octogenarian, having long considered all these defects, thinks it worthwhile to take some time away from the grandchildren to write this book. Indeed, you'll discern that writing this book is not really taking time from the grandchildren; it is being written for the sake of all our grandchildren—in whose number I naturally, even with a certain special pride, include my own. This book attempts the construction of a *better system of reason* for the grounding of constitutional human rights in this country.

We Americans first entered and still hold ourselves out on the stage of the world as a people and as a power dedicated to the securing of human rights; we were the very first people formally to make that seminal commitment. That is the way we have been thought of on Wenceslaus and Tienanmen Squares. Even more important, that is the way we want and need to think about ourselves.

We are also a people, and want to be looked on as a people, dedicated to the rule of law. It was, then, inevitable that we should use law for the delineation and the sustaining of human rights.

Given these two things, it came to us naturally to be the people, of all the peoples of the world, who some two hundred years ago invented the idea that the very *Constitution* of a country—the document, that is to say, that *constitutes*, that establishes, empowers and shapes the structures of government, and that declares itself to be the "supreme Law of the Land"—should set up, in the same breath of history, and with the same authority, certain affirmative guarantees of human rights under law.

Law is reasoning from commitment. Where do we find those *commitments* from which we may derive our reasoned constitutional law of human rights?

On the highest level and of fully general scope, there are just three such commitments: (1) the opening paragraphs of the Declaration of Independence (1776); (2) the Ninth Amendment to our Constitution (1791); and (3) the thirty words or so that are the "citizenship" and "privileges and immunities" clauses of Section 1 of the Fourteenth Amendment (1868) thereto.

That is all we have of first-water, unassailably authoritative commitment to substantive human rights in general. It is the keystone thesis of this book that these three utterances, unmatchable as they are in authority, constitute (each severally and the three in harmony) all the commitment we need from which to reason—*after the manner of law's reasoning*[2]—to an open-ended and open-textured series of human rights.

All we have, and all we need? It's worth taking another look at them.

First, the Declaration of Independence. That rather short document was not a Fourth of July oration (the first one of those was doubtless delivered on 4 July 1777) but a distinctly juristic act—the foundation of all later juristic acts in our territory. Near its beginning it declares, as foundation for all its claims:

"We hold these truth to be self-evident, that all Men are created equal, that they are endowed by their Creator with certain inalienable Rights, that among these are life, liberty, and the Pursuit of Happiness—that to secure these rights, Governments are instituted among men. . . ."

It was no less a man than Calvin Coolidge who said, "The business of America is business." I have always found this utterance repellently close to the even terser "Business is business," which is what you say when you propose to deal with somebody for your own gain alone, without alloy of humaneness, generosity, or charity. But Silent Cal was a century and a half late in defining the "business" of America. The Declaration of Independence did that, and will continue to do so until we decide formally to repudiate it. To paraphrase Dr. Johnson, the cultivation of human rights is the invaluable part of the business of our nation.

Should the Declaration be taken seriously? Well, the people who uttered this document pledged to its support their lives, their fortunes, and their sacred honor. That does sound a bit like a Fourth of July oration, doesn't it? But think again before you commit to that judgment. All the people who signed did so with the knowledge that they were taking a substantial chance of incurring the penalties of treason—hanging, drawing and quartering, forfeiture by attainder of all their property, and being branded with the hateful name "traitor." Some of them may have thought, or hoped, that the British government would not proceed to those extremes. But who, in the early days of a war, knows how bitter the events of that war, and the feelings engendered by those events, may become? These people were knowingly and literally endangering their lives, their fortunes, and their sacred honor. Can we think it right—in a time when the word "superpower" applies to us alone (Acton, thou shouldst be living at this hour!), while Britain is a mellowed old friend—to treat as mere bombast the words these people used formally to justify their great action to the world? How frivolous dare we be with our original and irreplaceable commitment?

When we consider the Declaration as a commitment of the new nation to a general system of human rights, to the "securing" of such rights as the ground for the very legitimacy of the nation, it is the short passage just quoted that must interest us. Since the organic connection of this passage with later commitments (the Ninth Amendment and the "citizenship" and "privileges and immunities" clauses of the Fourteenth Amendment) is the unifying concern of this whole book, I shall leave the matter at that, touching for now on only one collateral question about the Declaration. Since we are examining the commitment of this nation to a *law* of human rights, we must here consider the force, *in law*, of the doctrines of the Declaration of Independence.

It is my own view that the doctrines of the Declaration should be taken to have the force of *law*—the force *in law* of general commitments from which *particular* law can be derived. For a reason which will very soon appear, it is not necessary to insist upon this. But I will briefly give my grounds, because (whether my view be fully accepted or not) the rehearsal of these grounds, slightly transposed, can make clear the appropriateness of the Declaration as a *basis* for law, as a *nourisher* of law, whether or not it be taken to be law of its own unaided force.

The Declaration as a whole was an act of "constitution," a *juristic* act, an act of *law*, after the manner of law in all its fields, quite as surely as is a statute setting up a state police force. The operative passage is the penultimate sentence:

> We, therefore, the Representatives of the United States of America, in General Congress, Assembled, appealing to the Supreme Judge of the world for the rectitude of our intentions, do, in the Name, and by Authority of the good People of these Colonies, solemnly publish and declare, That these United Colonies are, and of Right ought to be Free and Independent States; that they are Absolved from all Allegiance to the British Crown, and that all political connection between them and the State of Great Britain, is and ought to be totally dissolved; and that as Free and Independent States, they have full Power to levy War, conclude Peace, contract Alliances, establish Commerce, and to do all other Acts and Things which Independent States may of right do.

These words demolish one legal authority and set up another. They are, then, constitutive words, constitutive of the authority of the United States, as that authority has been transmitted and developed down to now. The Declaration is the root of all political authority among us, of all legitimate exercise of power.

The Declaration (a short document) gives its reason for this new constitution of the basis of law, and the reason the signers consider themselves authorized to make such a change, in its second paragraph. That reason was that the British power had crossed the bounds *of legitimacy,* in that it had offended against the rights named in the second paragraph of the Declaration ("life, liberty, and the pursuit of happiness") and had thus failed in its radical and indispensable duty, as a "government"—the duty to "secure" these rights.

It is hard for me to believe that a political organism that places its right to life on these clearly expressed grounds can walk away from the legal obligation based on this constitutive fact of its history. The scope of such an obligation is bounded by prudent possibility, and we ought not be too hard on those who wobbled and trembled in the awful shadow of slavery. Let the dead past bury its pitiful dead— but we ought not to leave that past buried in darkness. Nor the obligations created by past acts at the beginning of our life. If we do that, what right do we have to spout the Declaration's words on the "Voice of America"? It may be too late to mend with full efficacy, but it's never too late to mend.

Should chattel slavery, now 130 years gone, continue to cast a killing shadow on our noblest and most fundamental utterance, as it continues to cast a cold shadow on the lives and hopes of black people?

Now you may not be convinced that the Declaration should be looked on as having permanent standing as a source of law. After all, this constitutive, supra-constitutional act, grounding all our later political acts, and in its expressive force binding all "governments"—national, state, county, city, drainage districts—doesn't *say* that it is "law." So let us pass on to the Ninth Amendment to our Constitution.

The Constitution of the United States—of which the Ninth Amendment is "to all Intents and Purposes" a part (Article V)—*does* declare itself to be law, part of "the supreme Law of the Land" (Article VI).

In 1791, less than fifteen years after the signing of the Declaration of Independence, this Ninth Amendment to the Constitution came into effect, having been drafted and submitted by the First Congress in 1789, thirteen years after the Declaration of Independence. The Ninth Amendment, pursuant to Article V of the Constitution, then became "valid *to all Intents and Purposes* as part of this Constitution" and therefore a coequal working part of "the supreme Law of the Land."

The Ninth Amendment provides: "The enumeration in the Constitution, of certain rights, shall not be construed to deny or disparage others retained by the people."

The academic writing on this Amendment seems to me to be in great part a multidirectional fluttering flight from the Amendment's rather plain meaning, a flight set going by the fact that the Ninth Amendment seems to be repudiating decisively the idea that no constitutional human right can be valid unless it is enumerated—or named—in the Constitution.

I can't make a better start here than by setting down my own earlier thoughts on this, with some slight clerical changes.[3]

This sentence stands at the end of a very short "enumeration" of rights—an "enumeration" nobody could possibly think anywhere near sufficient for guarding even the values it patchily and partially shields. The Ninth Amendment language was put where it is by people who believed they were enacting for an indefinite future. All sorts of other language may have been used around this language. But this was the language chosen to become "valid to all Intents and Purposes, as Part of [the] Constitution. . . ." What does it seem to be saying?

It could be read as saying that nobody really ought to deny, in discourse of a mixed moral and political tenor, that a number of rights exist, beyond the "enumerated" ones. But this is quite unbelievable. Virtually all of the Constitution, including the amendments preceding and later following this one, is *law*, sparely stated in the language of law. Attention here should be focused especially on the first eight amendments, together with which the Ninth Amendment entered the Constitution. These are austere, peremptory directions to lawmaking and law-enforcing officials, from Congress, through courts of law, down to magistrates issuing search warrants and military officers quartering troops. In the Constitution as a whole, and in this immediate context, the insertion of a precept of moral philosophy would not merely have changed the subject abruptly, but would have put the content of this Amendment in quite a different world than that of the Constitution, and of the "enumerated" rights just set out.

The Amendment could be read as saying no more than that

the bare fact of "enumeration" of other rights should not, in and of itself, give rise to the inference that no other rights exist, but that the forbidding of the drawing of this one inference in no way prejudices the question whether there really are, in addition to the enumerated rights, any "others retained by the people." I guess a computer, fed the words, would have to print this out as a logical possibility. I submit that it is not a serious psychological possibility that anyone quite neutral on the question of the existence of rights not "enumerated" would bother to set up this kind of directive as to what course the non-logic of *expressio unius* may take, leaving it quite open that the very same conclusion—no non-enumerated rights—may be reached by some other path of reasoning. The Ninth Amendment seems to be guarding something; such bother is not likely to be taken if the question is thought to be quite at large whether there is anything out there to be guarded.

It should be noted, in passing, that the most one could get out of even this computer printout is that the language of the Ninth Amendment does not *affirmatively imply* the existence of unenumerated rights; even a computer would have to print out that this language implies that such rights *may* exist—if you also fed into that computer the assumption, "The utterers of this language were not talking just to hear their heads rattle." This, while not strictly an existence-proof, would be a proof of the serious possibility of the existence of rights not enumerated; even this might be enough to legitimate a further quest. But the Constitution is not a computer program, and I submit that the preponderance of reason leaves us with the conclusion, about as well-supported as any we can reach in

law, that the Ninth Amendment declares as a matter of law—
of constitutional law, overriding other law—that some other
rights are "retained by the people," and that these shall be
treated as *on an equal footing* with rights enumerated.
This would have to mean that these rights "not enumerated"
may serve as the substantive basis for judicial review of governmental actions; any other conclusion would not only do
violence to expectations naturally shaped by the command
that these other rights not be "denied or disparaged" in respect
to the enumerated rights, but would also lead one back around
to the inadmissible theory, discussed above, that this Amendment, placed where it is, is merely a directive for the course
of moral philosophy or of purely political argument. Nor does
it make any difference whether the possibility of judicial review was immediately present to everyone's mind at just the
moment the Ninth Amendment passed Congress, or was ratified by the last necessary State. The idea that constitutional
rights were to form the substantive basis of such review was
so much in the air (and in the laws[4]) that it is unlikely that it
was overlooked. But in any case the direction of the Ninth
Amendment—that nonenumerated rights not be "denied or
disparaged," as against enumerated rights—was directed literally at the future, at the corpus of law-to-be, and affirmative settlement of the question (if, as I more than doubt, it was
a real question in 1790) of the rightness of judicial review, on
the basis of *any* right "enumerated" in the Constitution, would
settle the rightness of judicial review on the basis of those
rights not enumerated, though "retained by the people," because anything else would "deny or disparage" these latter, in
a quite efficacious way.

The only hitch is, in short, that the rights not enumerated are not enumerated. We are not told what they are. So the question is, "What do you do when you are solemnly told, by an authority to which you owe fidelity, to protect a generally designated set of things in a certain way, but are, in the very nature of the case, not told what particular things this set comprises?"

These are two possible courses to follow. One is to throw up your hands and say that no action is possible, because you haven't been told exactly how to act. The other is to take the Ninth Amendment as a command to use any rational methods available to the art of law, and with these in hand to set out to discover what it is you are to protect.

The first of these leads right back around, yet again, to a *practical* "denial and disparagement" of the rights not enumerated; it leads, indeed, to something a shade more imbecile than taking the Amendment as a direction of the course of moral philosophy, for it disclaims any power even to discover what rights are not to be "denied or disparaged" in out-of-court discourse. But at least you stay out of trouble.

The second course gets you into deep troubles. First is the trouble of deciding, by preponderance of reason, what methods are to be seen as legitimate, in our legal culture, for making out the shape of rights not named. Then there is the trouble—since no known legal method produces anything like certain results—of deciding where the preponderance of reason lies on the merits of *any particular claim of right,* when that claim is weighed by the methods you decide are legitimate. And the worst of it is that these troubles will never be done with, or even lessened. The methods of law are not a closed canon. The problems they must solve are infinite and unforeseeable. The solutions will never have the quality of the Pythagorean Theorem; time may even bring the conviction that some solutions, though confidently arrived at, were wrong, and must be revised.

Altogether, it's a pack of troubles. Maybe we ought to give up, and let the Ninth Amendment—and the priceless rights it refers to—keep gathering dust for a third century.

But there is one thing to note about the very real troubles that face us when we turn to the search that the Ninth Amendment seems to command. *These are the troubles not of the Ninth Amendment alone, but of law itself.* If they put one off the Ninth Amendment enterprise, maybe one ought to give up law altogether, try something else. But that course has its own problems. To turn to medicine, to music, to history—even to mathematical physics—is to accept the burden of troubles somewhat resembling those of Ninth Amendment law, or of law as a whole. For my part, too far superannuated to train for anything else, I would accept the challenge of Ninth Amendment law, as the same old (and forever new) challenge of law.

What methods are legitimate for finding and giving shape to the nonenumerated rights guarded by the Ninth Amendment?

Let me start with a rejection. Some people, faced with this question, would try to dig up every scrap of paper that happens to have survived since the eighteenth century; and to piece together some sort of "intent,"[5] with very little weight given to the transcendently relevant piece of paper, the one on which the duly enacted text of the Ninth Amendment was written.

I am one who thinks that, in a general way, our legal culture carries this sort of thing much too far. We sometimes treat statements made informally in one House of Congress as the exact equivalent, in everything but name, of formal statutory language; if it is right to do that, what are the formalities for? In the very teeth of Madison's quite sound and reiterated insistence that the confidential records of the 1787 Convention, not being publicly known until decades after the government was formed, ought not to be used to establish the public meaning of the Constitution's text, we sometimes seem to treat these records as all but superior in authority to that text itself. If we had to choose between our style of getting drunk on collateral and sometimes casual evidences of "intent," and leaving the stuff altogether alone, as the British did for centuries, I would choose the latter course—though I think sometimes a very cautious use of such material may be warranted.

But if there ever was a case where informal collateral evidence of "intent" must be useless, it is in regard to the finding of the rights that belong in the class of "others retained by the people." This language of the Ninth Amendment is apt for referring to things you haven't thought of or quite agreed upon; such language would be hopelessly inapt as a sort of coded-message reference to a closed class of "rights" you *have* thought of and agreed upon. If the decoded message read that rights A, B, C, D, *and no others*, were not to be "denied or disparaged," then the peculiar result would have been reached that we would have two kinds of "enumeration," the second kind being a coded enumeration, and that these *two* kinds of "enumeration" exhausted the class of rights to be protected, so that other rights, not thus "enumerated," *could* be "denied or disparaged." The informally arrived-at "enumeration" would thus be given an *expressio unius* force explicitly denied to the formal "enumeration" elsewhere. Something would have gone wrong here; doubtless the Greeks had a word for that kind of paradox. I am content to say that it seems to me to have no place in the robust common-sense world of the best work on American constitutional law.

Some pause might be given if we found a real consensus, uniting the *major pars* of the relevant eighteenth-century people, that some identifiable claim to a "right" was not to be looked on as guarded by the Ninth Amendment. But this would be a pause only. "Due process" is an evolving concept; "cruel and unusual punishment" is an evolving concept; the language of the Ninth Amendment seems even more apt than these to be mentioning an evolving set of rights, not to be bounded even by a negative eighteenth-century judgment based on eighteenth-century evaluations, and social facts as then seen.

I have treated this issue of collaterally evidenced "intent" quite abstractly; I don't know of any corpus of actual evidence that would enable or oblige one to treat it more concretely.

To me, the upshot is that we have to take this language as it comes to us. We are its inheritors; it "belongs in usufruct to the living," as Jefferson said of the earth. If we regard it (as I do) as directing us to do our best to discover for ourselves what unenumerated rights are to be given sanction, so that we may obey the Ninth Amendment command against their denial or disparagement, there is really no dearth of sound and well-tested methods for obeying this command, and so for moving in the direction of a rational and coherent *corpus juris* of human rights.

This statement gains a great deal of plausibility, or more, from the fact that that we have for a very long time been protecting unnamed (that is to say, "unenumerated") rights. We have done this, sometimes, under the guise of treating the language of the Constitution as highly metaphoric or otherwise figurative, as when we see "speech" in black armbands,[6] or see the making of a continuously disturbing noise near someone's land as a "taking."[7] But the *appropriateness* of any such metaphoric extension can be explored only by asking, for example, "Is the wearing of a black armband in the context in which we encounter it so similar, *in relevant respects*, to speech, that it ought to be treated as speech is treated?" And when we ask this question we recognize an old friend—the common-law method of arguing from the established to the not yet established, weighing similarities and differences, and deciding where the balance lies. Sometimes, as in the common law, this method creates a whole new heading, as with the "freedom of association" now generally recognized as arising, by the discernment of analogy, with the First Amendment rights literally "enumerated."[8] That is how we achieved the result of applying the double-jeopardy clause to cases where *imprisonment* or a fine is the penalty, though, if you read the Fifth Amendment, you find that, as named or "enumerated," this protection applies only to "jeopardy of life or limb." If, in time of peace, government attempted the "quartering" of sailors or government civilian employees in houses without the consent of the owners, or if consent was had from the "owner" of an apartment house but not from a tenant in possession under a lease, rational legal discourse could be addressed to the questions all these actions would raise in confrontation with the Third Amendment. You could, of course, talk as though the questions were whether a sailor is "really" a soldier, whether the tenant of an apartment is "really" the part-owner of a "house"—but even in this disguise these questions could be rationally addressed only by adding to them the phrase "in preponderantly relevant respects," or some such language. And this would lead right into the eternal question, "Is this a difference that *ought to make a difference?*" This question sounds familiar, because it is, first, a question repeated infinite times in the quest for rational justice, and, secondly, because it is the question continually asked—and answered in each case as best it may be—by the common law, the matrix of all our particular legal methodology. The issue is not whether the use of this method would be a bizarre innovation; the issue is whether any quest for decent law, with its parts rationally related, can possibly do without it.

Nor need this method of "analogy" be used only for small motions, like the supplying of the hiatus in the double jeopardy clause. The seeking of consistent rationality is a requirement of all good law, at every level of generality. If the central meaning of the equal protection clause is—as it pretty surely is—the forbidding of discrimination against *blacks*, then the propriety of applying that clause to discrimination against *women* can be reasoned about by marshaling the similarities (the genetic character of the trait, the maintenance of a grievously discriminatory regime by social stereotypes, and so on) in confrontation with the differences (for example, absence of whole-family discrimination stretching back through history). And since this is both a real and a complicated case, the Nineteenth Amendment would also serve as a starting-point for the eternal similarity-difference reasoning of law questing for justice: If women may not be excluded from voting, may they be excluded from office-holding? From jury service? And so on.

I must resist the impulse toward—and am really glad to eschew—any attempt here and now, or ever, to build a *corpus juris* of human rights on this basis, or on the others to be mentioned below. In a proper and profound sense, that corpus will never be built; it will always be building, like the common law at its best. If this method is not rational, then neither is the common law. And neither is any other attempt to give due effect to similarities and differences between already decided and newly presented cases and problems.

There is another generative principle in our legal system, the principle that law may be generated by due attention to the sound requirements arising out of social or political structures and relations.[9] This is how we got the warranty of fitness for human consumption in the sale of food; this is also how, in some States, we have recently gotten that warranty extended to bind the manufacturer and packager or canner, when structures and relationships changed in the food trade. This is how we got the insurers' right of subrogation, and the testimonial privilege for communications between penitent and priest. This is how we got the obligation of parents to care for their children. Our law—and, I venture to say, *all* law—has been and is continuously being shaped and reshaped by this generative principle. It is the principle from which we first derived the right, not literally "enumerated," to move from one State to another—a right established without reliance on any text.[10]

Now if we had only these two master methods—the method of similarity-difference reasoning from the committed to the not yet committed, and the method of reasoning from structures and relationships—we would have the means of building toward a rationally consistent, comprehensive, and fairly serviceable law of human rights. There is no question here of discerning that these rights are in any designative sense "mentioned" or "incorporated" in the Ninth Amendment, or that they derive from that Amendment; the language of the Amendment suggests—or commands, as I think—a quest outside itself, a quest for rights *nowhere* enumerated, not the mere tracing out of its references, which are, in the very nature of the case, of total vagueness.

But methods, in any mature and subtle legal culture, are never a closed class. Law ought to be seen to contain not only the means of striving toward rational consistency, not only the means of keeping the rules of legal decision in tune with the society's structures and relationships, but also the means—the methods—for reaching toward higher goals. Herein is the very best of what was so beautifully called, by Lon Fuller, "The Law in Quest of Itself." With the carefulness that is a condition of law's rationality, we may be able to discern and validate "other [rights] retained by the people" as latent in, and therefore susceptible of being drawn from, the noblest of concepts to which our nation is committed.

The earliest and best-attested source for such concepts is the Declaration of Independence. There is, then, a shorter way toward giving abundant application to the Ninth Amendment.

The Amendment speaks of rights *"retained* by the people,"— "retained," that is to say, in 1789—enacting that these are not to be "denied or disparaged" by virtue of their not having been "enumerated in the Constitution." The crucial word is "retained." What "rights" could "the people" be thought to have had *before* the enactment of this Amendment, though these are not "enumerated" in the Constitution?

Well, as we have seen, just thirteen years (a little more than two Reagan terms and one Bush term) before the Ninth Amendment went out to the States, the Declaration of Independence said that all mankind had certain rights—stated with great generality and with the greatest possible emphasis. Were these 1776 rights "retained by the people" in 1789? In 1791? They were said to be "inalienable." When and how were they "alienated" in that brief interval?

We are not talking about a time interval like the 450 years between Magna Carta and the coronation of William and Mary, wherein words may have undergone deep semantic metamorphoses. Let's make this a little more concrete. Of the fifty-five signers of the Declaration of Independence, thirty-four were still living when the Ninth Amendment became law. Some of them lived ten years longer; a few even longer than that. The Declaration of Independence and the Ninth Amendment are virtually contemporaneous documents. Their verbal fit is close: "All men" = "the people." The word "rights" is the key word in both. The word "retained," in the Ninth Amendment, exactly fits the situation, if you so much as take the Declaration at all seriously.

Thirteen years is a mighty short time for the vanishing of inalienable rights given by God—an attribution surely intimating some feel of permanency, even if you think that God is just a ghost story.

Let me move on to complete the tripartite canon. (It is well to put these things in place together and at the start; we will be continually returning to them throughout the book.) After our great Civil War, several Amendments to the Constitution were put in effect. (I remind you that the putting into effect of an Amendment to the Constitution is the most formal and authoritative act of which the American people are capable.) Slavery was abolished in 1865 (Amendment XIII) and in 1870 racial discrimination in voting eligibility was forbidden (Amendment XV). All who know the realistic history of Negro voting in the ninety-five years following this latter Amendment, in a large section of the country (a history graced by such monkey-tricks as the "white primary"), will smile. But the principle had been committed to in a form which could not be defaced by long years, and which at last shows signs of coming into and holding its own.

Between these two Amendments, in 1868, the Fourteenth Amendment came into the Constitution. It is a complicated thing, but we do not for our purposes need to go through it all. It will be enough to quote its first lines, the last words in our precious three-part charter of substantive human rights: "Section 1. All persons born or naturalized in the United States, and subject to the jurisdiction thereof, are citizens of the United States and of the State wherein they reside. No State shall make or enforce any law which shall abridge the privileges or immunities of citizens of the United States . . ."

It should be noted carefully that, under this provision, not only *national* citizenship but also *state* citizenship is conferred on certain vast classes of persons as a matter of *national* constitutional law. If you are born in the United States or are naturalized in the United States, then you are a citizen of the United States. If, being such a citizen of the United States, you live in Texas, then the *national* law of this Amendment ordains that you are a citizen of Texas; Texas has nothing to say about the matter. Not 10%, not 1%, just nothing. If you leave Texas and set up a residence in California instead, then you are, under the edict of *national* law, a citizen of California, again without California's consent. Each of these States might *add* (or try to add) a few citizens for sentimental or honorary reasons—say, all the living descendants of the Marquis de Lafayette or of John C. Calhoun—but these are a trace element. (The very validity of such romantic additions would be exceedingly doubtful, because it is Congress that is empowered to establish "an *uniform* rule of naturalization," and these words are probably to be seen as preemptive. The Fourteenth Amendment's confirmation of *state* citizenship should also be taken, I think, as preemptive.) In the overwhelming main the right to be and to call yourself a citizen of any State is not a right conferred by that State, but a right bindingly ordained as a matter of *national* constitutional law. It will be a more than satisfactory approximation to say (leaving out of account a handful of dubious oddities) that all citizens of Ohio are also citizens of the United States who reside in Ohio.

This denial to each of the States of the right to choose its own citizens might be looked on now as just another nail in the coffin of the theory that our States are "sovereign." That coffin can use all the nails it can get, because it yawns every now and then, on some inauspicious midnight, to give up

its undead, clad perhaps in the senatorial toga of Calhoun.
I will make more later of this national command of state citizenship; for now, it will do to say that unless this national
command is utterly vain, it must have imported more than the
right to put "citizen of North Carolina" after your name on a
calling-card.

I have written of the "harmony" of these three cardinal
commitments to a system of human rights. How does this last
of the three enter that harmony?

The "privileges and immunities" of national citizenship can
be, though they have not always been, treated in a straightforward manner. Even the lexicographer's art need not be
contemned, at least as a source of suggestion of possibilities.
As I write, I have turned around and looked up "privilege" in
the 1939 edition of the of the unabridged Merriam-Webster
that has somehow drifted through time into my possession
(1939 is not far from the middle distance between us and the
Fourteenth Amendment). I find that "privilege" can mean
many things—a grant of a special right to a particular person,
a grant of a patent or of a waiver, priority for a certain credi-
tor, "primage" in the sea-law of goods-carriage, congressional
or parliamentary immunity from arrest, the right of asylum or
sanctuary, a "call, spread or straddle" on the stock market or
the produce exchange. That's how words are—but none of
these seems at all the sort of thing you would put in the lead
clause of the treaty of peace after the greatest of wars up to
its time, or that you would think of imposing on a state government as a right of its citizens.

There is one definition in my trusty dictionary that seems
to rise to the great occasion:

3. Any of various fundamental or specially sacred rights considered as peculiarly guaranteed and secured to all persons by modern constitutional governments, such as the enjoyment of life, liberty, . . . [and] the right to pursue happiness.

The entry then quotes, as its example, the very words of Section 1 of the Fourteenth Amendment. I feel pretty safe in opting for that one, and not for a "straddle on the stock market," or "the grant of a manor," "ship's primage" or some concept of Roman law. Fortunately, lexicographers often have some common sense, and "3.", just quoted, fits a lot better in the context of this Amendment's language.

There is no reason for not taking the "privileges and immunities of citizens of the United States" (as our common-sense lexicographer evidently does) to include the "rights" set out in the Declaration of Independence. To avoid that conclusion, you have to drive a thin wedge between "citizens" and "people"; though in very much the greater part, they are the same human beings.

As this corpus of *national* "privileges and immunities" could have taken shape, under the deliberative processes of law, it could have become a visible and serviceable part of *national* law. In that character, it is a part of the supreme law of the land under Article VI, and under the view settled since *Marbury v. Madison,* a law superior to and controlling even Acts of Congress, as well as all other acts of the national government and any of its parts—and of course, under the express terms both of Article VI and of the Fourteenth Amendment, binding the States.

That would be all we really need on this score, but we have

just seen yet another way in which the States ought to be looked on as bound—a way not much attended to so far, but, I think, unquestionably solid. If you are a citizen of the United States and if you reside in Texas, then, by *national constitutional command* you are now a citizen of Texas as well. Texas has nothing to say about that. Now when a *status* is conferred on a person by the national Constitution, it can hardly be that the substantive consequences of that status do not go with the grant of the status, or that what these substantive consequences are to be is anything but a question of *national* law. The alternative would be that Texas citizenship, which the Fourteenth Amendment in its first sentence went to the trouble of granting and confirming to all citizens of the United States who live in Texas, is just an empty compliment, like a Kentucky colonelcy. (A compliment? Is it not rather an insult to the intelligence?)

The concept of "privileges and immunities" of citizenship "in the several States" is as old as the Constitution. (Article IV, Section 2.) But on the day the Fourteenth Amendment became law, every State entirely lost its right to exclude from its own citizenship, together (unless we live in Wonderland) with all "privileges and immunities" pertaining to such citizenship, *any United States citizen* residing within the State.

In one federal case, decided in 1825,[11] the "privileges and immunities" of *state citizenship* were very broadly defined. It may have been, though I think it unlikely, that a State might, before the great War and the Fourteenth Amendment, have been empowered to diminish these. But that's water that flowed under the bridge a long time ago. The Fourteenth Amendment quite surely redefined the position. The "privileges and immunities" of *state* citizenship are *now* fixed by national constitutional law unless the very clear grant, by that law, of state citizenship was a playful futility. Constitutional law does not deal in playful futilities.

So, as to the nation as a whole, there are "privileges and immunities." These are ample, and cannot be abridged or frustrated by any State. As to States, there are such "privileges and immunities" as go with citizenship in the State, this state citizenship being *mandated* by the national law of the Fourteenth Amendment.

Choice among these alternatives is perhaps not pressing; they take you to much the same place. Nor should this be surprising—since both are rooted in the same Declaration of Independence. One might add that there is nothing strange or paradoxical about the concurrence of national and state power in the securing of these "privileges and immunities." Such concurrence, on a much smaller scale, is a feature of the original Constitution, in the *ex post facto* and bill of attainder clauses of Sections 9 and 10 of Article I. As to all these concurrences of guarantees, large or small, the national interpretation has to prevail in case of conflict—because Article VI says so. Practically speaking, the guarantees of free speech and freedom of religion have been held to apply in just this way, though the Fourteenth Amendment "privileges and immunities" clause would furnish a much more acceptable basis than the one that has been used (see below, pp. 78–80 and 87–106).

Now many readers will know that these arguments were rejected or in part not reached, in the *Slaughterhouse Cases* in the Supreme Court in 1873,[12] ten years after Gettysburg. The arguments just above thus remain "visionary"—as "visionary" as

Abraham Lincoln's sacred prophecy at Gettysburg ("that this nation, under God, shall enjoy a new birth of freedom"). These first words of the Fourteenth Amendment brought his vision briefly into sight in law. The Supreme Court, by a five to four vote, waved that vision away—insofar as the newly confirmed moral unity of the American people, so dearly bought, could be waved away. Since we are examining the present real condition of American human-rights law, we have to take account of that destructive decision.

I will return to a detailed critical examination of the *Slaughterhouse Cases* in a later chapter. But for now we must not leave the matter altogether hanging. Let me deal as honestly and fairly as I can with the grounds on which the Court rejected any significant application of this clause of the Fourteenth Amendment to the States. I shall proceed schematically, assuring you that many questions will be dealt with later.

First, the Court points out that the Fourteenth Amendment recognizes that there is a *state* citizenship and a *national* citizenship, and that these two citizenships are distinctly referred to. So far, so good.

Secondly, says the Court, the mass of civil rights (before what we in the South used to call the "late unpleasantness"— that is to say, the greatest war, up to its time, in human history) was in the care of the States. One could argue very fundamentally about that, but that will be for later. Let us take it, *arguendo*, as broadly, though quite certainly not entirely, true. (This is not to concede, even *arguendo*, that there is anything anomalous or contradictory in the States' and the nation's both offering governmental protection for the same human rights.)

Finally, says the Court majority, it is incredible, whatever the bare words of the Fourteenth Amendment may seem to say, that the state power to define human rights within its territory "was intended" to be transferred to the care of the national government. Of this one may say, summarily:

1. The point is simply asserted, without any seeming attention to the awkward fact of the great Civil War. How can the Court pronounce, assertively and confidently, what "intention" may be projected as to the *future* adjustment of power in a national union that has been through such an experience?

2. The imputed "non-intent" sits more than uneasily in Section 1 of the Fourteenth Amendment, because the very next clause of that section, the famous "due process" clause, *does* radically and widely affect state independence as to *permissible procedures* for depriving any person of life, liberty or property. Of the procedures permissible, the States had been the sole judges. After the Fourteenth Amendment, there was and is an overriding national law of "due procedure." (There is also a new national law of "equal protection," touching and where needful changing every aspect of state law.)

How could the Court conclude that *these* fundamental changes were *thinkable*, on the same day that the extension of general protection to a wide set of *substantive* privileges and immunities was *incredible?*

3. In an attempt to avoid an impression that it has entirely nullified the "privileges and immunities" clause, the Court does list a pitiful handful of "privileges and immunities" of national citizenship which *already existed* before the Fourteenth

Amendment. This preposterous list is really a *reductio ad absurdum* of the Court's holding.

4. To close a loophole, the Court is most insistent on the point that each State, being the authority responsible for the securing of civil rights, in its territory, has full power to diminish these at will, not giving any weight to the change in authority as to the holding of the "privileges" of *state* citizenship, that occurred when, under the first sentence of Section 1 of the Fourteenth Amendment, state citizenship itself was now enjoyed by national constitutional command.

5. Finally, the Court makes no attempt to solve or even to face the problem of the "privileges and immunities" of those who are "citizens of the United States," having been "born or naturalized therein," but who are not residents of any State—because they reside in the District of Columbia, or in the territories, or in foreign countries.

To condense all this, the upshot of the decision is: (1) that United States citizens as such have only a scrappy handful of *preexisting* "privileges and immunities" of *national* origin, and (2) that each of the States in which these citizens reside has full power to alter or to abolish altogether whatever "privileges and immunities" they have as *state* citizens. What a result for the Civil War! What a crumbly basis for *national* political morality! What a complete moral triumph for John C. Calhoun, who never won a battle in his own life. The decision simply cannot be right. I will return to this point later.

he really doesn't like Calhoun

It takes some very picky reasoning not to read the words "citizens of the United States" to include "citizens of each of the States severally," and to claim for these directly the same privileges and immunities guaranteed to citizens of the United States. One dominant fact is that these are in very much the greater part the same people. "Law," as the late Jerome Michael liked to say, "is a practical subject."

The national privileges and immunities are set up not for ornament, but for a practical purpose—namely, that a valuable citizenship in a free society is to *prevail* in this nation throughout its territory. If this territory comes to consist (as it did not in 1868 or 1873) almost entirely of the territories of the States one by one, and the citizens of the States are the major part of the "people" of the United States one by one, these citizens will not be guaranteed the things promised in the charters of our freedom, if the States may whittle down those freedoms, those rights—free speech, immunity from religious coercion, the right to live with your orphaned grandchildren—contained in one hallowed phrase, "the pursuit of happiness."

The fact (an amazing one in view of the intervening great Civil War for national unity) is that, on the level of our highest values, this *Slaughterhouse* holding is a very close fit with the banefully "classic" doctrines of John C. Calhoun, the great heresiarch, on the relative importance and worth of *national* citizenship (not very much) and state citizenship (nearly everything). The South may be said to have surrendered to John Marshall but the Supreme Court, in *Slaughterhouse*, surrendered to Calhoun. As a bridge into the next chapter, let me remind you that most successful and unsuccessful claims of infringement of human rights are made *against the actions of state and local governments.* John Marshall, in an early opinion on a claim of right by Indians, remarked with certain truth that the people most likely to infringe against your human rights are not the people who are far away, but are those nearest you, where getting at you is easy. Attempts at book-banning, *de jure* or *de facto* racial segregation, the prohibition of the teaching of evolution, the censorship of movies, the regulation of clothing, are things mostly undertaken by state and local governments. (An apparent exception exists in cases where the national government runs, as in the District of Columbia, something that amounts to a local government.) If the national "privileges and immunities" are not good against the States and their subdivisions—if *that* is the Catch-22 of our boasted national regime of freedom—then we have set up nothing but a beeswax simulacrum of a free nation, for in fact and in truth, as to almost all its territory, and almost all its people, it is a *nation* that does *not*, as a *nation*, "secure" human rights. It was just that kind of result that the Civil War was in the deepest sense fought and won to prevent. Such a concept is death to Abraham Lincoln's sacred prophecy, at Gettysburg, that this nation, *as a nation*, might have a "new birth of freedom."

The purpose of this chapter has been the introduction, no more than that, of an important theme—that a satisfactory fundamental legitimization of an open-ended series of open-textured human rights, the only thing that can cover the lives of real people in their infinite variety and in their unforeseeable development, can be seen to flow from our basic commitments. A few collateral observations must be made.

First, I invite you to believe I have chosen the components of this foundation carefully, and that I have expounded each of them on the basis of the words chosen by the original expressers to be the words of authority. I think that is all we need, and that we proceed most safely if we respect those very words, following them as far as they may legitimately lead us. If the words of the applicable passages in the Declaration of Independence, of the Ninth Amendment, and of the Fourteenth Amendment's "citizenship" and "privileges and immunities" clauses were really dark, if they could be interpreted only by esoteric reason, or by "narrow verbal criticism," or by an attempt to piece together the informally and sometimes delphically expressed subjective "intents" of selected people long gone, in regard to questions they did not face, then we might have to enter on such inquiries—of proven inconclusiveness though they be. If the Ninth Amendment and the relevant parts of the Fourteenth Amendment were nothing but "ink-blots," giving no guidance to the mind, whether of judge or of citizen, then we might have to set, generation after generation, such Rorschach tests as recurrently anxious scholarship could devise *and of course endlessly revise.* But I earnestly invite you to consider whether the words actually agreed to as the actual operative words are really all that obscure in character.

The Declaration of Independence asserts that all people have the "right" to "the pursuit of happiness." What would make you think that that phrase—given its full generality of expression and its close collocation to "liberty," and considering that it is a "right" not to "happiness" (which no political organization can give) but to the *"pursuit"* of happiness— meant anything in 1776 that it does not mean now? It is true that some change has perforce taken place in judgments as to the chance that happiness will result from different choices of means of pursuit. Perhaps more people then than now thought that devotion to religion was the best road to happiness. But if the word "pursuit," directly associated with "liberty," implies the right to *choice of means*, there is no change in *general* meaning. Or how would you state the distinction, and support its reality?

Passing to the Ninth Amendment, the "ink-blot" Amendment *par excellence*, it evidently calls for expounding the meaning of the phrase "other ["rights"] retained by the people." But what could make you loath to turn, for this necessary purpose, to the most apt source, the Declaration, where the rights of humankind are set out, just thirteen years before the Ninth Amendment was proposed and transmitted to the States? Why keep worrying around with the same old ink-blot, when there is so ready (indeed so unavoidable) a route of escape.

When we get to the word "privileges" in the Fourteenth Amendment, we do have a somewhat technical word. So it was and is, as I have shown, in many curious contexts, none of them seeming to have anything to do with any imaginable objective of this Amendment. But as my lexicographic material has just showed, there is one accepted and common meaning of the word "privileges" which is judged (presumably by an apolitical lexicographer) to fit the *very words* of the Declaration of Independence—including the "right to the pursuit of happiness."

What in the world is the problem? The fit of the Declaration to the Ninth Amendment seems natural enough. The fit of the Declaration—as now incorporated in the Ninth Amendment—is the first thing that strikes the mind of a person whose professional job is to give apt illustration of the use of the word "privilege." What drives people to pass over all this? Particularly lawyers?

I am sure that the result of my line of thought—a generalized and endlessly productive system of human rights—is in itself repellent to many people, however clearly the authoritative words I have been addressing may seem, in their harmony, to lead to that result. There is a myth that lawyers must think small, even meanly, or lose the aura of professionalism. As in all other matters, we should think at the level of magnitude proportioned to the problem. Insistence on thinking small veils the largest facts from view. If we are to have a true *system*, a productive system of human rights, we have to commit ourselves to thinking large. If we are to take seriously the noble words of our past, we must pronounce them with emphasis and without apologetic hesitation. After all, in doing this we risk a good deal less than being hanged, drawn, and quartered.

Very closely akin to the fear of ampleness of thought is the fear of generality. Our national commitment to human rights starts with generalities: liberty, the pursuit of happiness. What is sometimes forgotten is that all law works from level to level, with commitment to great general principles that have to be worked into practice through insight and experience. This is true, for example, of our commitment to "freedom of speech." That concept must be worked into life at all levels of generality, and each level has its own problems of prudence and balance. The same pattern can be expected in our commitment to the right to the "pursuit of happiness," which must be worked out in the same way. But neither of these things means that neither "freedom of speech" nor "the right to the pursuit of happiness" can be worked into law.

This is a characteristic not just of one kind of law, or of one level of generalization in the commitment which we must go on trying to apply. Take the law of contract. It rests on acceptance of the vast principle that contracts are to be looked on as binding, and are to be enforced. The actual working-out of this principle is in lower levels of law. A contract to submit to a disfiguring mutilation would not be enforced. A contract by which a shipper by sea gives up his right to the carrier's diligence in care of the cargo is widely disallowed and treated as invalid, both by case law and by statute law. The definition and scope of a doctor's obligation to his patient cannot be freely manipulated by contract. And so on *ad infinitum*. None of this means that contract law is not law. It just means that the first principles of contract law are not absolutes. Of course no one could have an absolute right to enforcement of any contract, or to say anything regardless of its untruth or of the noxious character of the means used. But the general principle of contract law has serious reality and wide relevance in law. So does the principle of freedom of speech. Much flows from each of these principles, though not everything flows from them that they might seem to promise, if one didn't know how law of all kinds is developed on different levels.

This chapter can appropriately end with a condensed and rather assertive summary of my chief pattern of thought for this book and for the fresh beginning of our human-rights law. Each of the following points requires thought and discussion; each has already been given some of these things, and will get more—besides some collateral shoring-up. But the following list will serve to focus the mind on the points around which everything else in this book revolves:

1. The 1776 Declaration of Independence commits all the governments in our country to "securing" for its people certain human rights, "among which are life, liberty, and the pursuit of happiness." These are the certified cardinal values of our political morality. It is a separable question whether they are also "law," available as such. (I think they are.)

2. The 1791 Ninth Amendment to the Constitution is unmistakably "law," and unmistakably rejects the idea that a human right, to be valid in law, must be enumerated (or explicitly named). The Amendment does not say which rights are the "others retained by the people," and are therefore not to be disparaged or denied (what a silly thing to think it could have done that!). But the Declaration of Independence, uttered a mere thirteen years earlier, supplies this lack in major part: There is no apter reference than the Declaration for clearing up the words "retained by the people," whether the Declaration itself be "law" or not. The Ninth Amendment is certainly "law." All governmental units—the national government, and by the consequences of the rule of construction it sets in place, the state governments with their subdivisions—are bound by it, as a matter of law, and therefore ruled by its incorporation by reference of the Declaration's words.

3. The "citizenship" and the "privileges and immunities" clauses of Section 1 of the Fourteenth Amendment form a complex whole.

Citizens of the nation are, by national constitutional command, also citizens of those states, respectively, wherein they reside. Since state citizenship is a nationally commanded status, the substantive incidents of such citizenship are to be governed by national law. *To be a citizen of the State wherein you reside is a privilege annexed to and flowing from national citizenship.*

These two paths of thought lead to exactly the same result—the enjoyment, by the citizens of each State (as well as by the citizens of the nation, who are mostly the same people), of national rights founded on the Declaration of Independence and on the Ninth Amendment (see point 2, above). The national privileges and immunities of citizens of the United States are similarly derived and defined, and the States are forbidden to "abridge" them.

It is only against this point 3 that there is any precedent even technically binding—the *Slaughterhouse Cases* of 1873. This decision will be thoroughly discussed later. It is my own view that no sorrier opinion was ever written than the *Slaughterhouse* opinion, and that that case should be thrown into the rustiest trash-can of legal history. I have no illusion that this will be politically easy, but I think it is intellectually and technically easy.

Now the only thing I will add about this summary is that, if one shows it to a lawyer who, like most lawyers, has unequivocally (and often unthinkingly) accepted the most astounding, even scandalous, fact in American legal history— the consigning of the Declaration of Independence, the Ninth Amendment, and the Fourteenth Amendment "privileges and immunities" and "citizenship" clauses to virtual oblivion— that lawyer will smile tolerantly. This whole book is about the total unsupportability of these conventionally sanctioned attitudes of thought. I want you, now at least, to believe that I intend with the highest seriousness the proposals I am making for bringing home to their deserved positions of primacy these most precious of our treasures, forming our eternal tripartite and harmonious commitment, as a nation, to a *general* system of human rights.

And remember, as this lawyer you are talking to politely smiles, that the very great majority of "sound" lawyers, when I was already well grown and then some, thought that the segregation of blacks, by law, from cradle to grave, and the "white primary," were "perfectly legal."

Chapter 2

HUMAN RIGHTS
AND THE
STATES

In 1990, in a talk given in Moscow, I made what I commend to you now as the cardinal point about the place of our States in the American scheme of human rights. It is, incidentally, the most important general point about human rights in America. Because of the particular topic ("Glasnost") that had been assigned me, my thoughts were mainly directed to the human right of free speech, but the point is one of general applicability, and of a truth that goes to the life of real human rights in our country:

In the United States, the right to free expression, if it is to be enjoyed in the real world, has to be guarded (like all American human rights) within the structure of what some of us have called "federalism"—the *coexistence*, in as good as every square foot of the country, of two governments, the national government and the government of some State. Free expression might be guaranteed in the most absolute and efficient way against *infringement by national law*—and still, *if there were no national guarantee of free expression, good against the States*, you might live all your years in dread of going to prison for publishing or even owning a book dealing favorably with Socialism. The nation would not be in practical truth a free nation. There is no advantage, believe me, in going to a state prison rather than to a national prison.

Now it is true that most or all of our States have in their own constitutions some protection of free speech. But these protections—not being *national* law—would by themselves be subject to final interpretation by *state court* judges, most of whom are popularly elected, in decisions not reviewable in the *national* courts. And a state constitution is freely amendable by the legislature and the people of the State; in California, for one huge example, the state constitution may be and has often been amended by popular referendum.

If, as a last resort, you think this represents only a theoretical danger—that our States would never do bad things about free speech—then you ought to read a few of the hundreds of cases in which they have attempted suppression of speech of many kinds, including explicitly political speech. In this respect, they have shown far more imagination, over a much wider range, than the national Congress.

The upshot is that if we had no general national law binding the States to respect freedom of speech, enforced in the national courts, we would in no way be sure of at all enjoying this freedom, virtually anywhere. (This, I say again, is true of all human rights in the United States; without, for example, a national constitutional rule protecting the freedom of contraception against *state* violation we would not have freedom of birth-control and family planning, in this country, except where the States one by one chose to grant it from time to time. If none of them chose to grant it, we wouldn't have it anywhere—except in the District of Columbia, a national enclave—or I guess maybe in our Post Offices, where the question does not usually come up.)

It is well to mention, too, that "state law" means not only the formal law of a whole State, but also the ordinances put in place by cities and other subdivisions of the State, and the actions of State judges and other state officials, including police.[1]

Since this book is in large part about the force in law of "unnamed" human rights—such rights as are not specifically named in the Constitution or its Amendments—let me go a little deeper into the predicament we would be in, even as to free speech and freedom of religion, if we had to cite a constitutional text protecting these prime human rights *against the exercise of state and local power* before we could justify interposing a shield against such *state and local infringement.* Anybody who can read can read that the First Amendment to the Constitution, where these rights are "enumerated," forbids their infringement only by the *national Congress.* I shall later consider more fully the manner in which this very plain textual limitation of the force of the First Amendment guarantees has been surmounted (or clambered over in the dark) to bring them to bear against the States. At this point we can just note that, however you look at it, the explicit textual guarantees even of expressive and religious freedom do not, in their own terms, apply to the States at all.

But this illustration only dramatizes, perhaps just a little, the fix we would be in if *other* valuable human rights were not protected against *state* power, unless such rights were explicitly named. The pervasive and ineluctable fact is that power may be divisible, but the personhood of each human being is not so "divisible" as to make it possible for a person to be "free" when looked on as a subject of national power, while "unfree" as a subject of state power—free to marry at any age over eighteen, when considered as a subject of national power, but headed for a state prison if one marries before turning thirty-five, if a state law forbids that except when three solvent sureties make bond that no offspring of the proposed marriage shall become public charges. It is the same human being, and state prison is no better than national prison.

This is the lurking Cátch-22, the treacherous snake-in-the-grass, in "Our Federalism," with respect to our guaranteeing human rights as "one nation indivisible."

Following the plan of Chapter 1, that of setting out a fresh foundation for human rights in American law, I shall consider in this chapter the applicability of that course of thought to the judging of actions of the States. What has this three-part scheme to do with the States?

Are they not bound by the three commitments that I have invoked in the first chapter: the Declaration of Independence, the Ninth Amendment, and the privileges and immunities clause of the Fourteenth Amendment?

First, as to the Declaration: This transcendentally solemn act was assented to by the representatives of all the States then in being; States later joining are of course equally bound. The language of the Declaration speaks to the duty of *governments in general* to "secure" the rights to life, liberty, and the pursuit of happiness—with no distinction as to the levels or partitions of government so obligated.

I have already committed myself to the view that the Declaration should be looked on as embodying "law" in the full sense. If it does so, then of course it binds the States as a matter of law, and controls state law.

But, as I have said above, you need not make up your mind on that question, if you accept the closely related view that the most natural reference, by far, of the Ninth Amendment, in the phrase "others retained by the people," is to the "rights" already named (just thirteen years before) in the Declaration. I remind you of the close fit between the language of these two documents, of the propriety of taking the Ninth Amendment's words "others retained by the people" as referring in large part to rights to be looked on as "retained" because they had already been designated as rights of the people by the Declaration, and of the virtual contemporaneity of the documents—separated as they were by an interval less than the one that now divides us from the middle of Reagan's first administration. The acceptance of that equivalency makes the words of the Declaration in effect a part of or (more accurately speaking) a prime subject of the rule of construction commanded by the Ninth Amendment, and therefore a part of the Constitution, and so a part of the "supreme Law of the Land," binding within and upon every State.

In the present context, concerning as it does the human-rights obligation of the States, the only Ninth Amendment question remaining is whether that Amendment in its terms applies at all to the States. The radical reference-point in the Amendment is to the "enumeration in the Constitution of certain rights." "Enumeration" of *some* rights is not to be "construed to deny or disparage others retained by the people." It is the fact that a number of rights, good against the States, *are* "enumerated" in the Constitution. It is hard to calculate the firm number of these. Some of the rights are resultants from explicit rules, such as the Article I designation of voters in elections for Congress, as those qualified by state law to vote for the more numerous branch of the State legislature; by irresistible inference, this generates a correlative *right* in the same people to vote for Congressmen. Some are direct and specific, like those collected in Article I, Section 10. At least one provision (Article IV, Section 2, clause 1) in the original Constitution is in quite general terms. The right of citizens and others to enjoy the benefit of the "full faith and credit" rule of Article IV, Section 1, is binding on every State. There are certainly enough such protections of rights in the original Constitution and in the Amendments that are *to all intents and purposes* part thereof, to put to flight the idea that there were and are therein few or no rights good against the States, and therefore nothing of that sort to which the Ninth Amendment could apply.

Moreover, the Ninth Amendment, like all prior and subsequent Amendments, is "to *all* intents and purposes . . . part of the Constitution" (see Article V). There is no reason why a rule of construction, such as that commanded by the Ninth Amendment, should not be held to prevail as to the "enumeration . . . of rights" in those *later* amendments (such as the immunity from sexual discrimination *in voting* forbidden in Amendment XIX), which, under the Ninth Amendment "rule of construction," should not be held to be grounds for "denying or disparaging" other rights validly derivable by any legally proper method, whether from structure or analogy, or otherwise. (It would be interesting to rework the cases dealing with the post–Civil War Amendments [XIII to XV] with this in mind.)

To sum up on this point: In its literal terms, the Ninth Amendment rule of construction does apply to rights guaranteed against state infringement, because that is how the Ninth Amendment designates and defines its own application. If there is any overpowering structural reason to the contrary, I should like to hear what it is, but I'll let those who believe in it phrase it and support it. I don't believe in it, and I would be at a loss to construct support for it. We ought to stop quibbling about our seminal national commitments.

Let us go on, then, to the third of our master commitments to human rights—the applicable clauses of the (1868) Fourteenth Amendment. For convenience, I'll quote these again:

> Section 1. All persons born or naturalized in the United States, and subject to the jurisdiction thereof, are citizens of the United States and of the State wherein they reside. No State shall make or enforce any law which shall abridge the privileges or immunities of citizens of the United States; nor shall any State deprive any person of life, liberty, or property, without due process of law; nor deny to any person within its jurisdiction the equal protection of the laws. . . .
>
> ——
>
> Section 5. The Congress shall have the power to enforce, by appropriate legislation, the provisions of this article.

This Amendment was submitted to the States under the condition that no State that had joined the Confederacy could be readmitted to the Union without ratifying the Amendment. The parts of the Amendment of major and permanent significance are Sections 1 and 5, just quoted. Section 1 is therefore the salient, substantive provision of the treaty of peace ending our great Civil War. Its ratification was imposed on the rebelling States as the price of their reunion.

There are two distinct paths opened by the fact that not only *national* citizenship but *state* citizenship as well is enjoyed under the unmistakable command of this *national law.* I feel justified in stressing this again because conversations with students, even advanced law students, even some lawyers, have led me to think of it as a little-known fact. It's as though a cloud of unknowing covered these words in the Fourteenth Amendment. If I'm right on this, the cause may be that the States are so often referred to as "sovereign," and the power to designate the "sovereign's" own "citizens" is so commonly thought of as a power quintessentially pertaining to "sovereignty," that many people can't quite grasp what the Fourteenth Amendment so plainly decrees. Be that as it may, the fact is as it is: All or almost all the citizens of any State are (since 1868) citizens of that State because the Fourteenth Amendment says they are—not, I might add, that they "shall be," but that they "are."

So when we ask, at this turn in the argument, whether this command carries with it any human rights good against the so-commanded States, we are really asking whether the Fourteenth Amendment went to all that trouble just to do a vain thing—conferring a merely titular "status" of state citizenship that created no substantive consequences. If we cannot satisfy our minds with so lame and impotent a conclusion, then we have to ask what the rights of state citizenship are to be taken to be, *as a matter of national law.*

The first clue is that the phrase "privileges and immunities of citizens in the several States" occurs in the original Constitution:

> Article IV, Section 2. [1] The Citizens of each State shall be entitled to all Privileges and Immunities of Citizens in the several States.

In 1825, Supreme Court Justice Bushrod Washington, sitting as Circuit Justice, was confronted with a question about the meaning of the just-quoted provision. Did it render invalid a New Jersey law limiting to residents of that State the right to take oysters, clams, and shells in New Jersey waters? Rightly or wrongly, Mr. Justice Washington upheld the state statute, on the narrow ground that these mollusks in New Jersey waters were a part of the state's patrimony, in effect the "property" of its actual residents, the benefit of which the State might permissibly reserve to these residents. To reach a conclusion on this question, he had to consider what the phrase "privileges and immunities of citizens in the several states" *did* mean. His exposition, in *Corfield v. Coryell,* became classic:

> The next question is, whether this act infringes that section of the constitution which declares that "the citizens of each state shall be entitled to all the privileges and immunities of citizens in the several states?" The inquiry is, what are the privileges and immunities of citizens in the several states? *We feel no hesitation in confining these expressions to*

> *those privileges and immunities which are, in their nature, fundamental; which belong, of right, to the citizens of all free governments; and which have, at all times, been enjoyed by the citizens of the several states which compose this Union, from the time of their becoming free, independent, and sovereign.* What these fundamental principles are, it would perhaps be more tedious than difficult to enumerate. They may, however, be all comprehended under the following general heads: *Protection by the government; the enjoyment of life and liberty,* with the right to acquire and possess property of every kind, [*552] *and pursue and obtain happiness and safety;* subject nevertheless to such restraints as the government may justly prescribe for the general good of the whole.[2]

Now do you see anything familiar within that description of the "privileges and immunities of citizens in the several States"? Of course you do. Justice Washington has been looking at the Declaration of Independence, which was proclaimed when he was a lad of fourteen, and which, for that reason, he probably could not see as being lost in the remote cuneiform antiquity that shrouds the Code of Hammurabi. (After all, two of the signers of the Declaration, Jefferson and John Adams, were still living in 1825!) "Life, liberty, and the pursuit of happiness" are right there in the Corfield opinion.

That lexicographer we encountered in the first chapter of this book (see p. 26 above) knew his business. It was natural almost to the point of inevitability that the famous words of the Declaration should be taken as supremely suitable to fill out and explain the words "privileges and immunities of citizens." This is not a conclusion that we, or the lexicographer, or Mr. Justice Washington, must strain and struggle to reach. It is a conclusion we would have to strain and struggle to avoid reaching. Why should we want not to reach it? It says all that is best about our nation. It makes a unity of our story.

A crucial and fundamental clarification in authority occurred with the passage of the Fourteenth Amendment. The rather convoluted wording of the Article IV passage might seem to a "poring man" to leave it open whether a state might *subtract*, from the substantive content of its own citizenship, some of the rights mentioned in Mr. Justice Washington's words. True, he speaks of these rights as having "been enjoyed at all times" by citizens of every State, he calls them "fundamental," and he says "they belong of right to the citizens of all free governments." But the *permanency* of that enjoyment might *just barely* not be thought protected, as against some State that might want to diminish these rights. This was perhaps not a likelihood, and it certainly would be puzzling in the face of the holding that these are the rights enjoyed under "all free governments," but it was a rarefied theoretical possibility before the Fourteenth Amendment.

But since the Fourteenth Amendment was ratified, the *status of being a citizen of a State* has been something firmly based on the *national* law of the Constitution. How trivial and how paradoxical it would be to think that some State might alter the substantive content of this grandest and most comprehensive of its obligations to its own people, who are also *our* people, part of *us*, part of "The People of the United States"!

This may be a point at which you wonder why I am insisting so strongly—and at such length—on something so obviously right, if I have been right in my contentions that precede its statement.

I have given it as my view that (treating such a question as ever new), we ought now to recognize that the Declaration of Independence has the force of law, and that the States are bound by the law of the Declaration.

I have also put forward the view that the Ninth Amendment must properly be looked on as referring to the rights set out in the Declaration, and that the Ninth Amendment must be construed as applicable to the States.

I have just now put before you the view that the *state citizenship* subclause of Section 1 of the Fourteenth Amendment carries within itself the command that the States treat their citizens (bindingly made such by the Amendment) in accordance with the privileges and immunities *of the Declaration,* if "citizenship" in a political body within the American system implies that that body must treat you in accordance with the Declaration. On the other hand, nothing in this thought impairs the generality and reach of the "privileges and immunities of citizens of the United States" in the Amendment, which, by virtue of its part of being the "supreme Law of the Land," becomes binding in and upon every State.

Since these lines of thought end up at the same place, why go through them all? Nobody ought to be shocked by these conclusions, because that is just about what Justice Washington said in 1825, less than fifty years after the Declaration went out. I am doing this because I want to bring you to see that the commitment of the American republic to human rights has depth in time and in the harmony of its precious components.

Consideration of the "privileges and immunities" of the Fourteenth Amendment brings us to yet another instance in this harmony, in this case bridging the state and the national obligation to "secure" human rights.

The Fourteenth Amendment gives authority to the idea that these "privileges and immunities of citizens of the United States" exist. What are *these* to be taken to be? What, indeed, except the very "rights" named in the Declaration of Independence? Is not this conclusion powerfully confirmed by the Preamble, wherein the *purposes* of the ordainment of the Constitution are stated: "To form a more perfect Union, *establish justice*, insure domestic tranquillity, *provide for . . . the general welfare, and secure the blessings of liberty* to ourselves and our posterity"? How could it be thought that these great ends could be accomplished unless the privileges of citizenship were borne up by *national* law? How else would there exist a national power to "secure" the rights of the Declaration? The obvious answer to these questions is the key to the interpretation, the filling with content, of the *Fourteenth Amendment* phrases. If not, what was the situation of the residents of the District of Columbia, or of citizens of the United States living in the territories, or abroad?

From this conclusion may be drawn innumerable obligations binding the *national* government—an open set of obligations. And the key point here is that the privileges and immunities clause of the Fourteenth Amendment expresses an open-ended series of rights, set up and recognized in this clause as inherent in *national* citizenship. What is the duty of the States respecting this corpus of human rights, established by "the supreme Law of the Land"?

The States are not to use their general law-making power to "abridge" them. Are not "the Judges in every State bound thereby, anything in the Constitution and Laws of any State to the contrary notwithstanding"? "No State" may "make [a] law" that on its face or in its operation nullifies, or in some way curtails or abridges the force, the authority, the *efficacy in practice* of some privilege or immunity of national citizenship. When we get that far, we are really up against the Supremacy Clause of Article VI.

> This Constitution, and the Laws of the United States which shall be made in Pursuance thereof; and all Treaties made, or which shall be made, under the authority of the United States, shall be the supreme Law of the Land; and the Judges in every State shall be bound thereby, any Thing in the Constitution or Laws of any State to the Contrary notwithstanding.

Let's go back to the illustration I began this chapter with. I have noted that the First Amendment free-speech guarantee, on its face, applies only to Congress' actions. But if the enjoyment of "liberty" and the right to the "pursuit of happiness" are "privileges and immunities" of national citizenship, and if liberty to speak is a liberty, and if the acquisition of knowledge and ideas are, to many at least, vital ingredients in happiness, then the State that infringed on these rights would be "abridging" a privilege or immunity of national citizenship, now very clearly made law by the Fourteenth Amendment. That idea goes all the way back to *McCulloch v. Maryland*.[3] After all, the Maryland law in that case did not directly and in terms violate a national law; what it did was to *interfere with the functioning of a system created by national law.* The prime system the United States exists to secure is a national regime of human rights.

In the light of all this, what is to be taken as the reference of the phrase "privileges and immunities of citizens of the

United States"? Can you imagine anything more silly and trivial than the reduction of this largest of our generalizations about the consequences of our universal common citizenship to a scrappy half-dozen or so of minor and inchoate rights, excluding almost everything of weight, and shutting off the view that seemed so natural to Justice Bushrod Washington—Declaration of Independence and all? The all but astronomical sublimity of Mount Rainier, speaking silently out to the far streets of Seattle, is reduced to the merest dusty trace of a molehill.

Well, you don't have to answer the question whether you can "imagine" such a thing, for that is exactly, and without exaggeration, the holding of the *Slaughterhouse Cases*,[4] decided by the Supreme Court of the United States in 1873. In the exact etymological sense, the Court *annihilated* the privileges and immunities of national citizens, insofar as these were to be seen as ordained by the Fourteenth Amendment.

I am not going to put you through very much case-analysis in this book. But I have to do it this time, because you cannot really understand the shoddiness of the professional foundations of our current human-rights law, unless you have an understanding of what happened in the *Slaughterhouse Cases*. This is probably the worst holding, in its effect on human rights, ever uttered by the Supreme Court. It's worth (and requires) some effort to understand it.

The bare facts were simple; they can be most handily stated by quoting from the first headnote:

The legislature of Louisiana, on the 8th of March, 1869, passed an act granting to a corporation, created by it, the exclusive right, for twenty-five years to have and maintain slaughter-houses, landings for cattle, and yards for enclosing cattle intended for sale or slaughter within the parishes of Orleans, Jefferson, and St. Bernard, in that State (a territory which, it was said . . . contained 1154 square miles, including the city of New Orleans, and a population of between two and three hundred thousand people), and prohibiting all other persons from building, keeping, or having slaughter-houses, landings for cattle, and yards for cattle intended for sale or slaughter, within those limits; and requiring that all cattle and other animals intended for sale or slaughter in that district, should be brought to the yards and slaughter-houses of the corporation; and authorizing the corporation to exact certain prescribed fees for the use of its wharves and for each animal landed, and certain prescribed fees for each animal slaughtered, besides the head, feet, gore, and entrails, except of swine.

The interesting question, for us, is whether this statute violated the "privileges and immunities" clause of the Fourteenth Amendment, as quoted above: "no State shall make or enforce any law which shall abridge the privileges and immunities of citizens of the United States . . . "

Of this Louisiana law the Court majority says, early in its opinion:

It cannot be denied that the statute under consideration is aptly framed to remove from the more densely populated part of the city, the noxious slaughter-houses, and large and offensive collections of animals necessarily incident to the slaughtering business of a large city, and to locate them where the convenience, health, and comfort of the people require they shall be located. And it must be conceded that the means adopted by the act for this purpose are appropriate, are stringent, and effectual.

In common sense and in soundness of judicial practice, the case should have been decided and thrown out right there. If it was so clear, as it was to the Court and I think is to us, that this state law was a reasonable regulation of the practice of slaughtering—a practice necessarily and frequently regulated, and fraught with danger to the health and comfort of the people unless so located as to minimize these effects—then (if authority were needed for so obvious a thing), the statute would appear exactly to fit the passage the Court quotes with approval from *Corfield v. Coryell*, where Mr. Justice Washington added, to his comprehensive listing of the "privileges and immunities" enjoyed in our civilized society, the necessary words ". . . subject nevertheless to such restraints as the government may justly prescribe for the general good of the whole." The conclusion would be that, under the facts as stated and evaluated by this very Court, the reordering of the New Orleans slaughterhouse situation was well within the compass of this reservation. To claim that slaughtering animals wherever and however you want to is, under any tenable interpretation, a "privilege and immunity of citizens of the United States" was (I think it not too much to say) frivolous—even disgustingly so, if frivolity can disgust. No court faced with such facts has to jump at once to the most thin-aired level of discourse, and decide in a general way the most abstract question possible about this clause, settling silently and destructively such real questions as whether (for one among infinite such examples) the right to marry and have a family is a privilege or immunity guarded by national law against arbitrary state infringement. Or the right to have an important say as to the education of your children. Or the right to go to church or not to go to church, as you think best.

Questions like these, throbbing with real life, ought not to be decided when you have before you only a question about the right to unregulated slaughter of animals anywhere in town.

It would be interesting to know how this absurd case got selected as the one on which the *whole fate* of the solemn "privileges and immunities" clause was to be summarily (and mortally) settled. But the truth is that, having stated the facts of the case in such a way as to make any but one result inevitable, the Court yokes the subject of ubiquitous unregulated butchering to the whole lot of "privileges and immunities of citizens of the United States" and leads the clause itself lowing to the sacrifice.

The knife is sharp. On the basis of the most "narrow verbal criticism" in the literature, the majority of the Court dissects the wording of Section 1 of the Fourteenth Amendment:

> The next observation is more important in view of the arguments of counsel in the present case. It is, that the distinction between citizenship of the United States and citizenship of a State is clearly recognized and established. Not only may a man be a citizen of the United States without being a citizen of a State, but an important element is necessary to convert the former into the latter. He must reside within the State to make him a citizen of it, but it is only necessary that he should be born or naturalized in the United States to be a citizen of the Union.

> It is quite clear, then, that there is a citizenship of the United States, and a citizenship of a State, which are distinct from each other, and which depend on different characteristics or circumstances in the individual.

> We think this distinction and its explicit recognition in this Amendment of great weight in this argument, because the

next paragraph of this same section, which is the one mainly relied on by the plaintiffs in error, speaks only of privileges and immunities of citizens of the United States, and does not speak of the citizens of the several States. The argument, however, in favor of the plaintiffs rests wholly on the assumption that the citizenship is the same, and the privileges and immunities guaranteed by the clause are the same.

The language is, "No State shall make or enforce any law which shall abridge the privileges or immunities of citizens *of the United States.*" It is a little remarkable, if this clause was intended as a protection to the citizen of a State against the legislative power of his own State, that the word citizen of the State should be left out when it is so carefully used, and used in contradistinction to citizens of the United States, in the very sentence which precedes it. It is too clear for argument that the change in phraseology was adopted understandingly and with a purpose.

Of the privileges and immunities of the citizen of the United States, and of the privileges and immunities of the citizen of the State, and what they respectively are, we will presently consider; but we wish to state here that it is only the former which are placed by this clause under the protection of the Federal Constitution, and that the latter, whatever they may be, are not intended to have any additional protection by this paragraph of the amendment.

Perhaps the most striking thing about this analysis is that it skips past the fact that in overwhelming majority, the "citizens" of the States are the same human beings as "citizens of the United States." Everybody who is a citizen of the United States is a citizen of some State, unless that person does not reside in a State—and, even in that case, only so long as that person does not reside in any State, which he may at any time and without notice elect to do. With perhaps some bizarre exceptions, of more than doubtful legality, everybody who is a citizen of a State is a citizen of the United States. Every citizen of the United States who lives in a State is a citizen of that State as well. It's the same person in an only metaphysically different hat.

Another principal fallacy in the Court's reasoning is to be found in the word "transfer" in the following passage:

[With] few restrictions, the entire domain of the privileges and immunities of citizens of the States, as above defined, lay within the constitutional and legislative power of the States, and without that of the Federal government. Was it the purpose of the Fourteenth Amendment, by the simple *declaration that no State should make or enforce any law which shall abridge the privileges and immunities of citizens* of the United States, *to transfer the security and protection of all the civil rights which we have mentioned, from the States to the Federal government?* And where it is declared that Congress shall have the power to enforce that article, was it intended to bring within the power of Congress the entire domain of civil rights heretofore belonging exclusively to the States?

All this and more must follow, if the proposition of the plaintiffs in error be sound. For not only are these rights subject to the control of Congress whenever in its discretion any of them are supposed to be abridged by State legislation, but that body may also pass laws in advance, limiting and restricting the exercise of legislative power by the States, in their most ordinary and usual functions, as in its judgment it may think proper on all such subjects. And still further, such a construction followed by the reversal of the judgments of the Supreme Court of Louisiana in these cases, would constitute this court a perpetual cen-

sor upon all legislation of the States, on the civil rights of
their own citizens, with authority to nullify such as it did
not approve as consistent with those rights, as they existed
at the time of the adoption of this amendment. *The argument we admit is not always the most conclusive which is drawn from
the consequences urged against the adoption of a particular construction of an instrument.* But when, as in the case before us, these
consequences are so serious, so far reaching and pervading, so great a departure from the structure and spirit of our
institutions; when the effect is to fetter and degrade the
State governments by subjecting them to the control of
Congress, in the exercise of powers heretofore universally
conceded to them of the most ordinary and fundamental
character; when in fact it radically changes the whole theory of the relations of the State and Federal governments
to each other and of both these governments to the people; the argument has a force that is irresistible, in the absence of language which expresses such a purpose too
clearly to admit of doubt.

We are convinced that no such results were intended by
the Congress which proposed these amendments, nor by
the legislatures of the States which ratified them.

The discernment and enforcement of an ample substantive
corpus of national human rights under the name of "privileges
and immunities of citizens of the United States," would not
produce a general "transfer," but would simply bring into clear
view a *superior national law* of human rights, against which state
dispositions would be tested for their lawfulness. This would
be nothing but a new field for application of the Supremacy
Clause of Article VI of the Constitution, already thoroughly
familiar as the key provision in the structure of our nation.
Each of the States had, in 1787 and in 1873, and still has, a

law of contracts, prescribing in copious detail the requisites for the formation of contracts, and their remedial consequences. But all these state laws are subject, under the supremacy clause, to the Article I, Section 10 prohibition against any state law "impairing the obligation of a contract." Ordinarily state contract law goes its own way. But whether the "obligation of contract" clause is offended by some particular state law is a *national* question, finally decidable in the national courts. Similarly, real-estate transactions are generally subject to state law, but a national treaty may impose an outside rule on some classes of such transactions.[5] The escheat of property, on the death of the owner without heirs or a will, is ordinarily a state-law matter—but Congress has been upheld in imposing a different rule for a national purpose.[6] This is the normal balance of our so-called "federalism."

Moreover, the *Slaughterhouse* Court need have read only a few words further into Section 1 of the Fourteenth Amendment itself to see that general, sweeping changes in state-federal allocations of authority were *not* out of range on the very day the Fourteenth Amendment became law. No *State* is to deprive any person of life, liberty, or property *"without due process of law."* This command (leaving out of account for now its exceedingly questionable expansion into substantive fields, see pp. 90–93 below) is a *general* requirement of national law as to the fairness of procedure in state civil-and criminal cases and in other actions of the States. It has begotten and continues to beget illimitable questions as to the adequacy of state procedures. The application to the States of the Fourteenth Amendment "privileges and immunities" clause would similarly have begotten innumerable special problems. But there would have been no question in the latter case any more than in the former of turning over, "transferring," the normal tasks of the States in administering civil and criminal law. What was involved in the "due process" field, and what would be involved in the "privileges and immunities" field, is the subordination, wherever needful, of actions of the States to *national* standards—the Supremacy clause, as always, speaking the last word.

The same remarks may be made about the "equal protection" clause immediately following. Even if only applied to blacks (and it has been found impossible so to limit it) it states a general rule, applicable throughout the corpus of state law.

But the hugeness of these fallacies may well be thought to be no more shocking than the nose-thumbing insolence in the Court's final turning to the question, "All right, you may say that there *are* then no privileges and immunities of national citizenship? We'll show *you*":

> Having shown that the privileges and immunities relied on in the argument are those which belong to citizens of the States as such, and that they are left to the State governments for security and protection, and not by this article placed under the special care of the Federal government, we may hold ourselves excused from defining the privileges and immunities of citizens of the United States which no State can abridge, until some case involving those privileges may make it necessary to do so.

> But lest it should be said that no such privileges and immunities are to be found if those we have been considering are excluded, we venture to suggest some which owe their existence to the Federal government, its National character, its Constitution, or its laws.

One of these is well described in the case of *Crandall v. Nevada*. It is said to be the right of the citizen of this great country, protected by *implied guarantees* of its Constitution, "to come to the seat of government to assert any claim he may have upon that government, to transact any business he may have with it, to seek its protection, to share its offices, to engage in administering its functions. He has the right of free access to its seaports, through which all operations of foreign commerce are conducted, to the subtreasuries, land offices, and courts of justice in the several States." And quoting from the language of Chief Justice Taney in another case, it is said *"that for all the great purposes for which the Federal government was established, we are one people, one common country*, we are all citizens of the United States," and it is, as such citizens, that their rights are supported in this court in *Crandall v. Nevada*.

Another privilege of a citizen of the United States is to *demand the care and protection of the Federal government* over his life, liberty and property when on the high seas or within the jurisdiction of a foreign government. Of this there can be no doubt, nor that the right depends upon his character as a citizen of the United States. *The right to peaceably [sic] assemble and petition for redress of grievances, the privilege of the writ of* habeas corpus, *are rights of the citizen* guaranteed by the Federal Constitution. *The right to use the navigable waters of the United States*, however they may penetrate the territory of the several States, *all rights secured to our citizens by treaties with foreign nations*, are dependent upon citizenship of the United States, and not citizenship of a State. One of these privileges is conferred by the very article under consideration. It is that a citizen of the United States can, of his own volition, become a citizen of any State of the Union by a *bona fide* residence therein, with the same rights as other citizens of that State. To these may be added the rights secured by the thirteenth and fifteenth articles of amendment, and by the other clause of the fourteenth, next to be considered.

I summon the boldness to say that the Court would have done well to omit the final taunt of this "list." What the list painstakingly shows is that, in the Court's view, *nothing* was set up, or added, or created, or even newly recognized by the Fourteenth Amendment's "privileges and immunities" clause. Think in detail about that list of the rights taken by the Court to be given by the great words of the first sentence of this Amendment:

1. A citizen of the United States has a right to leave a State without paying an exit tax. That was the 1868 case of *Crandall v. Nevada,* which is perhaps somewhat tendentiously understated in the just-quoted passage. It stated the law, in any case, *before* the Fourteenth Amendment came into effect—and rests on *no* constitutional text.

2. Such citizens have the right to "demand" the protection of the national government when they or their property are in a foreign country or on the high seas. (To "demand"? I have called this an "inchoate" right, or perhaps I should have said an "imperfect" right. In any event, such as it was it existed and was immune from state "abridgment," before the Fourteenth Amendment.)

3. They have the right to use the navigable waters of the United States. This was well-recognized *before* the Fourteenth Amendment.

4. They have the right "to peaceably [sic] assemble and to petition the government for a redress of grievances." This right was set up in 1790, in the Bill of Rights. (But see below for the puzzle created.)

5. They possess all rights secured to "our citizens" by national statute, by the Constitution itself, and by treaties with

foreign nations, such as the rights secured by Amendments XIII (that is to say, the right not to be a slave) and XV (the right to immunity from exclusion from voting on the ground of race), by Article I, Section 9 (the right to the writ of *habeas corpus*), and by the "other clause of the fourteenth, next to be considered" (the reference seems to be to the "due process" clause and the "equal protection" clause, though this is not entirely clear). Of course, all those are "rights" already secured by national law, without reference to the Fourteenth Amendment "privileges and immunities" clause.

Four or six times zero is zero. In the Court's view, the "privileges and immunities" clause *had no operational meaning.*

On this list, the four Justices dissenting in the *Slaughterhouse* case commented:

> If this inhibition [the privileges and immunities of the Fourteenth Amendment] . . . only refers, as held by the majority of the court in their opinion, to such privileges and immunities as were before its adoption specially designated in the Constitution or necessarily implied as belonging to citizens of the United States, *it was a vain and idle enactment, which accomplished nothing, and most unnecessarily excited Congress and the people on its passage.* With privileges and immunities thus designated or implied no State could ever have interfered by its laws, and no new constitutional provision was required to inhibit such interference. The supremacy of the Constitution and the laws of the United States always controlled any State legislation of that character. But if the amendment refers to the natural and inalienable rights which belong to all citizens, the inhibition has a profound significance and consequence.

In the case of all these "privileges and immunities" so proudly deployed by the Justices in the majority, this is a just comment. I have lumped together, as my own number 5, those of which this appears on the very face of things—the Article I right to *habeas corpus*, treaty rights, and all other rights granted by other parts of the Constitution and its Amendments; these are available as a matter of law to all citizens of the United States (as well as, one should add, to all non-citizens, where their interests are implicated) under the Supremacy Clause of Article VI of the Constitution. (It is worth remarking that the first, second, and third of the rights in the list are *not* "enumerated" in the text of the Constitution, but had already been taken to be rights of national citizens.)

Brief comment may be made as to *Crandall v. Nevada*. That case, striking down a state law interfering (even by a small exit tax) with travel from one State to another, was decided without reliance on the Fourteenth Amendment, or indeed on any particular text, but on grounds of the very nature and structure of the Union. It was nevertheless a case *based on* the Constitution (for it could be based on nothing else), and in its timing, and its non-reliance on the Fourteenth Amendment, it is just another illustration of the fact that, as interpreted by the *Slaughterhouse* majority, the Fourteenth Amendment "privileges and immunities" added nothing, not even the *in extremis* stabilizing force of a fifth wheel.

The right of a citizen to demand the protection of the national government against foreign actions is an elusive and incomplete right. It's hard to see how Nevada could interfere with it, especially if the "demander's" "life, liberty and property" are at risk "in a foreign country or on the high seas." But if Nevada had, in 1866, passed and attempted to enforce a state law making it a crime for a citizen of the United States to make such a "demand," or even, God save us, for an official of the United States to accede to such a demand *while in Nevada*, I don't think any Court at any time, with or without a Fourteenth Amendment privileges and immunities clause, would have hesitated to declare such a law void.

The one other example given by the majority (the right peaceably to assemble and to petition for redress of grievance) poses special problems, since it is in one sense a "named right" in the First Amendment, but (because that Amendment limits only "Congress") is *not therein "named"* as binding the States. The 1833 case of *Barron v. Baltimore* had held that the first eight Amendments, the so-called Bill of Rights, were not binding on the States. The textual reason for this is particularly clear in the case of the First Amendment, because (as we have seen) only *Congress* is disabled by the Amendment. But the puzzle of this "illustration" develops in yet another direction. If the assembly and petition rights are privileges and immunities of national citizens, then what about freedom of speech? It is protected in the same Amendment, and like the right of petition (which is only a form of free speech), it is a foreseeable element in the operation of the national government. I don't think the *Slaughterhouse* Court would have wanted to explore that. On the main point, there is no suggestion that the Fourteenth Amendment added anything to the right of petition, or that a state could validly have interfered with that right before the Fourteenth Amendment. What is shown, once again, is that in the Court's view the privileges and immunities clause of the Fourteenth Amendment has added nothing, that this great resounding clause, which seemed, just after the Civil War, to be summing up the moral result of that war—"one nation indivisible"; "a new birth of freedom"— actually had no operative force whatsoever, and was to be a mere dead letter.

This *Slaughterhouse* court says that to give broad substantive force to this "privileges and immunities" clause would be violative of some well-understood postulates about what (with reverential capitals that can be heard even when they are not written) is sometimes called "Our Federalism." (It has to be interesting that neither the word "federal" nor any of its derivatives and cognates occur in the Constitution.) It would be my answer that sweeping change in "Our Federalism" would not have been really surprising, after the end of what was up to its time the greatest war in human history, a civil war fought initially, continuingly, and, above all, to establish forever the national supremacy, as expressed in the Supremacy Clause of Article VI, of which an ample substantive content for national citizenship would be, as I have shown, just one more application. There is a tottering absurdity, which ought to be repugnant to every lawyer's mind, of triumphantly performing a conjurer's trick with the clause, in order to *annihilate* (it must have been anticipated for all time) every hope that such great words, used at such a moment in history, should mean anything at all, except a few little things that were already there when this Amendment became law. You know, that just can't be right. Al Smith's 1928 words seem dead on center: "No matter how thin you slice it, it's still baloney." Nobody before or since ever sliced it quite as thin as the *Slaughterhouse* Court did!

How would you like to try to explain, to a foreign person, that these words, so placed, simply had and have *no* substantive meaning? That (as this spoiler Court construed them) they neither *confirmed nor set up a general and generous national regime* of the privileges and immunities of citizenship in the United States?

This kind of thing is not unexampled in the wider world. Some forty years ago, I heard an official visitor from the Soviet Union speak of the rights of the citizens of *that* Union— freedom of speech and political action, no arbitrary imprisonment, and so on. He read out all the statutes and constitutional provisions; he did not detail the ways they used to bring them to nothing—the "Catch-22s."

I wonder whether any American representative has anytime lately made a speech in Russia, quoting the words "privileges and immunities of citizens of the United States" (rather sonorously, one imagines) as summing up, in a felicitous phrase of comprehensive reach, the whole essence of the American regime of human rights as a matter of law. If so, the chances are that this representative doesn't know any better. I think maybe you really have to be a lawyer to get inside the *Slaughterhouse* case enough to see how shabby, how sorry, it really was.

I'll add a couple of grace notes.

The Court, as courts so often do, surveys the "history" of the post–Civil War Amendments. Then it goes on:

> We repeat, then, in the light of this recapitulation of events, almost too recent to be called history, but which are familiar to us all; and on the most casual examination of the

language of these amendments, no one can fail to be impressed with the one pervading purpose found in them all, lying at the foundation of each, and without which none of them would have been even suggested; we mean the freedom of the slave race, the security and firm establishment of that freedom, and the protection of the newly-made freeman and citizen from the oppressions of those who had formerly exercised unlimited dominion over him. It is true that only the fifteenth amendment, in terms, mentions the Negro by speaking of his color and his slavery. But it is just as true that each of the other articles was addressed to the grievances of that race, and designed to remedy them as the fifteenth.

Why in the world did the Court think such a set of considerations bore on its decision with respect to the "privileges and immunities" clause of the Fourteenth Amendment? Is there any suggestion, anywhere, that that clause applies to and benefits only *blacks?* Or benefits them especially? Was such a question presented in this case? No. This is just dust thrown in the eyes. Having at such length indicated a tender concern for black people, does or can the Court in any way apply that concern to this case? No, race is not in the "privileges and immunities" question in this case. Whatever else it is, *Slaughterhouse* is not a race case.

There was one real danger. The Congress that had passed the post–Civil War Amendments, and the early Civil Rights Acts, might at some time take some action under the national "privileges and immunities" clause that would be helpful to blacks, if that Clause were to grow real teeth.

The Court needn't have worried. Some such laws were actually passed. But the Supreme Court pretty well did them all to death. If you will look through the list (tabulated in the Annotated Constitution of the United States) of some twenty-five cases, in which the Court struck down Acts of Congress, through the whole nineteenth century—mostly after the Civil War ended—you will find that the majority of these decisions rested either on Bill of Rights procedural provisions or on infra-governmental grounds such as the relations of Congress to the Presidency or to the Judiciary. *Of the rest, the cases dealing with the affirmative powers of Congress, the only ones of any staying power or moment were the ones striking down or narrowing statutes protective of Negroes* (voting, lynching, public accommodations, and so forth). If you look over this list of cases, and consider that the latter half of the nineteenth century was a time of ample and Court-approved expansion of Congress' power (the last legal tender case, the proliferation of power in admiralty, the hanging of interstate commerce power on a mere "peg," as in *The Daniel Ball*), you are not going to be very far from the conclusion that the so-called "doctrine" of "strict construction" of Congressional power was generated by concern to keep black people in their place.

If you read Justice Bradley's opinion in the 1883 Civil Rights Cases, and some of his opinions on the "admiralty" power and on the power to make paper money legal tender, you will find it hard to believe that it's the same man. But his half-century well understood what "strict construction" was really meant to be in aid of. You may be surprised; the shade of John C. Calhoun would not be surprised—except perhaps pleasantly, on seeing that his doctrines as to national and state citizenship (that had seemed old crockery smashed to powder by the Civil War) had sprung up alive and well.[7]

The *Slaughterhouse* Court left another little message:

If any such restraint is supposed to exist in the constitution of the State, the Supreme Court of Louisiana having necessarily passed on that question, it would not be open to review in this court.

Remember the quotation from Mr. Justice Washington's opinion in *Corfield v. Coryell*[8] which the *Slaughterhouse* Court calls *"the first and leading case"* on the content of state "privileges and immunities." The *Slaughterhouse* majority *closes the quotation* and goes out on a track of its own, concerning the Article IV clause:

> The constitutional provision there alluded to did not create those rights, which it called privileges and immunities of citizens of the States. It threw around them in that clause no security for the citizen of the State in which they were claimed or exercised. Nor did it profess to control the power of the State governments over the rights of its own citizens.

> Its sole purpose was to declare to the several States, that whatever those rights, as you grant or establish them to your own citizens, or as you limit or qualify, or impose restrictions on their exercise, the same, neither more nor less, shall be the measure of the rights of citizens of other States within your jurisdiction.

> It would be the vainest show of learning to attempt to prove by citations of authority, that up to the adoption of the recent amendments, no claim or pretense was set up that those rights depended on the Federal government for their existence or protection, beyond the very few express limitations which the Federal Constitution imposed upon the States—such, for instance, as the prohibition of ex post facto laws, bills of attainder, and laws impairing the

obligation of contracts. But with the exception of these and a few other restrictions, the entire domain of the privileges and immunities of citizens of the States, as above defined, lay within the constitutional and legislative power of the States, and without that of the Federal government. Was it the purpose of the fourteenth amendment, by the simple declaration that no State should make or enforce any law which shall abridge the privileges and immunities of citizens of the United States, to transfer the security and protection of all the civil rights which we have mentioned, from the States to the Federal government?

If you look back and carefully read the *Corfield* quotation of 1825 (above, pp. 49–50), characterizing the "privileges and immunities" that Mr. Justice Washington believed to be those referred to by that phrase in Article IV, Section 2, you must at this point feel what may with extreme charity be called a jarring dissonance. Justice Washington (in what this *Slaughterhouse* court actually calls the "great and leading case" on the subject) sees the "privileges and immunities" of citizens in the several States as "fundamental"; they *"belong of right to the citizens of all free governments."* They include the Declaration of Independence, through to "the pursuit of happiness" *under that name.* Yet to the *Slaughterhouse* Court their "entire domain" lies all but totally within the power of the States.

That conclusion, once again, takes *no account* of the fact that, under the Fourteenth Amendment, the status of being a *state* citizen is a status bindingly ordained and confirmed by national constitutional law. If this *national* ordainment of *state* citizenship had just been omitted, then the "privileges and immunities of citizens of the United States" would have stood full-bodied and independent, because the conjurer's trick which made them antithetical to, and in a zero-sum game with, the "privileges and immunities" of the *very same people* as state citizens would have had nothing (instead of next to nothing) to rest on.

The lexicographers for the letter *P* would have had excellent illustrations for "paradox" and "perversity." They still do, as the text actually stands. The Court's conclusion goes directly against what would normally be thought the consequence of making the enjoyment of *state* citizenship a thing recognized and confirmed by *national constitutional law*.

Now look what they've done here—whether adroitly or by baneful instinct, who can say? The words I have emphasized stamp out any expectations formed around the just-quoted part of the *Corfield* opinion. In the face of the wholly new fact that state citizenship was confirmed as a national right by the Fourteenth Amendment, the Court may have thought (or at least dimly feared) that this new status of being a state citizen, having become something conferred by the national Constitution, might naturally be thought to be governed in its substance by national law. So they paused to step on *that!*

This was the *same* Court that at the same time was moving toward solidifying commitment to the idea, now black-letter law, that the mere grant to the federal judicial courts, in Article III of the Constitution, of judicial jurisdiction over "all cases of admiralty and maritime jurisdiction," was ground for the recognition and creation of a national system of substantive maritime law, binding on the States.[9] That was a much larger leap than it would be to conclude that the national *establishment* of state citizenship (not the mere mention thereof) implies a nationally protected *content* of such citizenship. The federal courts could have decided maritime cases on the basis of *state maritime law*, just as they do in many other cases governed by state law, though that would have been inconvenient. To read the command that huge numbers—nearly all—of national citizens should be citizens of the States wherein they reside, as leaving to the States the substantive consequences of one's being a state citizen, would be a total nullification of the very grant just given. The brighter sixth graders would have learned to smirk when reciting the words "one nation indivisible, with liberty and justice for all."

So that is *Slaughterhouse*. The opinion[10] blows a kiss at the recently freed slaves. That kiss was the kiss of death, as later cases that *did* affect black people showed, but it was blown from a long distance, since the *Slaughterhouse* case, in its central focus, had nothing to do with race, and the decision could therefore in no way be beneficial to black people. The only logical application of this historical stuff to the "privileges and immunities" question would have been to hold that, since the Court saw all these Amendments as being for the benefit of black people, and since the plaintiffs weren't black, the "privileges and immunities" clause, as to these plaintiffs, means nothing. But the Court did not choose to pursue this logic, because of course they were not saying, heaven forfend, that there is a special class of enforceable national "privileges and immunities" for black people. Instead, the *Slaughterhouse Cases annihilated* the "privileges and immunities" clause as a whole and in the most general terms, as to white and black alike. (It was a cost-free blown kiss.)

After this polite genuflection toward the black "beneficiaries," the Court gets on with its work of annihilation, even to the length, as I have just shown, of trying to cut off the reliability of benefit from the new *national* basis of *state* citizenship. It turns out nobody is to benefit incrementally in any way from the privileges and immunities clause. The Court even speaks with horrified disapproval of the possibility that Congress might deal with the substantive content of national citizenship—though, as a new matter, I should have thought that such Congressional action would be a power of sovereignty, as is the power to make paper money of the United States legal tender. We have here to deal with something more important than legal tender—the worth, one hundred cents on the dollar, of our own original national commitment to *secure* by law the human rights of us all and of those to come.

When all is said, the question whether the phrase "privileges and immunities of citizens of the United States" is full of meaning, or empty of meaning, has to be settled on the basis of competing alleged absurdities.

The Court puts the case wholly upon what it sees as the all but ineffable radicalness of the change proposed by those who would give this new and ample substantive meaning to this clause. That the case is put by the Court solely on that basis is clinched by the last sentence in a passage I have quoted: "We are convinced that no such results *were intended* by the Congress which proposed these Amendments, nor by the legislatures of the States which ratified them." This sentence states a bare conclusion about the minds of the members of both Houses of Congress and the members of the legislatures of three-quarters of the States. No evidence whatever is advanced, except the sheer implausibility of most people's wishing for so radical a change. (I am a bear on so-called legislative history, which, as I have sampled it, is chronically multifarious and inconclusive, particularly on such a great question as the present one, about which so much was said. But I believe that if I were going to announce my conviction on this "unthinkability," as the one and only ground for this savagely destructive decision, I would try to scare up something along the line of material evidence of the short statement of my conclusion as to the "did-not-intend" of a great many political people in far-flung places, and even of the people at large. I might even have tried to assemble some support, more than just my own assertion, that the settled institutions of the Republic would be shaken to their foundations if the meaningful privileges and immunities clause, rather than the meaningless one that was sanctified by the Court in *Slaughterhouse*, were to become law.)

The *Slaughterhouse* Court very greatly exaggerated the magnitude of the revolution it feared, or feigned to fear.

First, a real live "privileges and immunities" clause, as opposed to a waxen simulacrum, would be no more (and no less) than a new (or newly recognized) body of national law, developing as law develops, and binding as all national law is supremely binding. Such a body of law would fit in without any strategic theoretical change.

Secondly, such a body of human-rights law, national in authority, would not even in its content be wholly new. There were already explicit limitations on state power over human rights questions, culminating in the Thirteenth Amendment's abolition of slavery. Important and comprehensive *procedural* limitations ("due process of law") on the States were entering the Constitution in the very same section of the Fourteenth

Amendment, as was an "equal protection" clause applying, whatever its substantive scope might turn out to be, to *all* state laws. More such provisions, binding the States generally, were to come in the next few decades. A living "privileges and immunities" clause would have been just one in a series of national laws protecting human rights against any contrary dispositions in state law. Such a development, moreover, is deeply rooted in earlier history—in the commitments of the Declaration of Independence and of the Ninth Amendment. There is no question here of wrenching diastrophic change. The clause would have fitted into the structure of the Constitution, and into a process already in train within that structure.

Thirdly, but on a different and a larger view, what is all this astonishment about? The country had just been through one of the most painful and bloody wars in human history. As the Court recognizes, that war was rooted in that ultimate denial of human rights, slavery. About a third of the American States had fought not only to preserve this institution, but to spread it to the Western territories; in this grand design they had been abetted by the Supreme Court itself, in a decision handed down some dozen years before the Fourteenth Amendment passed. They also (and as a corollary) bitterly denied the national power over them, and gave their own last full measure of life's blood and property to make that denial stick. How strange would it really be that the victorious nation should address itself in a newly serious way to ensuring that human rights were to be thoroughly protected, in the future, by national power?[11]

This book is haunted by the heroic, brooding figure of

Abraham Lincoln. So much is plain from the Dedication and the Afterword, as well as throughout.

But as I have read and reread the manuscript of the whole book, and especially of this chapter, I have realized that in a quite different sense, another figure haunts the work—the figure of John C. Calhoun. I seem sometimes to have dated occurrences, even when they have nothing to do with him, by the date of his death. I have chosen to conjecture that the grisly undead corse of "states rights" rises ever and anon on midnight, wearing his senatorial toga. I have imagined him as a person all of whose descendants are sought to be given honorary citizenship in some imaginary State. All in all, it's a lot of Calhoun for such a short book.

I didn't start with the idea that Calhoun was a subject. The thought that he was playing a part came to me fully formed when I was rereading, for perhaps the tenth time, Mr. Justice Field's dissent in the *Slaughterhouse Cases*. This is the passage that arrested my attention, as I thought about its full implications:

> The first clause of this [i.e., the Fourteenth] amendment determines who are citizens of the United States, and how their citizenship is created. Before its enactment there was much diversity of opinion among jurists and statesmen whether there was any such citizenship independent of that of the State, and, if any existed, as to the manner in which it originated. With a great number the opinion prevailed that there was no such citizenship independent of the citizenship of the State. Such was the opinion of Mr. Calhoun and the class represented by him. In his celebrated speech in the Senate upon the Force Bill, in 1833, referring to the reliance expressed by a senator upon the fact that we are citizens of the United States, he said: "If

by citizen of the United States he means a citizen at large, one whose citizenship extends to the entire geographical limits of the country without having a local citizenship in some State or Territory, a sort of citizen of the world, all I have to say is that such a citizen would be a perfect non-descript; that not a single individual of this description can be found in the entire mass of our population. Notwithstanding all the pomp and display of eloquence on the occasion, every citizen is a citizen of some State or Territory, and as such, under an express provision of the Constitution, is entitled to all privileges and immunities of citizens in the several States; and it is in this and no other sense that we are citizens of the United States."

Until I placed the quoted words from Calhoun's 1833 Senate speech alongside the result in the *Slaughterhouse Cases,* I had thought Calhoun to be simply a rather unappealing antiquity. He believed human slavery was a positive good. The energies of his later years went into strengthening the position of the slave States, with a view to protecting slavery. He espoused and expounded the doctrine of nullification of national laws by individual States. He regarded secession from the Union as the right of any State. He thought that all the national powers set forth in the Constitution should be "strictly construed." He supported the refusal by the Senate to receive petitions aimed at slavery, and the closing of the United States mails to abolitionist literature.

I think all those are positions that have not stood the test of time and reflection. I never hear anybody speak well of slavery anymore, though I cannot vouch for other people's secret thoughts. If anything can be settled by history and acquiescence, it is settled that state secession is not lawful. Some heady talk of "nullification" did surface some decades ago, but it never really got anywhere. "Strict construction" of the national powers was far from prevailing or nearly prevailing even in Calhoun's own century,[12] and now has no *principled* constituency. (How does partial-birth abortion become a *national* subject?) Once in a couple of decades, the Supreme Court may test these waters, but nobody can think the oceanic mass of national legislation is going to be reduced to a "strict-construction" puddle; after all, the banks want a *national* law that protects due-on-sale clauses in mortgages, and everybody wants the national government to act against kidnapping and even crime in the streets, and so on, *ad infinitum.* We actually have a fully empowered *national* government; I don't know why some people think that is a bad thing.

It is astounding, then, that the views on the nature and relations of national and state citizenship respectively of such a figure as Calhoun should be the prevailing law in the United States as to the all-important matter of human rights. On all his other views—secession, nullification, slavery, "strict construction"—Calhoun is nowhere.

Yet the *Slaughterhouse Cases* held:

1. That the "privileges and immunities" of citizens of the United States as such are a derisory handful of "rights," construed *stricti juris,* which had existed *prior* to the Fourteenth Amendment.

2. That the fundamental "privileges and immunities" enjoyed in this country flow entirely from the States one by one, and that the States, one by one, may validly diminish these within their own borders, as far as they see fit.

This is a pretty close tracking of the implications of the Calhoun position in the passage quoted above, on the relative worth and importance of national and state citizenship respectively, with respect to human rights.

All that can be asked is that you weigh and consider critically the reasonings in the *Slaughterhouse Cases*, in the light of the fact that the result of that decision preserves in the amber of the United States Reports very much more than a trace of Calhoun's 1833 opinions on citizenship, as quoted above. The *Slaughterhouse Cases* have never been overruled. It is therefore, *today* technically the law of this country that the "privileges and immunities" of its national citizenship are just as they were when Calhoun spoke of that citizenship disparagingly in 1833, and that the *fundamental* privileges of citizenship are those of *state* citizenship, bestowed by the *States* one by one on their own citizens, and changeable or destructible at the will of each State.

I will permit myself to say again, as I have said of other aspects of the *Slaughterhouse* decision: "You know, that just can't be right!"

I must report one further irony: In one way the Slaughterhouse opinion may be more Calhounian than Calhoun in its *definite* and *express* holding that each of the States may diminish and devalue the "privileges and immunities" flowing from its citizenship. I would be surprised, in view of the general drift of Calhoun's "states rights" thoughts, if he would not agree, but in a passage aimed only at expressing his contempt for the concept "citizen of the United States," he had no occasion to reach that question.

At least he does *not* misquote the language of Article IV, Section 2 ("... *in* the several States"). The majority opinion in *Slaughterhouse* in quoting this section, substitutes the words "... *of* the several States" for "... *in* the several States," and repeats this change ("*of*" for "*in*") in the immediately following quotation from Mr. Justice Washington's opinion in *Corfield v. Coryell*, where the quote is correct. Let me be clear about this. Justice Washington, in his opinion, correctly quotes the Article IV phrase, but in quoting Washington's opinion, the Supreme Court opinion in *Slaughterhouse* takes some pains to alter the phrase to its own incorrect version! This misquotation does not occur in the dissenting opinion of Mr. Justice Field.

I don't know what we can make of these rather surprising misquotations of a very short and in the context quite crucial passage in the Constitution. Maybe they were just in a hurry to get the deed done: "... 'Twere well it were done quickly."

Early in November of 1863, Lincoln was invited to come to Gettysburg on November 19, and there to speak words of dedication of the new cemetery for the soldiers who had died in the great battle. Accepting, he worked nearly ten days on the short speech he finally gave, which became one of the most revered utterances in our history. He had no speechwriters; he read his draft to no one. His alone was the sublime expression of hope that "this nation, under God, shall have a new birth of freedom."

Only five years later, only three years after his death, the action of Congress and of the nation in proposing and ratifying the Fourteenth Amendment, gave an apt and authoritative form of words to this hope and prophecy, in the "privileges and immunities" clause linked to the status of national citizenship.

In the case we have been examining, the Supreme Court struck down the vision expressed in Lincoln's prophetic hope. But the words and the hope of Gettysburg are still there, for use when we are ready.

Chapter 3

THE TRANSITIONAL FUNCTION OF "SUBSTANTIVE DUE PROCESS"

In the first two chapters, I have proposed an approach to American human-rights law based upon three cardinal commitments and on their mutual harmony: the Declaration of Independence, the Ninth Amendment, and the "citizenship" and "privileges and immunities" clauses of the Fourteenth Amendment, Section 1.

The scandalous truth is, as we have seen, that the first two of these have been allowed to sleep, almost without stirring, for two centuries—to my mind quite long enough—while the third was forcibly drugged into coma by a 5–4 vote in the Supreme Court, in the *Slaughterhouse Cases* of 1873; I have examined, just above, the footless scramble to judgment of that decision, which I invite you to think of as one of the most outrageous actions of our Supreme Court. One way or another, all three of these cardinal commitments to human rights have so far played virtually no part in our human-rights jurisprudence. What have we used instead, and with what results? With no acknowledgment of the force of these commitments, we have been left with a miscellany of legal techniques, intellectually insufficient to support a *general* regime of human rights.

We are speaking here of a complex legal development that circles and meanders, and never stoops to simplicity. This material can be organized only in approximation. But I think I can give you a good approximation of the lines of question and context.

There has been an influential school of thought—on the Court and outside it—that has had respect only for the specific Constitutional texts protecting more or less specific rights. Some Justices on the Court have from time to time talked and acted as though these texts were the only proper source for human rights enforceable as a matter of constitutional law—though we must notice that as good as no Supreme Court Justice in either of our centuries has been consistent in this.

Since this narrow textualism has found and still finds recurrent strong expression—sometimes of a condescending or even a bullying tone—it is a good idea to look seriously and in detail, as we have done, at the constitutional texts available. Where would we be if we faithfully adhered to this austere canon? This inquiry cannot of course of itself tell us whether it is the right canon. But it will serve to suggest something about whether that is a dead-serious question. This purely textual material deserves a close reading.

Aren't you a little surprised that the *whole* of original Constitutional protection, by specific naming, of human rights—even counting the first eight amendments—is so skimpy? Perhaps we shouldn't be. The central understood purpose of the original Constitution was to *constitute* a structured and empowered government. The people who did this created—in the four months of a hot Philadelphia summer, about eight decades before the publication of *The Origin of Species* and three or four decades before the beginning of the steam railroad— a system that has lasted over two hundred years, surviving great changes and the greatest of shocks.

As to human rights, the 1787 framers of the Constitution did do one endlessly important thing. In the bill of attainder and *ex post facto* provisions they staked out the *principle* that the document that *constituted* the nation was also a fitting place in which to establish *human rights,* in the sheer interest of justice and in forthright contradiction to what might be the majority will at any one time. Sometimes, when I have read those two provisions, I have felt the back of my neck tingle; together they are the holdfast cell of the great and still growing American principle, now imitated throughout the world, that in the very *Constitution* of a national government there may be fixed provisions for human rights, binding on the constituted government by the very same authority as the one that structured and empowered it. This was one of the great American political inventions of our seed-time. I don't know whether that long step called for courage, or whether like some vastly creative things (perhaps the wheel) it may suddenly have seemed obvious to everybody. Reflecting on that very alternative ought to stir our courage to go on in the building of a wider and wider edifice of constitutional human rights.

After all, the greatness of the original Constitution, in the sphere of human rights, must be seen to consist in this dazzling invention, rather than in the uses to which it was put in the document. These were and I think can be said to remain very few. We are walking around in the spaces of a nation which we believe to guarantee "liberty and justice for all," as a matter of constitutional law. But outside the realms of free expression and religion (and in those only as to the national power), our original constitutional text, even including the Bill of Rights, protects only a few limited and special substantive human rights. That is not a slight variation on the principles of the Declaration of Independence, but very nearly antipodal to them.

In the wake of the *Slaughterhouse Cases,* wherein the Court chloroformed the "privileges and immunities" clause of the Fourteenth Amendment, it came slowly to be seen, or perhaps tacitly acknowledged, or perhaps something even vaguer and less visible, that common justice, and implementation of the shared national value of freedom, required more than this, not only for religion and speech but for other rights as well, such as the right to marry and to have a family. This, it seems clear, generated the development of "substantive due process."

An early Supreme Court case is illustrative of this—*Chicago v. Burlington & Quincy Railway Co.*[1] Just twenty-nine years after the passage of the Fourteenth Amendment, and twenty-four years after the *Slaughterhouse* decision, the Court held that the Fourteenth Amendment "due process" clause incorporated the Fifth Amendment right to be justly compensated if a State (or one of its subdivisions) took your property. In its effect, this case overruled or at least rendered null the 1833 holding, in *Barron v. Baltimore,*[2] that the Fifth Amendment guarantee of compensation for property taken by a government did not apply to the States.

Now if the government takes or desires to take your property, the right to "just compensation" (if it exists at all) is a *substantive* right, just as it is a substantive right in private law to be justly compensated if some other person unlawfully deprives you of your property. It is true that there is a *procedural* right annexed to the substantive right. That is true of every substantive right, if it really is a right. If someone wrongfully takes your property, you have a *procedural* right to sue for compensation. But the basic right is the substantive right not to have your property taken without just compensation.

This substantive right to just compensation for property taken would have fitted with comfort into the "privileges and immunities" clause that was clobbered by the *Slaughterhouse* Court. But clobbered the "privileges and immunities" clause had been. And yet some elementary sense of outrage made it impossible for the Court, in 1897, to let this "taking" go uncompensated. I don't know whether the phrase "substantive due process" had yet been uttered; I rather think it hadn't. But it was not to be borne that the city of Chicago could despoil the railroad of its property without even a theoretical nationally created obligation to pay just compensation. That would be one hell of a way to run the railroads. And so "substantive due process," perhaps still unbaptized under that name, entered the scene as a living thing.

Now when you say those words "substantive due process" over and over, you must see, if you have considered the examples I have given of the difference between substantive law and procedural law or "process," that the phrase is incorrigibly self-contradictory. I tell my students sometimes that it resembles a Zen Buddhist *Koan*, a saying that expresses or asks for the impossible, even the unimaginable, in order to tease or stimulate or press the mind or the spirit into awakening to a transcendent reality, to a transformation of the inmost self. I do not in the least intend to disparage that procedure when it is applied to the spiritual realm, though I am not therein an adept or anything like it, and must look at the matter through a glass darkly, from a great and much obstructed distance. But in all sooth that is not what we are talking about in law. I must repeat the words of my good friend and teacher, the late Jerome Michael, "Law is a practical subject." To generate a powerful and stable field of human-rights law, in a way that will import legitimacy to those who live within it, we need terms that seem to be saying something about human beings. For a thoroughly general system of human rights, we need *general* terms, though we must know that derivation of decisions and rules from those general terms will not exhibit the apodeictic quality of science or mathematics. But we get no aid at all from a term, like "substantive due process," that is simply a contradiction.

The imaginary "substantive due process" clause has nevertheless been used as the flickering imputed source of many substantive rights:

1. The right not to have your property taken without fair compensation (1897), as just seen.

2. The immunity from various forms of governmental activity impinging on economic practices—like the New York law struck down in *Lochner v. New York*,[3] limiting bakers' hours of work to ten a day and sixty a week. Many local, state, and national enactments of this type were successfully challenged. These cases, like the Lochner case itself, were for the most

part soon overruled, and the general line of judicial disapproval of much governmental "economic" regulation passed into disfavor.

3. The right to free speech, extended by liberal analogic and functional reasonings.[4]

4. Freedom of the press, similarly extended.[5]

5. The rights to free exercise of religion,[6] and not to live under rules "respecting" the establishment of religion.[7]

6. The right to teach and to learn foreign languages.[8]

7. The right of parents to send their children to schools other than the public schools (e.g., parochial or military schools)[9]—and a general right of parents to a large share in their children's training.

8. The right to practice contraception.[10]

9. The "fundamental right to marry."[11]

This partial list establishes that, under the phantomic "substantive due process" clause, the Supreme Court has validated many substantive rights. *The "due process" clause is being made to carry the load that would far more naturally have been assigned to the "privileges and immunities" clause of the Fourteenth Amendment, jointly with the two "citizenship" clauses in that Amendment.*

Probably the most important utterance celebrating and in an oblique way explaining this transfer of function (though not under that name) stands in Justice Cardozo's opinion in *Palko v. Connecticut.*[12]

Palko had been convicted, in a state court, of second-degree murder. The *State* appealed this judgment, on the ground that errors of law in the defendant's favor made by the lower state court had resulted in the conviction's being limited to this lower degree of murder, while correct rulings might have cleared the way for a first-degree murder conviction. The Connecticut appellate court agreed with the State, and remanded the case for a second trial, purged of these errors. On this new trial, Palko was convicted of first-degree murder. He applied, in turn, to the Supreme Court of the United States, on the ground that the second trial violated his right to not to be put "twice in jeopardy for the same offense"; he contended that the "due process" clause of the Fourteenth Amendment *incorporated* the Fifth Amendment guarantee against double jeopardy, which otherwise would not have bound the States. Addressing this contention, Cardozo insisted that there was "no such general rule" of total incorporation. He goes on:

> *On the other hand, the due process clause of the Fourteenth Amendment may make it unlawful for a state to abridge by its statutes the freedom of speech which the First Amendment safeguards against encroachment by the Congress, or like the freedom of the press, or the free exercise of religion,* or the right of peaceable assembly, without which speech would be unduly trammeled, or the right of one accused of crime to the benefit of counsel. In these and other situations immunities that are valid as against the federal government by force of the specific pledges of particular Amendments have been found to be *implicit in the concept of ordered liberty,* and thus, through the Fourteenth Amendment, become valid as against the states. . . . The line of division may seem to be wavering and broken if there is a hasty catalogue of the cases on the one side and on the other. Reflection and analysis will induce a different view. There emerges the perception of a rationalizing principle which gives to discrete instances a proper order and coherence. The right to trial by jury and the immunity

from prosecution except as the result of an indictment may have value and importance. Even so, they are not of the very essence of a scheme of ordered liberty. To abolish them *is not to violate a "principle of justice so rooted in the traditions and conscience of our people as to be ranked as fundamental."* We reach a different plane of social and moral values when we pass to the privileges and immunities that have been taken over from the earlier articles of the federal bill of rights and brought within the Fourteenth Amendment by a process of absorption. These in their origin were effective against the federal government alone. If the Fourteenth Amendment has absorbed them, *the process of absorption has had its source in the belief that neither liberty nor justice would exist if they were sacrificed.* This is true, for illustration, of freedom of thought, and speech. Of that freedom one may say that it is the matrix, the indispensable condition, of nearly every other form of freedom. With rare aberrations a pervasive recognition of that truth can be traced in our history, political and legal. So it has come about that the domain of liberty, withdrawn by the Fourteenth Amendment from encroachment by the states, has been enlarged by latter-day judgments to include liberty of the mind as well as liberty of action.

It is very important to note that this passage, which contains doubtless the most famous and most often-quoted justification for applying First Amendment law against the States, exhibits in just this regard a huge inconsequence of thought. (I raise this because it affords yet another insight into the dangers to clear thought that threaten, when the mind is directed to establishing that the phrase, "due process of law," incorporates *substantive* law.)

The *Palko* case itself concerned *criminal procedure.* Every single example given by Cardozo, on either side of his "line of division," concerns nothing but criminal *procedure*—jury trial, a real trial instead of a sham trial, and so on—except for the shield of free speech and religion. The latter is inserted in such a way as to suggest that the question as to "free speech" or "religious freedom" is much the same sort of question as the one about "right to counsel."

But the questions are not in the same part of the world. It is very clear that "due process of law" refers at a minimum to criminal procedure. The only question is then, "To what aspects and kinds of criminal procedure?" At the least, the original Bill of Rights guarantees, which are largely about criminal procedure, can suggest possibilities to the mind. And that is the use Cardozo makes of them—a very natural use, when one is asking, "What *procedure* is *due?*"

But the question as to "free speech" is not, "What *procedures*, named in the original Bill of Rights, are requisite to 'right *procedure*' under the Fourteenth Amendment 'due process'—that is to say, 'due procedure'—clause?" The question here could hardly be more different: "Does the 'due *process*' clause of the Fourteenth Amendment 'incorporate,' and so make applicable to the States, the *substantive law* of the free speech and religion guarantees?" It has to be added that this is a particularly difficult "incorporation," because the First Amendment addresses its prohibitions *only to Congress*. It is an enormous leap.

It ought to be added that the opinion's approval of the gathering-up of the First Amendment into the Fourteenth's "due process" clause was not strictly relevant to the issue in Palko. But in 1937, the same year, the Court had decided a handful of free-speech cases, in at least one of which this "incorporation" had seemingly been effected.[13] Cardozo's words in Palko became the next thing to definitive grounds for this "incorporation."

(As a matter of fact, the hard rule of the Palko case, as to double jeopardy itself, was expressly overruled in 1969;[14] *Benton v. Maryland*, 395 U.S. 784. I have not had the heart to research the grim question of whether Palko was meanwhile put to death.[15] If he was, perhaps his shade has by now convinced Cardozo's that the right not to be tried twice [or more] for the same alleged offense can be quite "fundamental.")

Now I've gone into all this to illustrate what zigzags and blurrings have to occur when you address yourself to the task of bringing to bear a clause about "due process," which means "due procedure," on matters of substantive law.

I think I ought to say just here that no better illustration ever occurred of the plain truth that "substantive due process" has taken over, though with its own innate feebleness, the function that naturally could and should have been the function of the "privileges and immunities" clause, as its content was foreshadowed by Mr. Justice Washington in *Corfield v. Coryell* (see pp. 49–50, above). Freedom to speak your mind is a precious *substantive right.* Jefferson said, "Knowledge is happiness"; one of the indispensable paths toward that happiness is to communicate with others and to have them free to communicate with you. Who will deny these assertions? Whether we read the Declaration of Independence, or Mr. Justice Washington's *Corfield* opinion, we know that we are committed as a nation to the idea that "the pursuit of happiness" is a basic right of humankind. Isn't that really a better way, a simpler way, a more convincing way to justify the protection of human speech and expression than all these twistings, these

misdirections, about "substantive due process"? Doesn't this
"pursuit of happiness" entail freedom of religion as well? These
"verdurous glooms and winding mossy ways" don't even ex-
emplify *elegantia, juris* or otherwise.

But let that for the moment be. The Cardozo dictum in the
Palko case has gone out even beyond free speech and religion,
and become an omnibus touchstone. One among the important uses of this sort occurs in the second Justice Harlan's dissent in the *Ullman* case:[16]

> However it is not the particular enumeration of rights
> in the first eight Amendments which spells out the reach
> of the Fourteenth Amendment due process, but rather, as
> was suggested in another context long before the adoption
> of that Amendment, those concepts which are considered
> to embrace those rights "which are . . . *fundamental;* which
> belong . . . to the citizens of all free governments," *Corfield
> v. Coryell,* for "the purposes [of securing] which men enter
> into society," *Calder v. Bull.* Due process has not been re-
> duced to any formula; its content cannot be determined by
> reference to any code. The best that can be said is that
> through the course of this Court's decisions it has repre-
> sented the balance which our nation, built upon the pos-
> tulates of respect for the liberty of the individual, has struck
> between that liberty and the demands of organized soci-
> ety. If the supplying of content to this Constitutional con-
> cept has of necessity been a rational process, it certainly
> has not been one where judges have felt free to roam where
> unguided speculation might take them. The balance of
> which I speak is the balance struck by this country, having
> regard to what history teaches are the traditions from
> which it developed as well as the traditions from which it
> broke. That tradition is a living thing. . . . And inasmuch
> as this context is one not of words, but of history and pur-

poses, the full scope of the liberty guaranteed by the Due Process Clause cannot be found in or limited by the precise terms of the specific guarantees elsewhere provided in the Constitution. This "liberty" is not a series of isolated points pricked out in terms of the taking of property; the freedom to bear arms; the freedom from unreasonable searches and seizures; and so on. *It is a rational continuum which, broadly speaking, includes a freedom from all substantial arbitrary impositions and purposeless restraints . . . and which also recognizes, what a reasonable and sensitive judgment must, that certain interests require particularly careful scrutiny of the state needs asserted to justify their abridgment.*

Justice Harlan five years later reiterated this position, linking it to Cardozo's *Palko* opinion:

> In my view, the proper constitutional inquiry in this case is whether this Connecticut statute infringes the Due Process Clause of the Fourteenth Amendment because the enactment violates basic values "implicit in the concept of ordered liberty," *Palko v. Connecticut,* 302 U.S. 319, 325. For reasons stated at length in my dissenting opinion in *Poe v. Ullman,* supra, I believe that it does. While the relevant inquiry may be aided by resort to one or more of the provisions of the Bill of Rights, it is not dependent on them or any of their radiations. The Due Process Clause of the Fourteenth Amendment stands, in my opinion, on its own bottom.[17]

Note what has happened here. The Cardozan gloss on "due process," "the scheme of ordered liberty," is functioning as an *independent* and *general criterion*, and not merely as a reason for preferring free speech to some of the other written Bill of Rights guarantees. The phrase has departed from the field of "incorporation" *vel non*, and taken on a general creative force.

There is no convincing way to find the substantive right to practice contraception in the specific texts of the original Bill of Rights—the first eight Amendments to the Constitution. In Douglas's opinion for the Court, in the *Griswold* case, an attempt is made to do this by recourse to concepts of "radiations" or "emanations," but that is not likely to seem solid except to those already convinced. It is true that getting convictions under the Connecticut anti-contraception statute might depend on the carrying out of "searches and seizures," some of which might not pass the "probable cause" test of the Fourth Amendment. But that need only mean that convictions might be difficult, as they are in many other cases. There undoubtedly would occur cases in which confession or admission, or very convincing circumstantial evidence, or the chance of a surreptitious third-party (or even a second-party) witness would be of avail. The real reason for the invalidation of the forbidding of contraception by married couples is that to tender to such couples the choice between sexual abstinence on the one hand, and on the other hand having at this time a child they do not want to have, or believe they should not have, is a savage blow to the pursuit of happiness, which is a vouchsafed right in the Declaration of Independence and was recognized as a fundamental "privilege and immunity" of citizenship a century and a half ago in *Corfield v. Coryell*. There could be no better case for exemplifying how much clarity can come from asking, at last, the *right* question.

Why not just stick with "substantive due process," irritating as it is? Because, quite visibly, this non-concept rests on insufficient commitment, and has too little firm meaning (if it has any at all) to beget the kind of confidence, in judges or in others, that ought to underlie the regime of human rights in the country that based and bases its life, and its claims to leadership, on its dedication to human rights.

The whole pattern of "substantive due process" cases exhibits this lack, and its dire results. But the problem is neatly illustrated in the case in which a grandmother faced the necessity of banishing from her house one of her two orphaned grandsons.[18]

I want here to make my point by considering this grandmother in a little more depth.

A Ms. Moore had taken in *two* small grandsons, respectively the sons of *two* different sons of hers who were absent. A zoning ordinance of the City of East Cleveland, by limiting residence in her "zone" to "single families" and by so defining a "single family" as to make it lawful for her to give a home to one of these first cousins but *not to both*, forbade this arrangement. The grandmother was convicted of violating this ordinance. (I still remember the shock that went through an audience in Iceland when I used and stressed this word, *"convicted,"* in stating this case.)

The City brought forward justifications for its law that were "marginal at best," and had "only a tenuous relation" to "overcrowding" and "traffic and parking congestion"—in the words of Mr. Justice Powell, writing the plurality opinion in the Supreme Court. Ms. Moore got to keep on giving a home to both her small grandsons. (My Icelandic audience was relieved, but still a little shocked.)

What would Ms. Moore have had to do to comply with this ordinance? To say merely that she would have had to stop giving family shelter to these two little grandsons is too shallow and hasty a description. Quite crucially, she would have had to *choose* between them, with all that would have entailed, as to them and as to her. She would have had to face consigning one of them to such publicly furnished care as might be available, while depriving the other one of close family association with his own first cousin—the nearest thing he had to a brother or a sister. There might have been available some arrangement for alternating them—say, a month apiece for each of them by turn, at her home and in an orphanage—but it is not clear that such an arrangement would have been possible, and, in any case, it would have had its own steady and recurrent agonies.

Now on the other hand, how much "traffic" and "overcrowding" did these little boys generate? We have to consider this comparatively. If the boys had been brothers, the ordinance would not have forbidden their both living under the same roof with their grandmother. Another grandmother, more abundantly blessed in fact and in law, might have, say, *six* grandchildren, all the children of *one* of her own children; *she* could keep all six of them under the one roof. How much "traffic" and "crowding" is prevented by applying a different rule to these first cousins? Mr. Justice Powell's words, "tenuous" and "marginal," seem restrained.

I will start with the plurality opinion of Mr. Justice Powell; it begins with what looks like an apology:

> Substantive due process has at times been a treacherous field for this Court. There *are* risks when the judicial branch gives enhanced protection to certain substantive liberties

without the guidance of the more specific provisions of the Bill of Rights. As the history of the *Lochner* era demonstrates, there is reason for concern lest the only limits to such judicial intervention become the predilections of those who happen at the time to be Members of this Court. That history counsels caution and restraint. But it does not counsel abandonment, nor does it require what the city urges here: cutting off any protection of family rights at the first convenient, if arbitrary boundary—the boundary of the nuclear family.

I am irresistibly reminded here of a reviewer's criticism of an actor appearing in *Hamlet:* "He played the King as though he were afraid somebody else might be about to play the ace." But Justice Powell is not to be faulted for this. "Substantive due process" is thin ice; you walk warily, and listen for cracks.

(As to the quotation about the King in *Hamlet,* I can't now put my hand on it. All I know is that I didn't make it up. How I wish I had!)

Let's look next at Mr. Justice White's *dissenting* opinion:

I cannot believe that the interest in residing with more than one set of grandchildren is one that calls for any kind of heightened protection under the Due Process Clause. To say that one has a personal right to live with all, rather than some, of one's grandchildren and that this right is implicit in ordered liberty is, as my Brother Stewart says, "to extend the limited substantive contours of the Due Process Clause beyond recognition." The present claim is hardly one of which it could be said that "neither liberty nor justice would exist if [it] were sacrificed."

Mr. Justice Powell would apparently construe the Due Process Clause to protect from all but quite important state regulatory interests any right or privilege that in his estimate is deeply rooted in the country's traditions. For me, this suggests a far too expansive character for this Court and a far less meaningful and less confining guiding principle than Mr. Justice Stewart would use for serious substantive due process review. What the deeply rooted traditions of the country are is arguable; which of them deserve the protection of the Due Process Clause is even more debatable. The suggested view would broaden enormously the horizons of the Clause; and, if the interest involved here is any measure of what the States would be forbidden to regulate, the courts would be substantively weighing and very likely invalidating a wide range of measures that Congress and State legislatures think appropriate to respond to a changing economic and social order.

These words rest on a hardly hidden major premise of virtually boundless *negative* judicial discretion as to the protection of human rights. What they say is, "I really think this is going rather too far, it will be a means of letting in upon us a flood of litigation." Such words should be impossible to write (or so I should hope) against any other background than that of perceived or felt fundamental weakness in the concept of "substantive due process." Justice White obviously thinks of that concept as one that may perhaps, now and then, justify a little protection of unnamed rights, but not very much—a concept easily malleable, not to solid distinctions of law, but to intuitive fears of judicial inconvenience. The passage is a perfect illustration of what you can lose when you rely on a highly vulnerable and totally puzzling general theory—such as "substantive due process."

I have stressed the intellectual hopelessness of "substantive due process." This pair of opinions—Powell's and White's—illustrates very well another baneful function of that non-concept. These opinions could prick out the start of a graph of the *feeble force* of this hapless concept, of the conviction and confidence it fails to beget. Mr. Justice Powell deals with it warily, as though it were about to fade away. Mr. Justice White treats it as of no force, properly speaking, at all.

Now that's no way for judges in a country that calls itself free to approach the problem whether a grandmother must turn one of her small grandsons out because somebody down at City Hall thinks, without any shown plausibility, that their both staying with her will create a parking problem and possible overcrowding.

What would happen if we finally threw the switch that would connect us with one of the most comprehensive and authoritative of our great commitments—"the right . . . to the pursuit of happiness?" A judge brought up in *that* tradition would come into court hardly able to contain his indignation that any governmental unit had dared try such a thing. I cannot think that Mr. Justice White would have felt differently. Each of the judges would have asked himself: (1) Is this a heart-crushing blow to the pursuit of happiness? and (2) Are the proffered justifications good enough to justify so heavy a blow? The answers are obvious. *This is what happens when you ask the right questions.*

There is nothing that should be of more anxious and perpetual concern to legal work, including above all the work of judgment, than "Are we asking the right questions?" And we will not be asking the right questions until we swear off "substantive due process" cold turkey and ask the questions authorized by our birthright commitments to human rights, and above all to the right to the "pursuit of happiness," as brought down to us by the Ninth Amendment, and by the Lincolnian "new birth of freedom," that ought to have been found, and ought still to be found, in the "citizenship" and "privileges and immunities" clauses of the Fourteenth Amendment.

Necessity, it is said, is the mother of invention. Sometimes the necessity is so pressing that it gives birth to an invention that doesn't work very well. That is how we got "substantive due process." It was unthinkable that in a supposedly free country the component States could at their own will suppress freedom of speech and religion, forbid contraception for no secular end, make a grandmother throw out one of her two little grandsons, impose unmeetable conditions on the right to marry, forbid people to learn a foreign language, and so on. As these and other examples show, "substantive due process" is an invention that now and then works a little bit in practice, but *does not work* intellectually. It has had perhaps a good transitional function, like the wood-frame support of an arch before you put the keystone in.

Chapter 4

JUDICIAL
REVIEW AND
MAJORITARIANISM

"The majority rules."

But the concept of "majority" is, at one and the same time, both essential and highly elusive. A majority of whom or what? How does or can one determine the existence of a majority?

In a democracy like ours, a stable and resolute national majority, if fairly evenly distributed throughout the country, can usually find a way to have its way. That is something like all the pure "democracy" that any country has or can have, perhaps all a country can bear and not oscillate into turbulence.

If what this perdurable and tenacious majority wants is the following of some policy not touching human-rights issues, its way to victory is relatively easy—though by no means always and absolutely so. If there arises instead a genuine issue of human rights under the Constitution, there may and often does arise the problem of judicial review.

Judicial review is not even arguably a "usurpation." It has as sure a warrant to existence under our Constitution and laws as does the power of the President (even in his second, lame-duck term) to name ambassadors, who must be confirmed by the population-skewed Senate. (Not just sort of skewed, but very heavily—one might say grotesquely—skewed.) I shall back up this claim a little later on. But for now let me just point out that this legitimacy of judicial review, like the legitimacy of the presidential power to nominate ambassadors, is ultimately subject to the will of a sufficiently strong and stable majority. Constitutional amendment could end either of these practices. It is vitally important, too, that neither of them is "counter-majoritarian" in its inception. Each became part of the Constitution (like everything else in the Constitution) by such "majoritarian" validation as was available in 1787–89. Judicial review of *State* actions for their *national* constitutionality is distinctly commanded by Article VI of the Constitution. Judicial review of Acts of Congress for their validity is unmistakably commanded by an Act of Congress that is itself still in full effect (see pp. 120–121 below) and its legitimacy is clearly assumed in many other acts of Congress. We have judicial review, after all, because the people of the whole country have not exhibited *in action* any desire to abolish it. It has been, by clear implication at the very least and sometimes by something stronger, affirmatively approved and facilitated by Congress. This is because nothing like a steady majority of the people really disapproves of "judicial review" as a whole and in principle, but only of some of the outcomes of judicial review. Try to find somebody who thinks the "contraception" case was forbiddenly "anti-majoritarian" and who has the same view as to the application to the States,[1] through the dubious use of the Fourteenth Amendment, of the doctrine that private property may not be taken by a State for public use without just compensation, whatever a "majority" in that State may want. Opposition to judicial review is a sometime thing—like "strict construction" of the powers of Congress.

How purely "majoritarian" are our other government institutions, in theory or in practice?

On the national constitutional level, the answer has to be "not obsessively so." Let us look over the ways in which the people who form our highest policies are to be chosen. There is the small fact that the presidential electoral system makes it clearly possible (though, as time has shown, not likely) that a President may be elected by the votes of a minority of the voters—by fewer popular votes than his nearest competitor. (I believe this has happened in about one case out of twenty.)

Under the Twelfth Amendment, if there is no majority of electors (a thing that might easily occur), the choice of one among the *three* highest candidates (in electoral votes) goes into the House of Representatives. In the vote there, *each State has one vote*. There can here be no thought of "popular majority," as essential to legitimacy, when Alaska and California have the same voting strength. Since, moreover, it takes *twenty-six States*—each casting *one vote*—("a majority of the whole number of States") to select a President in the House, and since Alaska and California each cast *one* vote, and since any State with an even number of Congressmen may be tied, or by some other distribution fail to create a majority within itself for any candidate, and so be unable to cast a vote, it may be very difficult to muster these twenty-six votes. 25–12–13 would not do it; 23–10–8 would not do it. Not even 25–25 would do it. Since the twenty-five may contain a total population much greater or much smaller than that of either the twelve or the thirteen or both together (58% of the American people live in nine states), that might mean that a choice would be impossible, though a popular majority choice existed. Very likely, there would be a "deal," but each low-population State would negotiate with just the same strength as the highest-population State—and one Congressman whose vote might change the vote of a State would negotiate with a strength all but entirely unrelated to numbers of people. A result that reflects the wishes of a majority of the American people, however that may be ascertained, is one possibility, but it is far from a certainty or even a very strong likelihood.

If no result is reached in the House of Representatives, by a "deal" or otherwise, nobody is elected President, and then the person elected Vice President takes over the Presidency. Since the Constitution calls for a vote in the Senate between the two vice-presidential candidates with the highest electoral votes, and since the Senate chooses by a general vote, a tie is unlikely. But if it occurs it would necessarily have to be resolved by the Vice President (outgoing, though he may be a just-defeated lame duck). Remember again that the Senate does not represent the American people *per capita* with even approximate equality. A strong "majority" in the Senate need not express the desires of a majority of the whole people. A bare majority in the Senate may easily be representative of a decided minority of the people. A tie in the Senate does not imply a tie in popular desire.

Now that's the way the President may be chosen. If the spread of votes is fairly even geographically, and if the popular majority is fairly large, *and if no strong third candidate* draws enough strength in a few States to capture their electoral votes, the system works tolerably well. But this tolerable working is not something that inheres in the system as set up in the Constitution. It has been attained only by the growing up of extra-constitutional devices that strongly tend toward drastically narrowing down the number of candidates who can be taken seriously. If there are only two candidates then it is near certain that one of them will get an electoral majority. The system does not work to secure the election as President of the person a majority of the people want as their first choice, for the good reason that, since George Washington, no such person normally, if ever, has existed. It is a system that conduces to a tolerable compromise.

The person who may be put in the presidential office by this process, flawed as it is from the pure majoritarian point of view, appoints all the principal officers of the United States, guides and sets our foreign policy, makes through appointees immeasurable administrative decisions affecting millions of people, holds the commandership-in-chief of the armed forces, and exercises the awesome power of the veto of Congressional actions, including statutes (a power, by the way, very problematic from the majoritarian point of view). What is the "majority" to be taken to want, when the Houses of Congress pass a bill, then the President vetoes it, and the override vote *just barely* falls short of two thirds in one house? As I believe I have demonstrated above, the person who has this veto power is not selected by a single popular-majority vote, nor by processes that guarantee its equivalence. In his second term of four years, the President need not think of a *future* election.

It is not my purpose here to express disapproval of all this. Straight-out popular election would have its own grave problems; it has been often considered and never in the final result adopted even by the Congress that would have to propose the necessary constitutional amendment. My purpose is a different one—to show that, as to choice of the President, our system, as originally set up and as it now stands, is not anything like committed to pure majoritarian principle.

The limitation of the President to two terms, like all "term limits" so-called, is radically anti-majoritarian, even "undemocratic." No matter how strong the popular desire may be, no matter what special circumstances produce this desire, the second-term President is ineligible to reelection.

How about the procedures for choosing people to serve in Congress? One needn't say much about the Senate! But one ought to think a lot about it. The population ratio between California and Alaska is about sixty-five to one, yet each of these States sends two Senators, and this huge disparity is guaranteed even against constitutional amendment, unless *Alaska* consents (see Article V). The naked constitutional theory is, then, that when it comes to the Senate, States with widely different populations are to have the same representation in that upper House, even though almost all the States and nearly all the people might agree that this apportionment of Senatorial power ought to be changed. How "majoritarian" is that?

Nor is it to the point that the Senate's function in passing laws can be looked on as "merely" negative. (This might equally be said of the more nearly "majoritarian" House of Representatives.) The power to refuse passage is the power to influence and sometimes to dictate content. It is also the power to refuse to repeal any law now in force.

This Senate, constituted without any attention at all to proportion of population, or to any "majoritarian" theory, has the power to confirm or deny confirmation to presidential appointments to federal offices, including ambassadorships and judgeships, and to *refuse* ratification to treaties by *one third plus one* of its membership. A treaty disfavored by that many Senators may be strongly favored by a majority of the American people. On the other hand, two-thirds of the Senate may (though it is not very likely) represent less than a majority of the American people, since 58% of the people live in nine states, sending eighteen Senators.

The House of Representatives, the "democratic" branch, is elected by *districts*. Since every State must under Article I have an integral number of Congressmen of its own, there is some disproportion between States in the number of persons represented by Congressman—because some States have populations just large enough to have, say, three Congressmen, and some States have almost but not quite enough to get four. You can check this out in any World Almanac. Perhaps these, and similar disparities, are not very serious. What is serious is the possibility, one that exists in any geographical-district system, that a majority of the House may have had the votes of a minority of the national aggregate of voters. This was not far from happening in 1994; what was hailed as a Republican congressional "landslide" was actually a fairly close-run thing in popular vote, nationwide, as between the two parties. Such a thing's not happening depends on the near-even geographical distribution of party-vote. More districts like the New

York City district of Charles Rangel could easily swamp, in aggregate popular vote, five or six districts where candidates of the opposite party ran ahead in close elections.

I am not against this either. On balance of all factors, I think the district system very likely works better, for us, than proportional representation would; in any case our Constitution would make it just about impossible to institute proportional representation.

But the entrance on the national scene of even one strong third party, let alone more than one, can change things. This is dramatically shown in the last election Margaret Thatcher won in Britain. She governed, both before and after that election, more or less in the tone and style of Napoleon I. But she had won only 43% of the popular vote. That election shows that third parties, which benevolent folk often hail as "giving the people a wider choice," may end up giving those people whose choice of policies is more or less alike, the choice of putting in power the crowd they like least of the three. Isn't this what probably happened in Lady Thatcher's 43% "triumph?"

Now I am not attacking the national government as "undemocratic." I am simply showing that it is *not* majoritarian in strict theory or practice. It balances doctrinaire "majoritarianism" with other values and devices. (The only one of these devices I cannot approve of or understand is the one that comes into play when and if nobody gets a majority of the presidential electoral votes. To set up for such a case a "one-state-one-vote" rule in the House of Representatives radically changes the constituency base of the original group of "electors" and adopts the "Senatorial" model of equal votes for each State. I can't think why this was considered a good change, particularly when the first electoral ratio to population might have been roughly replicated by giving the choice to a joint session of the House and the Senate with each member having one vote, thus taking a "second opinion" from a new Congress constituted on the same mixed basis as the original Presidential electors, and incidentally eliminating the problems that can arise if a State's delegation is tied up and so unable to vote at all.)

I've devoted this much analysis to the national government because that, for a reason I'll shortly give, seems to me to be the only one that matters, the only one whose quasi-"majoritarianism" really needs to be brought into view. But I think one can validly though summarily say that all the States exhibit in one way or another this falling from the dubious grace of doctrinaire majoritarianism, yet, like the national government, have that mediated and sometimes slow response to stable and persistent popular will that is probably the best kind of "democracy." The alternative is government by very frequent single-issue plebiscite; I have elsewhere explored what I believe to be the fatal flaws in one fairly recent and relatively modest proposal for national legislation by referendum.[2]

Why have I said that anything like a real clash with majoritarianism does not exist in the case of the collision of state laws and other local actions with *nationally* established human rights? I would think that the question answers itself in Article VI: "This Constitution, and the Laws of the United States which shall be made in Pursuance thereof; and all Treaties made, or which shall be made, under the Authority of the

United States, shall be the supreme Law of the Land; and the Judges in every state shall be bound thereby, any Thing in the Constitution or Laws of any State to the Contrary notwithstanding."

Since the great majority of cases that strike down state laws, in first instance or on final appeal, in the interest of human rights, rest on claimed *national* rights, and since the action rests on the interpretation of *national* law by *national* judicial officers, commissioned for this work by the *national* government, and since, though Article VI appears to answer this "majoritarian" objection as to state law cases, it continues to be urged as to them, one might say a bit more.

The uniform interpretation of national human-rights law, *as of all national law*, is a prime national function, guarding our practical and moral unity as a nation. Without such interpretation the national law of human rights is at the mercy of the States, one by one. No one would think of contending that a valid Act of Congress or a treaty must be interpreted and applied with a generous measure of deference to "majority" sentiment in one or more States. I can't think why this should not be true as to human-rights positions. Again, one wonders about the result of the Civil War. This "majoritarian" stuff, treating state majorities as a factor for weakening national law, is just "nullification" in clumsily stitched sheep's clothing.

Now so far I've dealt only with the principal power-expressions of national and state governments and with the organs that are given responsibility for these. But another point needs making, on this "majoritarian" matter: Many claimed infractions of national human rights occur through the actions of delegees or sub-delegees or sub-sub-delegees of national *or* state power.

There are indefinitely many such cases. A Secretary of State decides to withhold a passport. A police chief decides how to interrogate a suspect, and a state judge upholds him. A local zoning board votes to keep a grandmother from taking in more than one of her grandchildren to live with her, unless they are all offspring of the same one of her children. A mother claims that her retarded, involuntarily committed son suffered deprivation of constitutional rights when officials of a state institution gave him inadequate protection against injury. The Board of Regents of a state university does not rehire a teacher, allegedly because of things he has said.

In all such cases, the issue of "majoritarianism" is fictive, because *neither* a national *nor* a state "majority" is in any showable way implicated—unless you mean something like "the majority of the zoning board," or the one to zero "majority" of the police chief.

(None of this is said in derogation of "delegation." The non-rule that a delegated power may not be delegated is absolutely impossible of application to any complicated enterprise, including government. That is the way most decisions have to be made. But the delegation to a zoning board of the decision whether the rule in the grandmother case is a good one on *local* considerations [see above] cannot *be held to foreclose from fullest inquiry the question whether such a rule violates a national constitutional human right or any other national law*. And the concept of majoritarianism, proffered as a ground for such deference, is the substitution of fiction for fact.)

There is only one authority that even faintly arguably expresses the will of a national majority on a question of *national* rights—that is, the Congress, by the passage (and including either the signature of the President or passage over his veto) of a national law. That is the closest thing to the expression of the national majority will and understanding that we have. As I have shown, it has both major and minor imperfections. It also may have a rather short half-life, if viewed non-fictively, because, first, anybody who knows anything about Congress knows that repeal is not something that rather automatically occurs as soon as only 45% of the people now are of the same mind as they were, and secondly, because the President, perhaps in his second "lame duck" term, may veto the repeal, and thereby subject repeal to a ⅔ + ⅔ double bottleneck with no possibility that he will ever face the voters. All this gets rather fanciful, but that is because the idea of "majoritarianism" is itself rather fanciful. And (one cannot too often repeat) let us not lose sight of its entire unsuitability when applied to the States (against which most human-rights complaints are made) and to subordinate officials, both national and state.

The only thing not quite meaningless, as an application of the idea of majoritarianism to *nationally vouchsafed* human rights, would be the committing of such decisions to formal actions of *Congress*. But that is hardly even imaginable. Congress has neither time nor resources for going into a question about the zoning law as applied in East Cleveland, or the constitutional fairness of the treatment accorded to a retarded institutional inmate. Congress is a non-starter here. The alternative would be an elaborate administrative network with investigative and adjudicative power to move in and deal with this legion of unforeseeably occurring cases, trying the facts and deciding the law. Why would anybody think that to be a better way than using the judges to do this job of judging? The *individuation* of law, including constitutional law, is the business of judges.

Until the question can be answered, maybe we should, as it were, tentatively be thankful that we *have* the judges, and that their work in this regard is of well-attested legitimacy, both as to state law and practices and as to national law and practices, when either of these confronts a national human-rights claim.

As to the States, ponder Article VI yet again: It says that whatever the Constitution (as amended) may be held to provide is part of "the supreme Law of the Land; and the Judges in every State shall be bound thereby . . ."

Now I don't know how you'd get around that. The only way is to gloss the words *"Judges in every State"* as meaning "the state judiciaries." But note that (aside from the absurdity of a Constitution that would oblige state judges but not federal judges to obey it) it ignores the listing of the Constitution as a component in "the supreme Law of the Land." How could federal judges fail to apply that law, part of the "supreme Law," to state actions? And federal judges *do* sit "in" the States.

What about Congressional actions? Here the argument is slightly more complicated, though I think not of doubtful outcome. The Supremacy clause of Article VI establishes that the Constitution is *law*, inviolable in all courts. The remaining question is whether it has the status of a *superior* law, as against the two other kinds of "supreme Law of the Land"—Acts of Congress and treaties. To establish that it does, you have to read the Constitution itself. It speaks with unmistakable authority. Article I, Section 9, categorically forbids the passage by Congress of *ex post facto* laws and bills of attainder. The First Amendment rests on the same assumption: "Congress shall make no law . . ." But one need not stop there. If you start reading Article I, you find that the composition of Congress, and the qualifications for voting in congressional elections, are set out with the voice of authority; no action by a body not constituted in this way is even to be looked on as taken by "Congress." The intervals between elections for the House and the Senate are prescribed (contrasting with the British conception of an omnipotent Parliament that, regardless of prior agreement or understanding, could and did change the corresponding interval from three years to seven, in the famous Septennial Act of Queen Anne's day). Congress is peremptorily forbidden to lay a tax or duty on articles exported from any State. But why go on with this? If ever a document spoke with the claim, literally irresistibly and repeatedly implied, of being a law that is to govern and control another kind of law, our whole Constitution speaks that way. Of course this applies *a fortiori* to the actions of other federal officials.

The pure constitutional case for judicial review of the constitutionality of state and federal laws and other governmental acts of power seems clear. When somebody at a cocktail party says that such review is not mandated or contemplated by the Constitution, I tell them to take two aspirin and get a good night's sleep. One more piece of evidence might make that sleep more tranquil—Section 25 of the Judiciary Act of 1789, passed by the first Congress, in which sat a good many people who had been at the Constitutional Convention just two years before:

> That a final judgment or decree in any suit, in the highest
> court of law or equity of a State in which a decision in the
> suit could be had, where is drawn in question the validity
> of a treaty or statute of, or an authority exercised under the

United States, and the decision is against their validity; or where is drawn in question the validity of a statute, or an authority exercised under any state, on the ground of their being repugnant to the Constitution, treaties or laws of the United States, and the decision is in favor of such their validity, or where is drawn in question the construction of any clause of the Constitution, or of a treaty, or statute of, or commission held under the United States, and the decision is against the title, right, privilege or exemption specially set up or claimed by either party, under such clause of the said Constitution, treaty, statute or commission, may be reexamined and reversed or affirmed in the Supreme Court of the United States upon a writ of error. [But] no other error shall be assigned or regarded as a ground of reversal in any such case aforesaid, than such as appears on the face of the record, and immediately respects the beforementioned questions of validity or construction of the said Constitution, treaties, statutes, commissions, or authorities in dispute.

This language names certain kinds of cases which are expected to come up, how frequently no one could have guessed, in the state courts, and to have been decided there in certain ways. It then confers appellate jurisdiction on the Supreme Court of the United States, in certain of these decided state cases.

There are three classes of cases named, somewhat simplified:

(1) One in which the state court has finally decided *against* the validity of a statute passed by Congress (such decisions could only be on the ground of the repugnance of the Act of Congress to the Constitution).

(2) One in which the state court has finally decided *in favor* of the validity of a *state statute* as against the claim that it is repugnant to the national Constitution or laws.

(3) One in which a claim of right has been based in the national Constitution and laws, and the state court decision has been *against* the party claiming such national right.

In all three cases, the decision of the state court denying the federal claim (including a federal constitutional claim) is to be taken up by the national Supreme Court as a court of appeal, *and reversed or affirmed.*

The first thing to notice about this statute is that it simply *assumes* that such cases, turning on the validity of federal constitutional claims, will be an expectable part of state judicial business. Remember that this Act was passed in 1789 and signed by George Washington. This is very persuasive evidence of the "original understanding;" they didn't feel any need to *grant* state courts jurisdiction over such cases; the occurrence of such cases was simply assumed, as a part of the normal judicial work of all courts. You must note that (except for a brief interval) there did not exist, until 1875, any lower federal courts with *general* jurisdiction over cases "arising under" the national Constitution and laws. It resulted that such cases could and would be brought in the state courts, and this section's provision for appeal to the Supreme Court was to be, until after the Civil War, the only sure guard for the national interest in the prevalence of national law.

Back to Section 25 of the 1789 Act: One crucial thing is yet to be added. As to all three of the categories of cases set out as subject to Supreme Court review, *each such case is to be "re-examined and reversed or affirmed"* by that Court. *Now what does it mean to affirm a case that has held an Act of Congress invalid? Does it not mean to approve of and to confirm the state court's holding that the Act is invalid? This statute of 1789 therefore contains a congressional direction that the Supreme Court, as a court of appeal, take up and re-examine a case that has held an Act of Congress to be invalid, and affirm that holding if the Supreme Court thinks it correct.*

I am spelling this all out because it is one of the "Tested Sentences That Sell," in the Two Hundred Years War about "judicial review," that Congress has not even acquiesced in the Court's claims to the power to invalidate Acts of Congress. How can people go on saying that? Here, in an Act of Congress passed in 1789, and signed into law by George Washington, Congress goes further than merely acquiescing; it directs the Court to perform this function, where it is, as a court of law, free to reverse, or to *affirm* if convinced that validation of the Act of Congress is required. That 1789 Act, George Washington's signature and all, is as "majoritarian" as you can get. It also sets up a kind of paradox. In order to hold this Act, the 1789 act, not binding on itself in this particular, the Court would have had to hold this section of the 1789 Act of Congress unconstitutional.

This particular "majoritarian" action has not been receded from. This Act of 1789 has been codified over and over again, and modified only in ways completely consistent with its main message. Matthew Arnold encountered, in a Celtic epic poem, a warrior of whom it was said, "He went forth to battle, but he always fell." There has been fulmination and agitation in Congress, through the decades, against judicial review. But Congress, *as Congress,* acts only through the passage of bills, orders, resolutions, and votes. And Congress in two hundred years has never passed any law, or any "Bill, Order, Resolution or Vote," impairing the force of this direction to *reverse or affirm*. Instead, Congress has often facilitated the performance of this function. Judicial review is as lawful and as thoroughly legitimated as anything in the government of the United States. It is one with apple pie.

I think it is well that this should be so. To the reasons I gave thirty-five years ago, in my first book on constitutional law, I would add another. An active federal judiciary is the most hopeful bulwark for human rights in the United States. Or, one may fear in hours of dark thought, the least hopeless— but even that is something.

If human rights are to be kept ever-refreshed, seeing to this must be the business of somebody. The attainment and keeping of this good requires vigilance and effort. Every articulated set of values in government must have its executors.[3]

The federal judiciary is very, very far from perfect for this purpose. But at last we have to face the question, "Who is better?"

This is not a new question; the most decided early answer to it was given by Madison in Congress as he argued for the adoption of the Bill of Rights, in 1789:

> If they are incorporated into the Constitution, independent tribunals of justice will consider themselves in a peculiar manner the guardians of those rights; They will be an impenetrable bulwark against every assumption of power in the Legislative or the Executive; They will be naturally *led* to resist every encroachment upon rights expressly stipulated for in the Constitution . . .[4]

It is true that Madison, with Jefferson, came to have deep disagreements with some actions of the federal judiciary, particularly in the case of the Circuit Judges who upheld the Alien and Sedition Acts, but that disagreement was based on those judges' not being *vigorous enough* in protecting human rights by the judicial power. Jefferson was dissatisfied, in that controversy, with the federal judges' not having been sufficiently "activist," not "anti-majoritarian" enough. (The Alien and Sedition law was, after all, an Act of Congress, the nearest thing to an embodiment of a national majority.)

I am not out to eulogize our federal judges one by one. No indeed. But as an organized corps, the federal judiciary has at least a *capacity* for serving the cause of human rights, in time to come, that seems to me the most hopeful—or, perhaps, the least hopeless. There is given a way; whether the will can grow time alone must tell—from time to time. Of what *other* corps, now in existence or foreseeably likely to come into existence, is even this much true? Can anything better be said, *mutatis mutandis*, of Congress or the Presidency?

Particular suitors can approach a *court* with a claim of right that must be heard, and expect to hear *reasons* for the disposition made of the claim. These reasons are openly given, and those having any general interest are normally published; this gives full scope to criticism, and at the same time may augment the understanding of other judges working the same decisional process. Human-rights claims are made *in the name of the law,* as the outcome of *reasoning from commitment;* judges are practiced in this kind of reasoning, and some of them are expert at it.

What corps of persons, institutionally capable of giving actual life to human rights, has these indispensable characteristics, even though the realization of these characteristics to their fullest is far from universal? An elaborate *administrative* framework, charged with the same responsibilities? Whereof would we form such a corps, or find the will to do so? Would not the people who formed that corps, at their best, simply resemble judges?

Our history has provided us with such a corps. The price we pay for it is life tenure, which in practice means an average of say fifteen to twenty years. (I haven't done the simple arithmetic on this lately; when I wrote my earlier book on judicial review, in 1959, the average term of all the Supreme Court justices since the Civil War was about thirteen years.) In any case it is a price that Madison knew all about when he made the statement I have quoted above; it is the price successive Congresses have been willing to pay; it is the price of immunity from pressure, necessary as a condition to steadfast commitment to human rights—though not, alas, always a sufficient one.

I would conclude now, as Madison did over two centuries ago, that the commitment of judges to this (among other) work is a good deal. Not a perfect one. But what would be a perfect deal?

I would make one final point about this. It is a good deal even as to the invalidation of Acts of Congress, the only actions even close to "majoritarian" on the national level—on the same level, that is to say, as nationally ordained human rights. But it is a deal indispensably necessary in a nation having the structure of ours, with fifty local authorities, the States, competent to handle virtually any subject matter unless na-

tional law conflicts. There has to be some corps of authorities acting for the national interest as to any claimed violation of any kind, by the States, of nationally established
human rights, just as there is in regard to claimed violations
of any other national law. Without that, something we call
"Our Federalism" (however reverentially capitalized) can be
the Catch-22 of human-rights guarantees virtually anywhere
in our national territory.

If you doubt the concrete realism of this, read a lot of cases,
in the Reports, of what the States have tried to do, and been
kept from doing by the federal courts.

Let us thankfully use these judges for what could be—and
in some cases already is—their highest function: the making
good of our claims, before the world, of being a country
wherein governments are restrained from acting in ways destructive to liberty and obstructive to the pursuit of happiness.
That is exactly what we promised on the first day of our nationhood. If the ongoing fulfillment of that promise cannot
be kept in process by a corps of judges, it can't be accomplished at all, at least in our political system as it stands and
will in enormous likelihood continue to stand. The national
Congress is not so structured as to be up to the vigorous and
vigilant detail work, or to the steadfastness of vision that is
required. As to the States, the prospect for *moral* nationhood
would be not merely bleak but absolutely hopeless.

Perhaps a nation is fated to play out again and again the
same plot. Today we seem to be fighting the Civil War all over
again. The myth of state sovereignty ought to have been seen
to be obsolete when the first State was admitted to the Union
out of territory already belonging to the nation. (Kentucky,

↳ only
police dedicated
to corps of
human rights

I make it, in 1792, over two hundred years ago.) Wisconsin, among about thirty-three other states, never had any arguable sovereignty to give up. Indeed, from 1788 on, state "sovereignty" was a paradoxical puzzle, for the States were subjected to the supremacy of national law by the Constitution (which, unlike the Articles of Confederation, contains no reservation of their "sovereignty"). They were from the beginning denied powers pertaining to "sovereignty"—coining money, making treaties, engaging in war except in case of imminent danger that would not admit delay. The core power of "sovereignty"—stating what shall be the "supreme Law of the Land"—was denied them by Article VI of the Constitution of 1788. It's been downhill for state "sovereignty" ever since. In the Fourteenth Amendment, each State was stripped of the power to say who should be its citizens, and in later Amendments, who among these citizens were not to be denied the right to vote in state elections. I put it to you that nobody reading the Constitution and all its Amendments would think they were dealing with "sovereign" States. That's just Civil War talk, though we're still hearing it a hundred and thirty years after the end of that war.

And we need to be aware of it in this context. If national judicial review of state actions ceased to occur, there would be no security for any human rights to which the nation, as a nation, is committed. Justice Oliver Wendell Holmes, Jr., and Justice Joseph Story clearly saw that the only security for the law of the Union was in the national judicial power to review state actions. This is still true, and always will be true, unless the world utterly changes. Absent such review, the *obligatory* law of human rights in general would be the law of the State

or States least oriented, for the time being, toward the d
of human rights.

Why don't we face reality? We actually have only one corps of people—the federal judges—who have *in principle* the job of policing human rights in the name of national political morality, and who are so organized as to be able to do this work. The work of the federal judges in this regard sometimes does not have a quality pleasing to someone devoted to the ideal of a comprehensive regime of human rights. But at least there is hope of it.

Chapter 5

THE CONSTITUTIONAL JUSTICE OF LIVELIHOOD

You will not be surprised when I tell you that I think the place to start on this is the Declaration of Independence. The Declaration indelibly identifies the rights to life, liberty, and the pursuit of happiness as the inalienable endowment of humankind, and tells us that the quintessential purpose of government is to "secure these rights," and so (as the basic condition of its tenure), to "effect," so far as may be, the safety and the happiness of its people.

Well, many do die, quickly sometimes, sometimes more slowly, of poverty. "Liberty" is pervasively deadened by poverty into a dumb simulacrum, clean-shorn of "the blessings of liberty." But the "right to the pursuit of happiness" is enough to go on with. As a *right*, and not as a mere mocking truism (like a slave's "right" to choose whether his "happiness" will be better served by picking cotton all day "in the summertime, when the living is easy," or by being whipped), the right to the pursuit of happiness is the right to be in a situation where that pursuit has some reasonable and continually refreshed chance of moving toward its goal. The duty of government to *secure* this right is a duty to *act* affirmatively, just as the duty to effect the safety of the people is a duty so to act. (I have said enough elsewhere of the carrying-over of such claims into constitutional law, in the strictest sense, by the Ninth Amendment rule of construction—though it is my view that the Declaration itself is sufficient.)

There is more to be said as to the textual basis for a *constitutional* "justice of livelihood." Let us go to the Preamble to the Constitution, and to an echoing of the preamble in Article I, Section 8. The Preamble declares that a purpose of the Constitution is to "promote the general Welfare." Then, in a phrasal echo that can hardly be accidental, means are bestowed on Congress to tax and spend "for the . . . general Welfare." Do not these twinned phrases pick up and carry forward the very themes of the pursuit of happiness and of the duty of government to aim at maximizing happiness, that are found in the Declaration? And does not the possession of the *power* to seek and to support the general welfare generate a resulting *duty* to do these very things—even without the Declaration, strongly corroboratory though that document be?

There will always be some vagueness about the boundaries of the concept "general welfare." But our American generation is in the happy or unhappy position of not having to worry about these peripheral and penumbral problems. It is a clarifying metaphor, pointing to ugly truth, to say that our country contains two countries—a country wherein everybody has plenty of good food to eat, and a country wherein nutrition is a bad problem and getting worse. In one country, infant mortality, the death of babies, is going down; in the other country it is going up. A house thus divided against itself is not a penumbral case on the general-welfare question. It is a classic case—I ought better to say *the* classic case—of welfare diffused not *generally* but instead *partially*—with, as we all know, scandalously strong streaks of racism and sexism in that partiality.

The possession of a decent material basis for life is an indispensable condition, for almost all people at all times, to the pursuit of happiness. The lack of this basis—the lack we call "poverty"—is overwhelmingly, in the whole human world, the commonest, the grimmest, the stubbornest obstacle we know to the pursuit of happiness. I have suggested that poverty may be the leading cause of death; it is pretty certain that it is the leading cause, at least among material causes, of despair in life. Of course some few people, through extraordinary talent or rabbit's foot luck (and talent is itself a gift of luck), do clamber over the obstacle. But the right to the pursuit of happiness is going to be, for all but a small minority of those in poverty, the pale sardonically grinning ghost of a right.

I go on now to the *affirmative constitutional duty* of Congress diligently to devise and prudently to apply the means necessary to ensure, humanly speaking, a decent livelihood for all.

We have, to be sure, tended to conceive of constitutional human-rights law as a set of limitations on governmental power. This concept contains much health, but not the whole truth. A serious thirst for human rights, and so for human-rights law, cannot be slaked with no more than a canon of "thou shalt not's." Sins against human rights are not only those of commission, but those of omission as well. Other nations seem to have gotten ahead of us in explicit and detailed recognition of this; herein the papacy has outstripped the United States of America. We started the conversation about human rights, but we seem reluctant to carry it forward.

Since we are talking about human rights as a part of our own constitutional law, it might be useful to ask whether affirmative duties are exotic in that law. We closely study the *empowerments* of the national government, the *distributions* of power within it, and the *limitations* that lie upon it, devoting little attention to the Constitution as a source of affirmative obligation to act, sometimes on a very grand scale.

If we start at the beginning, we read no further than Article I, Section 2, before we find that Congress lies under the mandated duty of providing for a decennial census. Failure to make such provision would be plainly unconstitutional. Congress is under an affirmative duty to assemble at least once a year. Each House must keep a journal. Congress must provide compensation for its own members, for the President, and for the Article III judges.

The President lies under the highly general affirmative duty to "take Care that the Laws be faithfully executed." Here is exhibited a characteristic that may be seen to inhere to some degree in all the most important duties mandated by the Constitution. The President cannot do everything imaginable to bring it about that the laws be faithfully executed; he is limited by his own physical and mental powers, by other claims on these, and by the amplitude of the means put at his disposal by Congress. The duty has to be a duty to act prudently within these limits, without ulterior motive, sensitive to the force of the powerful conscience-stirring word "faithfully." It cannot be any more—or, I should think, any less—than that. But is it not a duty?

Let us take a closer look at some of the mandated congressional duties. The duty to provide for the mandated decennial census must be a duty to commit reasonable resources, in a prudent way, to producing a serviceably accurate count; to produce a wholly accurate count, as of any given moment, is rigorously impossible, and the fact that the census is mandated to be taken only once in ten years shows that a reasonable approximation is enough to satisfy the needful equality in constituency basis. Yet who can doubt that a congressman who openly refused to vote for any census, or who would vote only for a census-taking procedure which he knew to be grossly flawed, would be in flagitious breach of constitutional duty?

Something like this must be said of Congress' affirmative duty to appropriate money for compensating the President, the judges, and indeed its own members. The words "a compensation" must mean an amount of money that reasonably compensates; otherwise the provision, thrice carefully set out in respect of all the principal officers of the United States, could easily be brought to nothing. Nobody can say, with any show of reason, that there is a duty to compensate the President at one exactly calculable level. Yet a Congress that refused to appropriate a reasonable compensation for the President would be in breach of constitutional duty.

All or most of the constitutional duties expressly set out combine two characteristics. They are, on one hand, real duties. But they can be fulfilled by good-faith action, over a pretty wide and not distinctly bounded range. In these respects they resemble a very great many duties in law outside constitutional law, as well as many duties outside law.

There is then nothing exotic to the Constitution in the

proposition that a constitutional justice of livelihood should
be recognized, and should be felt by the President and Congress as laying upon them serious constitutional duty. In the
early phases of this work, I find I am most often asked the
question, "How much?" or "Where will you draw the line?" (So
many people are more anxious about "drawing the line" than
about getting food out to hungry children.) I will come back
to this later, but I think it well to try to suggest, at the beginning, that the establishment of duty is one thing, while the
specification of prudent quantities and means is another—
though it must be remembered as well that the decently eli-
gible *range* of means and measures is one thing when you are
under no duty at all to act, and quite another when you are
under a serious duty to act effectively.

This characteristic continues to be visible as we consider
constitutional duties that arise from structural and relational
considerations. Congress is literally commanded, as to the
support of the other branches of government, only to provide
for compensation for the President and the judges. Yet the
Constitution assigns to the President duties and functions
that require the spending of money—for people, space, travel,
supplies, and so on. Since, under Article I, Section 9, clause
7, Congress alone can provide this money, Congress must lie
under a constitutional duty to appropriate such money in reasonable amounts, and even to levy taxes, or to borrow, so that
the money will be there to appropriate. The same duty is dis-
cernible, *mutatis mutandis*, as to the judiciary. These duties, derived from the Constitution, cannot be given arithmetical
exactness, but are nonetheless constitutional duties.

That is exactly what I want to see recognized as the duty

of Congress: continually to move, by the general diffusion of welfare, to give life to a constitutional justice of livelihood, and so to prepare the way for the tasks allotted to government in the Declaration of Independence: "to effect their Safety *and* Happiness."

When we are faced with difficulties of "how much," it is often helpful to step back and think small, and to ask not, "What is the whole extent of what we are bound to do?" but rather, "What is the clearest thing we ought to do first?" When we descend to that level, one reasonable answer occurs. Somebody's count has been that a million and a half people in the State of New York are undernourished. About half our black children under six live in poverty, which very commonly entails malnutrition. Some helpless old people have been known to eat dog food when they could get it; it is not recorded that any Cabinet member has yet tried this out on elderly persons in his own extended family. Now you can bog down in a discussion about the exact perimeter of "decent livelihood," or you can cease for a moment from that commonly diversionary tactic and note that, wherever the penumbra may be, malnourished people are not enjoying a decent livelihood. In a constitutional universe admitting serious attention to the Declaration of Independence, a malnourished child is not enjoying a "right to the pursuit of happiness."

And we are very lucky herein, because we know pretty much what to do. We suffered chronic, endemic malnutrition in America. We got rid of it, humanly speaking, by food stamps, school lunches, and a few smaller measures. Then we just let it come back, because we wanted our own taxes lowered. The Great Society nutrition programs, widely spoken

of as having somehow failed, in fact succeeded brilliantly—
a thing that often happens when you attack a problem directly
and simply, giving hungry people more food the shortest way,
instead of feeding them rancid metaphors about boats that rise
when the tide rises. All we have to do is reinstitute adequate
funding for these programs. Then we can think about the next
thing. We won't have to think very long.

I know that I'm talking into the political wind. The country is now infatuated with an idol called "the economy," which
the high priests seem to agree is doing real well, though millions of children are not getting enough to eat, and millions
of adults who want work cannot find it. But winds change;
they always have, and doubtless they always will. A period of
no power is a period for the reformation of thought, to the
end that when power returns it may be more skillfully, more
fittingly, used. The way I want to see thought reformed is by
our ceasing to view the elimination of poverty as a sentimental matter, as a matter of compassion, and our starting to
look on it as a matter of justice, of constitutional right.

Let me say summarily that I know the judicial courts probably cannot handle this matter comprehensively—though
perhaps we may come to know more about that later, through
trial. But Congress too is bound by the Constitution; that instrument imposes many affirmative duties in Congress. Of
course no exhaustive and detailed blueprint of constitutionally required action can be laid out, now or ever, but this fact
is not to be taken to make impossible the discernment of legal
duty, including constitutional duty.

It's hard to know when to stop this argument—or even why
one had to start it. The facts speak: infant mortality, hungry

children, poverty-blighted lives, and the exceedingly plain relation of these and other such things to the pursuit of happiness. The person who needs a lot of argument on this probably won't be convinced by argument. But this chapter, as its title shows, at least aims at shifting the issue, in some minds, from an issue of sentiment, an issue of compassion, to an issue of constitutional justice. This kind of justice must be done, or we will never attain to any other kind of justice. The general diffusion of material welfare is an indispensable part in the general diffusion of the right to the pursuit of happiness.

Chapter 6

OF TIME
AND THE
CONSTITUTION

I have always been haunted—both troubled and nour-
ished in spirit—by certain words near the close of the Preamble to our Constitution: ". . . and our Posterity. . . ."
These words start up the music of human time—the music
of memory, and the vastly open-chorded music of hope. It
is in that time—in its fleetingness and in its infinity—that
our Constitution has its ordainment and establishment.

How much time are we talking about, thus far?

I spoke those words to a meeting of a then-new Board of
Governors of the American Bar Association, in 1991.

"Time" (in a different sense from the meaning of the legal
phrase) "is of the essence" in considering the Constitution, and
therefore its system of human rights. The thesis of this book—
that our constitutional law of human rights is underdevel-
oped, unnourished by the three basic commitments that still
might be called on to form it—generates questions of time.

1. First, has time made these unredeemed commitments
stale? Are we merely stirring the dust of an irrelevant and perhaps not serviceably accessible antiquity?

2. Is it too late, because of abundant and well-reasoned precedent, to effect the change I have advocated here, and so to bring down into modern and future years the epochal commitments of the Declaration of Independence, the life-giving rule of construction of the hitherto virtually unused Ninth Amendment, and the fulfillment of Lincoln's sacred prophecy: that "new birth of freedom" that I think ought to be taken to have been provided for in the "citizenship" and "privileges and immunities" clauses of the first words of the Fourteenth Amendment?

3. As to the future, what kind of development dare we dream of for time to come, if these ideas make their way?

To answer the pervading question I posed myself when addressing the American Bar people—"How much time are we talking about?"—everything depends on the purpose of that general question, and the frame within which it is asked. For purposes of these questions I have just deployed, human life is the right standard of measure. The Protagorean saying, "Man is the measure of all things," is not helpful for intergalactic distances, or for the mass of an electron. But for a history as long or as short as that of the United States Constitution and its laws, it will serve; indeed no other is really imaginable. I cannot, I believe, improve on what I said to the Bar Governors in 1991, just a few days before yesterday:[1]

James Madison was born in 1751; he died in 1836. Benjamin Harrison was born in 1833, and died in 1901; he was thus a child of three when Madison died. Dwight Eisenhower, born in 1890, was eleven when Benjamin Harrison died. Eisenhower died in 1969, 218 years after the birth of Madison. I start with Madison because of his indelible character as chief among the makers of the Constitution. The other two names were suggested by closeness of fit. Altogether, what is shown is that just three human lives, none phenomenally long, can comfortably cover 218 years. Measured backward from now, that count of years takes us back to 1779, three years after the Declaration of Independence, and eight years before the Constitution was sent out from the Philadelphia Convention.

I like to take the reach of my country's time in more personal ways.

Long, long ago, I had a strong friendship with a black man who had been raised to the age of fifteen as a slave. He was freed when the Union troops reached Texas. He had, then, been born about 1850, the year John C. Calhoun died. In Austin there were in my early days Confederate veterans in some number. I used to play with the gold watch one of them carried. Many of the old black people in Austin, when I was, say, ten, had personally been slaves; for that, they needed only to be a little over sixty.

Looking the other way in time, I've been teaching law, these years, to people in their early twenties; I've gotten to know many of them rather well, in quite recent days as well as in the longer past. On any conservative estimate, a good number of these people will live at least sixty years more, some a good deal more—very senior alumni and alumnae of the Yale and Columbia Law Schools, leaders in the mid–twenty-first century.

I won't squeeze these figures, for they can be only approximate; I hope all my students will reach one hundred. But the figures show that one person of my age can have talked about slavery with a man who had been a slave for fifteen years, and then can talk about the American future with young people supremely fitted to face and to shape that future for long into the twenty-first century. My reach in time—through voices I have heard and ears that have heard my voice—is something over two hundred years, not so far from the entire span, up to now, of our nation's life under our Constitution.

Let me take my own mother. She was born in 1885. As a young woman, she saw the first automobiles come to Hillsboro, Texas—though I believe an old buggy remained, until about the time of my own birth, the Hillsboro family's only means of transportation. She died in 1975, two months short of ninety—five or six years after she watched television coverage of the first landing on the moon.

Now people were born in 1816, and some of these lived to the same age my mother attained. As far as time is concerned, one of them, when about sixteen, could have had a talk with the aged James Madison, who remained amenable to conversation until near the very end. This eager young interlocutor would have lived until 1905; before that time my mother, an alert young woman of nearly nineteen, who had taken all the prizes in school, might have listened, eager in her turn, to such a person's reminiscences of that meeting with Madison. Then their talk might have turned to the newfangled automobile.

Time is little understood, hard to measure in the mind. But the obvious calculations I have played with lead me to believe that our feelings about our problems of today, and about the bearing of the past upon these problems, are too much colored by an illusory projection of a very considerable antiquity.

With these measures in mind, let us turn to the first of these questions I have put: Are we so far from the Declaration of Independence, from the Ninth Amendment, from the Fourteenth Amendment that we cannot understand what they are saying?

"Speech community" is a flexible term. I know that I talked with my maternal grandmother with no sense of any difficulty in our understanding one another. I know I would and could have talked with the same ease with my maternal grandfather, had he not died quite early. She was born in about 1860 in Norway, and he in 1857 (the year of the Dred Scott case) in Louisiana. This intimate familial speech community covers an eighty-one-year-old man, and people born about sixty years before he was born. Either of these two grandparents, so far as time goes, could have talked with an old person who in childhood had looked up at the face of George Washington. My grandmother's life and mine, even in their considerable overlap, cover about 155 years, more than all the time from the Fourteenth Amendment to now, and considerably more than half the time back to the Declaration of Independence.

Of course there are slow changes in language, even in such a short period. George Washington would not have understood the word "automobile," just as we do not perhaps understand the word "civility" quite as he did. But the assumption that eighteenth-century language is something like as incomprehensible to our minds as the language of King Alfred, or even of Chaucer, is unbased. We have to ask in each instance what makes us think the language under our eyes was used with important difference in those times.

The terms we have to deal with in our present context are few, and I do think not in the least problematic: the phrase "life, liberty, and the pursuit of happiness" in the Declaration, the rule of construction in the Ninth Amendment, and the "privileges and immunities" formula in the Fourteenth Amendment.

Let us take the naming of the "pursuit of happiness" in the Declaration. This is a thoroughly general phrase. What is it that would make you think that the meaning of these words has changed in their generality? "Happiness" may be a problematic concept, in these days as in those of the Declaration— and in much the same ways then as now. But words are bound by context, and by the nature of life, and it is not "happiness" that is guaranteed, but the right to "pursue," to seek, to try to attain it. In immediate context it stands side by side with the word "liberty." What did the *"pursuit* of happiness," by the exercise of one's "liberty," mean in 1776 that it doesn't now mean?

The Ninth Amendment rule of construction is of at least as obvious a meaning. The designation of the rights that are to be preserved from denial or disparagement, as those *"retained* by the people," makes just as clear a reference to the rights claimed in the Declaration (some thirteen years earlier) as it can now be seen to make.

The phrase "privileges and immunities," uttered in the Fourteenth Amendment when my maternal grandfather was about eleven, cannot be shown to have suffered any sea-change since at the latest 1825, when, in the leading case on the Ar-

ticle IV "privileges and immunities," Justice Washington explicitly and verbatim included the "right to the pursuit of happiness" as one of them. The only change has been the putting
onto the Fourteenth Amendment phrase (in the *Slaughterhouse
Cases*) of a jerry-built gloss that deprived it of *all* operational
meaning. Whether we can leapfrog over that barrier depends
on our assessment of what warrant the Court had for setting
it up—a question I have already fully discussed, in Chapter 2
above (I remind you that this gloss was so bizarre that knowledge of it did not become diffused in lexicographic circles—
see p. 26 above).

On the whole, to answer the first of my questions, there is
just no reason at all for seeing in the passage of time any
ground for treating any of these terms as puzzling. It is interesting to note that this is true of most of the expressions
in the Constitution and in the statutes contemporary with that
document—and later with the Fourteenth Amendment. To
confirm this, start reading Article I. You will hit a few terms
not now colloquial (like the use of "electors" rather than "voters" in Article I, Section 2), but almost all these tiny difficulties are resolved by the immediate context. In the very
example just given, you have no difficulty in understanding
who are to be the "voters" who elect the "Congressmen"—also
a term that does not occur in the Constitution. The great Document contains some legal terms of art, which are not more
than momentarily problematic, some words of indefinite semantic borders, then as now (e.g., "emolument"), a couple of
shamed concealments ("other persons" equals "slaves"), and a
few imaginable syntactic problems, such as in Article IV, Section 2 (see above, p. 49), but on the whole the Constitution

is a piece of language still easily readable as language—though seeing into its deep structure and implications may have been and still may be a much more complicated matter. What I am asserting is that these three prime commitments to human rights are similarly readable by anybody who has a good command of English, and (in case the word "privilege" gives momentary difficulty) a 1939 Unabridged Merriam-Webster dictionary or its equivalent.

It's worth saying, too, that even if we found a letter dated 1788 from Edward Rutledge to Roger Sherman, setting out that Rutledge, in signing the Declaration of Independence, had thought it the general view on that occasion that "the pursuit of happiness" meant only the "pursuit of material prosperity," such a letter could scarcely be looked on as an authority; it is only one cup of water scooped out of the vast ocean of eighteenth-century English. It is my very strong view that we ought to make what we can out of the language *chosen by authority to be the language of authority*, and resort to other language, which is all we can possibly have, only when the authoritative language is of real obscurity. None of the language in our three principal commitments to human rights comes close to such obscurity—except possibly the word "privilege," which we can look up in the Merriam-Webster, where we find it to be highly multivocal, but with one meaning right on the money (see above, p. 26).

I have labored this point pretty hard. That is because I have discovered how prevalent the practice is of turning at once to general considerations and to scraps of collateral discourse, while never giving the chosen authoritative language itself a chance. I have more than once been in an argument about the lawfulness of judicial review of state laws, for their conformity to the national Constitution, with people who turned out never to have read Article VI of the Constitution, or, if they had read it, showed no sign of remembering it. If you read it for yourself you'll see that it completely settles the question whether judges *in their courts* are duty-bound to treat the national Constitution as a superior law to any state law, and to decide the cases before them accordingly.

I could multiply such instances. There is very serious truth underlying the (quite apocryphal) story that a celebrated Supreme Court Justice once remarked, in a case concerning the meaning of an Act of Congress, "There is no legislative history available on this point, so we shall have to look at the text of the statute."

But let me not dwell on this any more, lest I convey the impression that I know of some unofficial, unauthoritative "history" of such weight as mortally to impeach what I think is the clear meaning of our three commitments to human rights. I do not know of any such material, and I believe if it were at all strong I would by this time have heard of it. But, as I formerly wrote in this general context, "Those who believe in astrology always know more about astrology than do those who do not believe in astrology."

The second question I have asked is whether it's too late to effect the change I am advocating.

This is just a question we are asking about ourselves, about our political society. It is obviously not too late, if we are in sufficient numbers resolved that it not be too late.

But while this remains true, the question can be made a little more malleable by asking two sub-questions. First, has so much authority, very high and carefully considered authority, accrued around the rejection of the force in law of these three great commitments that it becomes too hard to fight against that authority? Secondly, how characteristic is it of our constitutional law system that it can change its direction, even after a long time?

As to the accrual of authority, it will be helpful to take up each of these three prime commitments one by one.

The Declaration of Independence was appealed to in some very early cases, though not, I think, in the Supreme Court. These cases would not in themselves suffice to establish a consensus that the Declaration could not play a part in law. Neither do they establish the contrary. As to this, I think we cannot pass over the fact that the Declaration became a rallying-point for anti-slavery forces, and was an embarrassment to the pro-slavery people. It was inextricably caught up in the most terrible of American questions—the question that woke Jefferson like a firebell in the night, and that at last caused our great Civil War. That judges should treat it gingerly is more than understandable—such diffidence was something like inevitable. The best we can say is that no Supreme Court ever rejected in terms the idea that the Declaration was a primary source of legal right. The 1825 *Corfield* case, discussed above (see pp. 49–50), goes a long way toward upholding the authority of the Declaration in establishing the principal human rights protected in law, and the authority of that case, though in a perverse way, was recognized in the *Slaughterhouse Cases* opinion.

The same is true (and I am not historian enough to think I fully know why) of the Ninth Amendment, whether considered as a thing in itself or as an entitlement of the entry of the Declaration into constitutional law. I would conjecture—and no more than that, for I have no right—that the Ninth Amendment lay sleeping because it too might open up a path for additional rights antagonistic to slavery. Certainly it would have done so if connected to the Declaration. There was an anti-slavery thesis, floated by how many people I don't know, that the national government was barred by the Fifth Amendment from giving any affirmative aid to slavery, such as the drastic Fugitive Slave law of 1850. Perhaps it was feared that the creative freedom seemingly promised by the Ninth Amendment might give rise to other anti-slavery contentions not textually originating in Amendment V or other Bill of Rights provisions. Or perhaps there was heard by others than Jefferson the clang of a firebell in the night.

Finally (and most paradoxically!) we come to the *Slaughterhouse Cases*, wherein the third of our great commitments was dirked. That case is all the top authority we have against us.

I have written the immediately foregoing paragraphs with great diffidence, because I control neither the broad historical source-material nor the instructed concepts needful to explain why it was that the early record of the Declaration and the Ninth Amendment was so dreary. I strongly feel that it had to do with slavery. I even think it probable that the *Slaughterhouse* Court, realizing that giving full scope to a set of "privileges and immunities of citizens of the United States," which now definitely included blacks, would open a wide door for claims of right by the freedmen, paled at what it saw as the unmanageableness of this. In a country founded largely on slavery, a country wherein slavery continued for nearly ninety years after it was proclaimed that all men were "created equal," with the God-given right to liberty, a country wherein (after the great War and the abolition of chattel slavery) every sleazy trick concoctable was brought into play, and blessed by the Court, to keep black people in their place—in such a country, I think it not really far-fetched to suspect strong subjacent connections between anti-black racism and many puzzling judicial or political actions or inactions. Certainly the underdevelopment of a system of human rights, the election not to make very much of our three great commitments to comprehensive human rights, is not really astounding, just as it is no great breach of probability to guess that the failure of the United States now to go to a civilized system of general medical care results, in part at least, from a reluctance to give that much to poor black people.

The upshot is that, while there is little venerable authority availing of the Declaration and the Ninth Amendment, there is no daunting authority against their use. The problem-case is of course *Slaughterhouse*, but that case is not even a good piece of legal shoddy, fit to stuff mattresses.

Now, granting that the strong affirmative use of the great commitments on which I rely has not occurred up to now, is that enough to discourage you? How characteristic is it of our constitutional legal system to change its direction?

In 1842, the Supreme Court, in *Swift v. Tyson,*[2] held that the federal courts, hearing cases involving issues of state common (that is to say, non-statutory) law, were to apply their own version of this common law, rather than the version the relevant State followed in its own courts.

In 1938, in *Erie R.R. v. Tompkins,*[3] after nearly one hundred years and hundreds of cases in which the rule of *Swift v. Tyson* had been distinctly followed and applied, the Supreme Court overruled *Swift v. Tyson.*

In 1922, the Supreme Court said in *Prudential Insurance Co. v. Cheek:*

> But, as we have stated, neither the Fourteenth Amendment nor any other provision of the Constitution of the United States imposes upon the States any restrictions about "freedom of speech" or the "liberty of silence"; nor, we may add, does it confer any right of privacy upon either persons or corporations.[4]

When those words were uttered in the opinion of the Court, the Fourteenth Amendment had been around for fifty-four years—"substantive due process" had already strongly come on the scene.

By 1937, just fifteen years after *Cheek*, Mr. Justice Cardozo, in the famous *Palko* passage I have quoted (see above pp. 94–95), was explaining, after the fact, why it was that the free-speech guaranty of the First Amendment was held binding on the States. For it was in 1937—only fifteen years after *Cheek*— that the Supreme Court for the first time reversed a state court conviction on substantive free-speech grounds.[5] In the sixty years since, it has been almost forgotten that the Supreme Court said what it did in *Prudential Insurance Co. v. Cheek.* It just doesn't fit in.

In 1867, the Supreme Court first decided a case *in favor* of a person who claimed that a State's action impermissibly "burdened inter-state commerce."[6] (The doctrine relied on had been spoken of with qualified approval, in principle, in

1851[7]—but even that is sixty-four years after the adoption of the Constitution.)

In 1949, in *Hood & Sons v. Du Mond*,[8] Mr. Justice Jackson, for the Court, wrote:

> The Commerce Clause is one of the most prolific sources of national power and an equally prolific source of conflict with legislation of the states. While the Constitution vests in Congress the power to regulate commerce among the states, it does not say what the states may or may not do in the absence of congressional action. . . . Perhaps even more than by interpretation of its written words, this Court has advanced the solidarity and prosperity of this Nation by the meaning it has given to these great silences of the Constitution. . . . [The] principle that our economic unit is the Nation, which alone has the gamut of powers necessary to control the economy, including the vital power of erecting customs barriers against foreign competition, has as its corollary that the states are not separable economic units.

Never has the absence of a constitutional text supporting a line of constitutional decisions (a very common thing) been dealt with more elegantly than in the phrase "these great silences of the Constitution"! Still, the decisions are there, issuing in steady progression from "the great silence."

Since 1871, before and after 1949, formulation has succeeded metaphysical formulation, line after tormented line has been drawn, to mark out the areas of permissible and impermissible state interference with interstate commerce—as to migration, as to mudguards, as to milk, as to minnows. This regime, which the Court in *Hood* sees as so important to national economic unity, and which takes up a lot of space in modern constitutional-law casebooks, is virtually a post–Civil War creation.

There is an even more profound point to be made here. If a "great silence" is all that is needed to generate this mass of decisions—striking down, year after year, laws of the States, and thus frustrating the state "majorities" that made them— why should it be disturbing if the same thing should occur, with equal prolificity of doctrine, in regard to human rights? As it is, there isn't even a "great silence" here; we have the three texts I have been exploring. In the passage I have just quoted from Mr. Justice Jackson, he said, "[The] principle that our economic unit is the Nation . . . has as its corollary that the states are not separable economic units." When we at last face up to the Declaration of Independence, can we think that it generates no "corollary" that the States are not to be separable moral units, with respect to their "securing human rights"?

Then interference by the States with the practical enjoyment of the national "privileges and immunities" is an interference with values that the national government was established to maintain and is obligated to maintain—to "secure." It is state interference with the effective working of a comprehensive national plan of human rights. Even if the national bestowal of *state* citizenship were thought not to import a command that the human rights of the designated persons be binding directly on the States, the States would nevertheless be forbidden to frustrate, to bring to nothing, the obligation and therefore the *function* of the national government in regard to human rights, or to the values the national plan aims at advancing. The symmetry with the "economic nationhood" cases is striking.

It is in the implications of this symmetry that understanding of the high political place of general national protection of human rights, against the States, is to be found. The "economic nationhood" cases rest not on any particular words of the Constitution, but rather on a general concept of our national "union." Can there not also be made out a legitimate and authenticated concept of the nation, the "union," as one *wherein* human rights are to *prevail?* To deny this is total denial of the authority of the Declaration of Independence, even as a statement of the nation's goals and reason for being. Why would we do that, while at the same time finding in the "great silences" of the Constitution a concept of *economic* nationhood, capable of generating innumerable judicial decisions? What would support this choice of free trade over the claim of the American people to enjoy, as a whole people, the benefits of living under a unified regime of human rights? This would be inexplicably perverse, particularly since we do not have to choose between these concepts of nationhood, but can live by both of them. Indeed, they overlap in a penumbra between them, as in the right-to-travel cases. But can a nation live by trade alone?

Can we really bear to say, even (and above all) to ourselves, that the unity of this Union is a unity only in governmental power and in economic exchange, but is not a moral unity in the observance of human rights? Even the Preamble of the Constitution strongly speaks against this: "to . . . secure the Blessings of Liberty to ourselves and our Posterity." We betray this very statement of purpose, uttered at the creation of the national government, if we accept and act upon the view that there is no national law of general human rights binding on the States.

These structural considerations lead to the same conclusion as the texts: "Our Federalism" need not be, must not be, the "Catch-22" of human rights—the fine-print "catch" that fatally undermines the boldface guarantee.

In 1791, the Sixth Amendment guaranteed the accused person in criminal proceedings the right "to have the assistance of counsel for his defense." In *Johnson v. Zerbst* (1938),[9] the Supreme Court first held that this was an absolute Constitutional right *in the Federal courts*, and that assigned counsel must be provided. There had been differences in practice in the federal courts, but (*148 years* after the adoption of the Sixth Amendment) the Supreme Court held the right to be an absolute one for all federal criminal defendants.

Meanwhile, in 1930, the Supreme Court held that the Fourteenth Amendment "due process" clause required the appointment of counsel in state *capital* cases, where the defendants were indigent and ignorant.[10] Thirty-two years later, and a total of nearly one hundred years after the passage of the Fourteenth Amendment, in *Gideon v. Wainwright* the Supreme Court *first held* that counsel must be furnished in all serious state criminal cases.[11]

Now I could go on and on with this. Consider carefully the elapsed times—it took sixty-two years after the passage of the Fourteenth Amendment for the Court to hold, even in a state case involving eight death sentences, that "due process" required the appointment of counsel for indigent state defendants! It took nearly a hundred years (after much backing and

forthing) for the Court to decide that counsel must be furnished in *all* serious state criminal cases.

On the other hand, the Court can reverse itself fairly
quickly. There elapsed fifteen years between its 1922 *Prudential* statement that there was no free speech as against the
States, and the holding (soon greatly proliferated) that such
protection did exist.

Now in the matters covered in this book, we are dealing
with something that, if it is a mistake, was one of the biggest
mistakes, doubtless the very biggest, in our history—the failure to make any use, in law, of our unique and irreplaceable
national heritage of committed principles.

On the other hand, as to the Declaration and the Ninth
Amendment, there is not even any impressive decisional authority to "overrule." The overruling of *Slaughterhouse* ought to
occur, but even that is not seriously out of the time-scale I have
set before you. *Slaughterhouse* came down just a decade before
my father was born. It has encouraged no "reliance" of a de-
sirable kind. Indeed, the correction of such a huge mistake
could really never be "too late."

I put it to you that there is nothing in the history of these
great utterances in our charter of human rights, or in our dealings with hugely consequential precedents, to make us hesi-
tate to move toward the righting of this hugely consequential
mistake—the failure to use these precious utterances, "in their
spirit and in their entirety."

How would such a resolve come about? How would it
work out in the future?

One can safely say: "Very slowly."

The first step would be the formation of a new professional and public opinion around the idea that we have gone astray in rejecting, perhaps in rather silently repudiating, the mighty force of these commitments (the Declaration, the Ninth Amendment, and the citizenship and privileges and immunities clauses of the Fourteenth Amendment) as the ultimate complex source of our commitment to an open-ended, open-textured system of human rights in general. This book is obviously an attempt to push that process along. If such a movement of professional and public opinion does at last go forward, nobody will ever know what "caused" it; it will be the result of the efforts of many, many people.

When we turn to the chief and most easily invocable holders of power over the subject, our Supreme Court, we can expect nothing very dramatic, no definite "event." You have sampled, just above, illustrations of the slowness with which our judiciary moves toward change, toward the ultimate acceptance of the obvious—such as the obvious truth that no defendant can have a reliable expectation of getting "due process of law" if he has no right to a lawyer. It took fifty-eight years for the Court to accept and act on the obvious truth that segregation by law is pervadingly an insult to black people, an "assertion of their inferiority," and is massively harmful to them in more concrete ways.

But these cases, like some of the other examples I have given you of judicial glaciality, involved relatively simple issues. A dramatic event, like the *Brown* case, was clear in outline, though far from free of subsidiary problems of implementation.

By contrast, what I am advocating in this book is a change in method, in total outlook, in perception of the very foundations of our human-rights law. If it comes, my own thought is that it will come as a *realization* not only of the power of our cardinal commitments to human rights, but even a realization that all along, particularly in this century, we have lamely and inarticulately decided a miscellany of cases that can confidently be justified only on the grounds of our three *general* commitments:

People have a right to practice contraception.[12]

Persons have a right to marry, even people who cannot prove that the children of a prior marriage will not become public charges.[13]

A grandmother has a right to bring her own grandchildren to live with her.[14]

People have a right to send their children to a religious or military school,[15] or if their religion so commands, to send them to no high school at all.[16]

There is a general law of free speech and of freedom of religion *applicable to the States*, though the First Amendment, in its terms, applies only to the national *Congress*. (As to this enormous body of federal constitutional restriction on the States, see Cardozo's justification in the *Palko* case, and my critique of that justification, above pp. 93–97)

A prisoner has a constitutional right not to be sterilized for having three felony convictions.[17]

A person has a right to teach and learn a foreign language.[18]

A person has a right to travel from one State to another, without being taxed for doing so.[19]

A female naval lieutenant has a right to have her husband's medical expenses covered, as those of the wife of a male lieutenant would be—though the equal protection clause of the Fourteenth Amendment does not, as a matter of text, apply against the national government.[20]

The cases in this miscellany (and it could be augmented) rest on different constitutional grounds. Most of them either invoke or seem silently to rely on "substantive due process." Some seem to have sidestepped reference to any constitutional formula, and proceed directly to invoke the very solemnity and importance of the right sustained, without bothering to link it up with any specific textual material. The explanation of the right to travel from one State to another has been explained in a number of ways, so many and varied that in *U.S. v. Guest*[21] Justice Stewart says that it doesn't matter what the correct explanation is, because "all agree" that the right exists; he later adds, in a footnote, that this right is "in the Constitution," without saying just where.

Our constitutional law has never been wholly "textual." But these miscellaneous recent cases on human rights seem to me to evidence, as to human rights, an uneasiness, a discomfort, with the conventional (though erroneous) dogma that human-rights protections must be found in specific texts. With Stewart's treatment of the right to travel from State to State (see the last paragraph), compare his dissent in the contraceptive case, *Griswold v. Connecticut*,[22] where he says he has looked at the Constitution and could "find nothing" in it that invalidated a state law making contraception a crime. And in turn compare the ease with which the same justice accepted the derivation of a general substantive maritime law from a mere grant of *judicial jurisdiction* in admiralty cases to the federal courts.[23]

There's something out of joint here, some major dislocation of method.

Now look over the cases I have just listed, above. *Would not each and all of them* fit comfortably into the explanatory category of "the right to the pursuit of happiness"?

 He's picking pretty easy ones.

Is not the prohibition of contraception a deadly blow at the pursuit of happiness, forcing on couples the alternative of abstaining from intercourse, or having children they do not now, perhaps for very good reason, want to have, or think it right to have?

Is not the denial of the freedom to "exercise" one's religion a cold obstruction to the "pursuit of happiness," in the very way that countless millions have tried to pursue it?

Jefferson said "knowledge is happiness." Is not interference with freedom of communication an efficacious way of choking off this kind of the "pursuit of happiness"?

Are not a grandmother and her orphan grandchildren barred from one kind of the "pursuit of happiness," when they are told they may not live together in her house?

Are not parents who are forbidden by law to send their child to a military school being balked in the pursuit of that high happiness that comes when you think you are doing the best thing for your child?

Let me put it another way: In the "substantive due process" discourse, extended as it is, the phrase *"fundamental values"* plays a big part. But how do you proceed even to a first approximation to knowing what values are "fundamental"? Are they not the "values" of the Declaration, the values that people pursue in their pursuit of happiness?

What I am suggesting to you is that the frank acceptance of the "right to the pursuit of happiness" as a prime—probably *the* prime—foundation for a law of human rights would have a refreshing, clarifying effect on the feeling of legitimacy in most—if not all—constitutional human-rights material: It would easily explain most of the cases as to which difficulty has been felt.

It would do more than that. Just to take one random sample, it would make an open-and-shut case of the claim that the right to listen to music of your choice and to produce music of your choice are constitutionally protected. How could anybody conclude that producing and listening to music is not an indispensable, a very precious part, of "the pursuit of happiness"? You could skip all that part about how music resembles speech in some ways but not in others, and about the question whether "freedom of speech" includes "freedom to listen," and "freedom not to speak," and go right to the heart of the matter. Nietzsche said that "life without music would be a mistake"; some people feel that way, others are less intense. To some, of course, music is just a bore; they can "pursue happiness" in their own way. Bless us all!

This kind of simple insight could bring the "right to the pursuit of happiness" as a touchstone to any human activity or concern. It would reach out to every field of human rights. It would make plain the wrong in every kind of discrimination hurtful to women. It goes to the essence of the wrongs done by the law and outside the law to those having homosexual preferences. It could clarify the ultimate grounds of the banning of racial discrimination against blacks and other racial minorities. I won't try to go through all the other applications; they are as wide as human pursuit of happiness, to which our Declaration is not embarrassed to refer.

This most important factor that would be certain to make the application of, say, the "pursuit of happiness" criterion a slow, and indeed never finished, process is the ongoing assessment, over the range of law and life, of *justification* for governmental activity.

I am not on the enterprise here of searching the largely unknowable psychic content of eighteenth-century people. The matter is much simpler than that. Anybody, in the eighteenth century or the twentieth, who anticipated that a solemn guarantee, like this of the "right to the pursuit of happiness," was an "absolute"—that this "pursuit" was guaranteed to be absolutely immune from the interposition of any obstacles— would be certifiably insane. This plain fact has been clouded, sometimes, by the insistence (now, I believe, shelved under the impact of the facts) that the "freedom of speech" and "free exercise of religion" are "absolute." That view is one that can be held only by somebody who hasn't thought about a sound truck turning up at midnight and bawling obscene curses near a sleeper's window, or a radio broadcast that promises that a mixture of witch hazel and thrice-distilled water will cure skin cancer, or a fundamentalist snake-worshipper driving a herd of rattlesnakes down a street. The interesting and encouraging thing for us is that *although* free speech and freedom of religious exercise are *not* "absolutes," but must over a wide and never quite predictable range yield to suitable justification, these freedoms are very serious constitutional values, having ponderous weight in law, and that justification must, to be of avail, take into account the weight and seriousness of the constitutional commitment, given its source. If this course is followed—and it usually is—in law, then the free speech and freedom of religion guarantees are a part of law, and have real meaning as law.

I am sure that that is all we can ask for the Declaration's "right to the pursuit of happiness," carried down into and through the Ninth and Fourteenth Amendments. What I am asking is that that right be accepted as a pervading part of law, be accorded in law the great weight given by the authority of the Declaration and its sequels, and that, in view of that authority, all claims of justification should have to be judged, if they are to prevail, on a scale suitable for allowing such justification.

That is all we have now as to free speech, which, under this rule, has worked its way into the whole legal fabric, and plays a vital, serious part therein.

We do not have this now as to "the pursuit of happiness." If later on we do, the result will be a thoroughgoing and never-ending working-over of the regime of law, with attention to the enormous weight of the principle, given its origin.

An Afterword

"Now he belongs to the ages."

These words of Stanton, on his learning that Lincoln had died, assure us that we are not done with Lincoln. He belongs to our ages, gone and to come. His is the spirit I have invoked to quicken this book.

In one of his best-remembered sayings, he hopefully foretold that "this nation, under God, shall enjoy a new birth of freedom." Lincoln did not use such words lightly.

The distinct event after his death that seemed to announce this "new birth of freedom" was the opening passage of the Fourteenth Amendment, recognizing, as clearly and as broadly as words could do, the "privileges and immunities of the citizens of the United States." These "privileges and immunities" are set out, with becoming breadth, in the Declaration of Independence, the lode-star of Lincoln's life.

Eight years after Lincoln died, our Supreme Court, in the *Slaughterhouse Cases*, did its best to bring to nothing his sacred prophecy. The country has accepted, for a time, that terrible deed.

But the spirit and mind of Lincoln belong to the ages, out of the power of any Court, or of one short period of history, to bring them to nothing. When we are ready, we can take up the work to which he continues to beckon us. If we do so, we will be treading the ways of his journey, from his reverence for the Declaration of Independence to his vision at Gettysburg.

When you will, you can join the supreme company of his great soul.

Notes

CHAPTER 1

[1]*On Reading and Using the Ninth Amendment,* in *Power and Policy in Quest of Law: Essays in Honor of Eugene Rostow* 187 (M. McDougal & W.M. Reisman, eds. [Norwell, MA: Kluwer Academic Publishers, 1984]), reprinted in C. Black, *The Humane Imagination* 186 (Woodbridge, CT: Ox Bow Press, 1987) and in *The Rights Retained by the People* 337 (R. Barnett, ed. [Lanham, MD: University Press of America, 1989]); *Further Reflections on the Constitutional Justice of Livelihood,* 86 Columbia Law Review 1103, 1104 (1986); *"One Nation Indivisible": Unnamed Human Rights Against the States,* 65 St. John's Law Review 17 (1991).

[2]In this connection, consider with special care the points on p. 37, below.

[3]See Black, *On Reading and Using the Ninth Amendment,* Chapter 1, footnote 1.

[4]See especially Section 25 of the Judiciary Act of 1789, quoted and discussed below at pp. 120–121.

[5]Usually of a negative and constricting kind, often exploiting, perhaps unconsciously, the inbuilt ambiguity and even

multivocality of the phrase "did not intend," and whistling past the evident *generality* of the language under interpretation.

[6]*Tinker v. Des Moines School District*, 393 U.S. Reports 503 (1969). Of course, the analogic and functional extension of the freedoms of "speech" and "press" is far wider than a single example can do more than hint at.

[7]*U.S. v. Causby*, 328 U.S. 256 (1946).

[8]*Citizens Against Rent Control v. Berkeley*, 454 U.S. 290 (1981).

[9]See my *Structure and Relationship in Constitutional Law* (Baton Rouge: Louisiana State University Press, 1969), *passim*.

[10]*Crandall v. Nevada*, 73 U.S. 35 (1867).

[11]*Corfield v. Coryell*, 4 Wash. CC. 371, 6 F. Cas. 546 (C.C.E.D Pa 1825). The most important passage in this opinion is quoted on pp. 49–50 below.

[12]16 Wallace (88 U.S.) 36 (1873).

Chapter 2

[1]Black, *The American Law of Free Speech as Applied Against the States*, in *Faculty Presentations, Moscow Conference on Law and Economic Cooperation* 177 (1990). Paper presented to panel in Moscow on "Glasnost."

[2]*Corfield v. Coryell*, cited above, Chapter 1, footnote 11.

[3]17 U.S. 316 (1819).

[4]The *Slaughterhouse Cases*, 83 U.S. 36 (1871).

[5]*Havenstein v. Lynham*, 100 U.S. 483 (1880).

[6]*Oregon v. U.S.*, 366 U.S. 643 (1961).

[7]I will be back at you once more (pp. 80–84) with a sum-up of the exceedingly profound relations of the anti-nationalist doctrines of Calhoun to the holding in *Slaughterhouse*.

[8]See above pp. 49–50. You should reread this much-quoted passage with care at this time.

[9]*Butler v. Boston & Savannah S.S. Co.*, 130 & S. 527 (1889); *Kermarec v. Cie Gen. Transatlantique*, 358 U.S. 625 (1959).

[10]I am reminded here of a story told me by the late Allison Dunham, who for a few years, some forty years ago, occupied an office next to mine at Columbia Law School. He was stopped for speeding by an officer in Pennsylvania. Taken immediately before a Justice of the Peace, he was fined some $10. Fresh from his clerkship with Mr. Justice Stone, and recalling a recent Supreme Court case, he said, "Your Honor, may I respectfully inquire whether a part of this fine will go to you personally?" "Yes," the J.P. answered. "In that case," replied Dunham, "the recent case of *Toomey v. Ohio* says that you may not judge this case." "Wait a minute," answered the "judge." "What court was that case decided by?" Dunham played his trump: "Your Honor, by the Supreme Court of the United States!" "Well," said the "judge," "that's all right. This is Pennsylvania." I think Dunham paid the fine.

[11]It is a venerable hypothesis that the strange disappearance of the Fourteenth Amendment "privileges and immunities" clause from the toolkit of working law is to be explained by a reluctance in the Court to exclude aliens from benefits that the clause might otherwise have been held to engender. I find this ridiculous, in regard to a century covering the infamous Chinese Exclusion Case, *Chae Chan Ping v. United States*, 130 U.S. 581 (1889), the somewhat later cases that denied anything like even procedural due process to persons of oriental appearance who were so much as *alleged* not to be citizens, and the deplorable anti-alien cases of even later

decades. But in any event, the fear of harm to aliens is easily exorcised.

In *Truax v. Raich*, 239 U.S. 33 (1915), the Court indicated that Congress' decision to admit certain aliens was tantamount to a direction that they be allowed by the States to live amongst us and to enjoy in a substantial sense the privileges conferred by this admission. (The *Truax* Court, perhaps redundantly but very significantly, held that aliens' admission brought them under the "equal protection" clause, as to discriminations between them and citizens.) The Court has flip-flopped on the detailed application of this principle, but the principle is evidently sound, and serviceable as a complete answer to the fear, doubtless in many cases a pretended fear, that it would be dangerous to give much scope to the "privileges and immunities" clause, because the benefits extended thereby to citizens could not be thought to be within the reach of aliens. For a somewhat fuller discussion of the *Truax v. Raich* principle, and its few legitimate qualifications (as to voting and eligibility for major policy-forming public office), see C. Black, *Decision According to Law* 56–62 (New York: W. W. Norton, 1981).

[12]See my *On Worrying About the Constitution*, 55 *University of Colorado Law Review* 469 (1984), reprinted in Black, *The Humane Imagination* 118 (1986).

CHAPTER 3

[1]166 U.S. 226 (1897).

[2]32 U.S. 243 (1833).

[3]198 U.S. 45 (1905).

[4]See Chapter 1 at notes 5 and 7.

[5]*New York Times v. Sullivan*, 376 U.S. 254 (1964).

[6]*Sherbert v. Wiener,* 374 U.S. 398 (1963).

[7]*Edwards v. Aguillard,* 482 U.S. 578 (1987). (The "religion" cases, both as to "establishment" and "free exercise," are confusing and decision oscillates.)

[8]*Meyer v. Nebraska,* 262 U.S. 390 (1923).

[9]*Pierce v. Society of Sisters,* 268 U.S. 510 (1925).

[10]*Griswold v. Connecticut,* 381 U.S. 479 (1965).

[11]*Zablocki v. Redhail,* 434 U.S. 374 (1978).

[12]302 U.S. 319 (1937).

[13]*De Jonge v. Oregon,* 353 (1937).

[14]*Benton v. Maryland,* 395 U.S. 784.

[15]See my *Capital Punishment: The Inevitability of Caprice and Mistake* (New York: W. W. Norton, 2d ed., 1980), Chapter 9, pp. 85–93.

[16]367 U.S. 497 (1961).

[17]381 U.S. 479 (1965).

[18]*Moore v. East Cleveland,* 431 U.S. 494 (1977).

CHAPTER 4

[1]*Griswold v. Connecticut,* cited above, Chapter 3, footnote 10.

[2]Black, *National Lawmaking By Initiative? Let's Think Twice,* in *Human Rights,* Fall 1979, at 28; reprinted in Black, *The Humane Imagination* 66 (1986).

[3]I would like to think that John Marshall had something like this in mind when he strongly spoke in favor of judicial review, in the Virginia ratifying convention: *"To what quarter will you look for protection from an infringement on the constitution, if you will not give the power to the judiciary? There is no other body that can afford such a protection."* (Emphasis added.)

[4]1 Annals of Congress 457 (1789).

CHAPTER 6

[1]The entire address was later published as "And Our Posterity . . . ," 102 *Yale Law Journal* 1527 (1993). I borrow from it freely here, but with some omissions and changes.

[2]41 U.S. 1 (1842).

[3]304 U.S. 64 (1938).

[4]*Prudential Insurance Co. v. Cheek,* 259 U.S. 530 (1922).

[5]*De Jonge v. Oregon,* 299 U.S. 353.

[6]*Steamship Co. v. Portwardens,* 73 U.S. 31 (1867).

[7]*Cooley v. Board of Wardens,* 53 U.S. 299 (1851).

[8]*H.P. Hood & Sons v. Du Mond,* 336 U.S. 525 (1949).

[9]304 U.S. 458 (1938).

[10]*Powell v. Alabama* 287 U.S. 45 (1932).

[11]370 U.S. 335 (1963).

[12]*Griswold v. Connecticut*, cited above, Chapter 3, footnote 10.

[13]*Zablocki v. Redhail*, cited above, Chapter 3, footnote 11.

[14]*Moore v. East Cleveland*, cited above, Chapter 3, footnote 18.

[15]*Pierce v. Society of Sisters*, cited above, Chapter 3, footnote 9.

[16]*Wisconsin v. Yoder*, 406 U.S. 205 (1972)

[17]*Skinner v. Oklahoma*, 316 U.S. 535 (1942)

[18]*Meyer v. Nebraska*, cited above, Chapter 3, footnote 8.

[19]*Crandell v. Nevada*, cited above, Chapter 1, footnote 10.

[20]*Frontiero v. Richardson*, 411 U.S. 677 (1973)

[21]383 U.S. 745 (1966)

[22]*Griswold v. Connecticut*, cited above, Chapter 3, footnote 10.

[23]See the *Kermarec Case*, cited above, Chapter 2, footnote 9.

FOREWORD

[1]As for example in Magna Carta or the French Declaration of the Rights of Man.

[2]As described by Thomas Hobbes and Jean Bodin, among others.

[3]The Lawfulness of the Segregation Decisions, 69 Yale L.J. 421 (1960).

⁴Philip Bobbitt, *Constitutional Interpretation* (Blackwell, 1991).

⁵These six modalities are: *historical* (relying on the intentions of the framers and ratifiers of the Constitution); *textual* (looking to the meaning of the words of the Constitution alone, as they would be applied by the average contemporary "man on the street" today; *structural* (inferring rules from the relationships that the Constitution mandates among the structures it sets up); *doctrinal* (applying rules generated by precedent); *ethical* (deriving rules from the commitments that reflect the American ethos of limited government); and *prudential* (seeking to balance the cost and benefits of a particular rule. Ibid.

⁶I was recently accosted by a lawyer on the street—a former government official, I think—who asked me if I thought a sitting president could be indicted. When I said I doubted this, he demanded to know "where in the Constitution it says 'the President is above the law.'" I suppose we saw the same sort of thing at the time of Watergate.

About the Author

Charles L. Black, Jr., Sterling Professor Emeritus at Yale Law School and Adjunct Professor of Law at Columbia Law School, is the author of many distinguished works on the Constitution and the co-author of a treatise on Admiralty, and has published three volumes of poetry.